Research Methods in Biomechanics

D. Gordon E. Robertson
University of Ottawa

Graham E. Caldwell
University of Massachusetts, Amherst

Joseph Hamill
University of Massachusetts, Amherst

Gary Kamen
University of Massachusetts, Amherst

Saunders N. Whittlesey
University of Massachusetts, Amherst

Human Kinetics

D0068821

Library of Congress Cataloging-in-Publication Data

Research methods in biomechanics / D. Gordon E. Robertson
. . . [et al.].
    p. ; cm.
  Includes bibliographical references and index.
  ISBN 0-7360-3966-X (hard cover)
  1. Biomechanics--Research--Methodology. 2. Biomechanics--Mathematics.
  [DNLM: 1. Biomechanics--methods. WE 103 R432 2004] I. Robertson, D.
Gordon E., 1950-
  QP303.R47 2004
  612.7'6'072--dc22

                           2004002355

ISBN: 0-7360-3966-X

The Web addresses cited in this text were current as of January 14, 2004, unless otherwise noted.

Acquisitions Editor: Loarn D. Robertson, PhD; **Developmental Editor:** Anne Rogers; **Assistant Editor:** Amanda S. Ewing; **Copyeditor:** Nancy Elgin; **Proofreader:** Erin Cler; **Indexer:** Robert Howerton; **Permission Manager:** Dalene Reeder; **Graphic Designer:** Andrew Tietz; **Graphic Artist:** Denise Lowry; **Cover:** Cover images provided by novel GmbH, www.novel.de. Copyright novel GmbH.; **Art Manager:** Kelly Hendren; **Illustrator:** Mic Greenberg; **Printer:** Edwards Brothers

Part-opening images created by the authors using Visual3D software from C-Motion, Inc.

Printed in the United States of America   10 9 8 7 6 5 4 3 2 1

Human Kinetics
Web site: www.HumanKinetics.com

United States: Human Kinetics, P.O. Box 5076, Champaign, IL 61825-5076
800-747-4457
e-mail: humank@hkusa.com

Canada: Human Kinetics, 475 Devonshire Road Unit 100, Windsor, ON N8Y 2L5
800-465-7301 (in Canada only)
e-mail: orders@hkcanada.com

Europe: Human Kinetics, 107 Bradford Road, Stanningley, Leeds LS28 6AT, United Kingdom
+44 (0) 113 255 5665
e-mail: hk@hkeurope.com

Australia: Human Kinetics, 57A Price Avenue, Lower Mitcham, South Australia 5062
08 8277 1555
e-mail: liaw@hkaustralia.com

New Zealand: Human Kinetics, Division of Sports Distributors NZ Ltd., P.O. Box 300 226 Albany, North Shore City, Auckland
0064 9 448 1207
e-mail: blairc@hknewz.com

To Dr. James (Jim) G. Hay, 1936-2002,
for his inspiring leadership in sport biomechanics education and research,
and to all our graduate and undergraduate students
who assisted and inspired us.

CONTENTS

Part III Additional Techniques

PREFACE

This book was developed with biomechanics, biomedical engineering, and kinesiology students and laboratory researchers in mind. The purpose of this book is to outline concisely and extensively the mathematical and technical tools necessary for investigating human and animal motion. In the past, such information had to be gleaned from disparate sources, including the periodical literature, or through hands-on demonstrations by professors or seasoned researchers. Our text provides students and researchers with the tools necessary to collect and analyze the mechanical characteristics of human movements using current biomechanical technologies.

The authors assume that the readers have taken an introductory course in biomechanics or Newtonian or engineering mechanics. Readers should have an understanding of vectors and elementary vector algebra and be familiar with the International System of Units (SI), although these areas are reviewed. Furthermore, readers should know the fundamental laws of mechanics, namely Newton's laws, and basic human musculoskeletal anatomy. The text examines how these laws apply to complex human motions, including the analysis of a human motion segment by segment and combining segments for limb or total body measures. Although knowledge of human anatomy is desirable, it is not essential for in-depth understanding of the analytical tools described.

The text is divided into 11 chapters in three parts. Part I describes the area called *kinematics,* which is concerned with motion description without regard to its causes. This section and part II include chapters specifically concerned with two-dimensional (2-D) and three-dimensional (3-D) analyses. Therefore, the text can be used for both intermediate and advanced courses in biomechanics, ensuring continuity of terminology from year to year. In an intermediate-level course, it is often unnecessary to perform complex 3-D methods to answer particular biomechanical questions. Therefore, suitable methods are outlined using two dimensions alone. If the motion under study is not planar, appropriate 3-D methodologies are included.

Part II pertains to the kinetic analysis of human motion—*kinetics* being the study of causes of motion.

In general, this means the quantification of forces and the work, impulse, and power produced by forces. One chapter describes how to obtain various body segment parameters, such as mass, center of gravity, and moment of inertia. Not only are segmental parameters derived, but also methods for determining the total body's center of gravity and moment of inertia. As in part I, part II contains chapters on 2-D and 3-D kinetic analyses. Furthermore, methods for measuring forces and moments of force both directly and indirectly are presented. In biomechanics, it is rarely possible to directly measure forces in muscles and ligaments. These forces are estimated using a process called inverse dynamics. A method to perform inverse dynamics is outlined systematically, with explanations about its limitations and interpretation.

Part III contains four chapters about electromyographic kinesiology, muscle modeling, computer simulation, and signal processing. Electromyography (EMG) is the field of study that records and interprets the electrical signals produced by skeletal muscles when they are recruited to produce force. This discipline offers a direct way of determining the sequence of muscle activities and, therefore, explains how the brain and peripheral nervous system act together to coordinate human movements. Muscle modeling is a method of indirectly estimating the forces produced by individual muscles during a movement sequence. This is important, because it is not possible to measure these forces directly without using surgical intervention. Computer simulation involves the process of forward dynamics, in which a model of the musculoskeletal system predicts the kinematic motion from a set of initial conditions and prescribed kinetic patterns. Simulation allows researchers to explore optimal movement patterns, the effects of possible surgical interventions, and the role of specific muscles to a motion sequence. The final chapter, on signal processing, outlines technologies used when noise is present and needs to be removed. This technological area is of special interest when smoothing displacement signals prior to double differentiation (to obtain acceleration) or to analyze the frequency characteristics of EMG signals from muscle contractions.

The text concludes with a summary of SI units, conversion factors for converting measurements from American to SI units, an outline of basic electronics, outlines of vector and matrix mathematics, derivations and numerical integrations of double pendulum equations, and two computer subroutines for signal analysis. The book's glossary defines biomechanical terminology, and the index provides a quick reference for finding essential biomechanical concepts outlined in the text.

Biomechanics Analysis Techniques: A Primer

Gary Kamen

Every scientific discipline uses a unique set of tools that new scientists must master before they can contribute knowledge to that discipline. A molecular biologist could not begin to conduct research without knowing how to use and interpret information from a spectrophotometer or a gas chromatograph. A geologist intent on studying a volcano must select appropriate oscilloscopes, amplifiers, and seismographs, store the signals these instruments record, and later analyze those signals using signal-processing techniques.

Just as knowledge of basic Newtonian physics and mastery of instrumentation and analytical techniques are prerequisites for conducting research in molecular biology and geology, so too are they required in the study of biomechanics. This book is a comprehensive resource on the tools needed to conduct research in biomechanics.

WHAT TOOLS ARE NEEDED IN BIOMECHANICS?

Some prerequisite knowledge is necessary to begin applying the principles of biomechanics in research. This text assumes that readers have a basic understanding of geometry, trigonometry, and algebra, including elementary vector algebra. Knowledge of basic mechanics according to Newton's laws is also necessary, although this text provides many examples of their application in the field of biomechanics. A good working knowledge of human anatomy is also

important. This text includes examples that apply to the human musculoskeletal and neuromuscular systems, and knowledge of the constraints imposed by the anatomical system is essential to acquiring a comprehensive understanding of the mechanics involved. Readers seeking additional information on these topics should consult the many textbooks that provide good reviews. A list of suitable readings and texts is included at the end of each chapter.

APPLICATIONS OF THE PRINCIPLES OF BIOMECHANICS: AN EXAMPLE

Of course, thoroughly understanding two- and three-dimensional kinematics and kinetics, anthropometrics, muscle modeling, and electromyography is useless without good research ideas or biomechanical problems to be solved. Consequently, before we begin detailing applied biomechanical principles, let us consider an example that illustrates how we can apply the knowledge to be gained from this text.

Locomotion is the hallmark that distinguishes organisms in the Animal Kingdom from plants, and animals have devised myriad methods to enable movement. Problems related to locomotion constitute a major area of focus for many biomechanists. Ants, despite their small body size, move quickly by moving each leg at the correct velocity. Fish propel themselves efficiently even though they are subject to the large drag force of water. The anatomy of horses constrains their movement, yet they manage to select the right gait for the environmental conditions and the velocity of locomotion. Because the bipedal gait of humans and solutions involving human postural control are a primary focus of this text, let us consider a locomotion and postural control problem that requires biomechanical tools to solve.

On returning home from a difficult day at the office, our subject must ascend the front steps to her home, open the door, and walk inside before setting her briefcase on a table. After climbing the stairs, she reaches for the doorknob, turns it, and finds it locked. She must retrieve her keys from her pocket while juggling her briefcase and not lose her balance. After entering the house, she places the briefcase on a table.

These seemingly simple tasks actually require the successful interplay of a complex system of musculoskeletal design and neuromuscular control. Consider some of the many subproblems to be solved:

- How does our subject climb the stairs? How can we describe the characteristics of the movement at each joint?

- How much muscular force is needed during the transition forward to the next step?

- Where does she decide to stop and reach for the doorknob?

- How does she maintain her balance while reaching for the door?

- How does she first turn the doorknob, then retrieve her keys and unlock the door?

- How does our subject walk into the house with her briefcase? At what point, for example, does the leg planted on the floor begin to move forward again?

- How is the briefcase placed on the table? What prevents it from being placed either too far forward or too far back?

These problems all require solutions gleaned using biomechanical tools. For example, we need to know about the displacements and forces produced at various joints. How do we refer to units of displacement or force so that other scientists can understand us? In the scientific literature, the preferred units of measure are part of an international system—the metric system—whose use has been agreed upon by all scientists. Virtually all biomechanical conferences and journals require the use of this system, so this text uses it exclusively. The system consists of seven fundamental quantities from which all other measurements are derived; for biomechanists, the most

important ones are the kilogram, meter, second, and ampere. The basics of this system are outlined in appendix A.

Kinematic analysis describes the motions we see. As a person approaches stairs, we observe repetitive flexion and extension movements at the hip, knee, and ankle joints. These angular displacements change as the person begins to climb the stairs. Patients with a lower-limb injury may use a different pattern of angular joint displacement to perform these activities. We use displacement information to compute velocities and accelerations during performance of these tasks. The motions we observe may include linear as well as angular motion. Monitoring a point on the person's trunk allows us to describe the instant-to-instant linear displacement and the resulting velocity of walking.

We measure kinematic variables using an imaging system such as a film or video camera or instruments attached to the joint to measure displacement. Some motions may be too complex to describe using simple planar (two-dimensional; 2-D) coordinates. Overuse knee injuries are sometimes ascribed to inappropriate motions of the knee joint in the frontal or sagittal plane. Thus, more complex, three-dimensional (3-D) analysis may be required to adequately describe these motions. Methods for acquiring kinematic data and computing planar kinematics are covered in chapter 1, and chapter 2 considers 3-D kinematics.

Special computers capture the images and compute the trajectories of reflective markers placed over a subject's joint centers and then analyze the motion patterns. These data are then processed to derive various kinematic measures, such as the range of motion of each joint, the velocity and acceleration of each segment, and the path of the center of gravity. Chapter 3 details methods for determining the body's center of gravity. These data can then be synchronized with the ground reaction forces to enable inverse dynamics analysis.

Kinematics describes the movements we see, but to understand why the motions occur as they do we must examine the kinetics, or underlying linear forces and rotational torques, that dictate the kinematic motion. External forces are those caused by interaction of a person with the environment. As our subject walks toward the stairs with her briefcase, she subconsciously forms a plan for how to transition from level surface walking to stair climbing. Each footfall generates a ground reaction force (GRF) that can be measured with appropriately placed instruments, such as force platforms. Force platforms are embedded wherever the researcher wants to measure the mechanical cost of performing the movement task. For example, a force platform embedded in a step will show GRFs that are slightly higher than those generated during walking. The peak vertical forces will be slightly lower than two times the subject's body weight while she is ascending the stairs, but only 25 to 50% higher than her body weight when walking. As our subject opens the door, a force platform will show that her center of gravity has shifted to prevent imbalance. Chapter 4 describes methods of recording and analyzing forces, including how to use force platforms to measure ground reaction forces.

The internal forces and torques generated at each joint in the body can also be approximated using GRFs. GRF patterns change as people ascend stairs because they must overcome the force of gravity (attraction by the Earth) to raise their center of mass with each step. Measuring GRF patterns and segmental kinematics, in conjunction with inverse dynamics, can elucidate the strategies people use to maintain balance and keep internal forces within acceptable levels during movement.

To aid in performing the task of climbing and to help maintain balance, our subject might partially support herself with the handrail. Force transducers mounted between the handrail and its attachment to the wall or ground can quantify the amount of that support. Interaction with the door also requires force, this time in the form of rotational torque applied to the doorknob. Torque is measured by instrumenting the doorknob assembly with strain gauges. Our subject must also apply torque to the unlocked door to make it swing open on its hinges. Finally, placing the briefcase on the table requires of our subject both balance and gradual changes in her application of force to put it in the desired location. Appendix C sets out the basics of electricity and electronic instrumentation with strain gauges and other devices.

After measuring GRFs and computing 2- or 3-D motion patterns, inverse dynamics analysis is used to calculate for each joint the smallest possible force that is necessary to complete each action. It uses Newton's second and third laws to determine what forces and moments of forces exist at each joint. Chapter 5 introduces the theory and methods for performing inverse dynamics analyses for planar motions, whereas chapter 7 develops the techniques used for spatial (3-D) analyses.

The biomechanist can also compute the mechanical cost of the work done and the mechanical power required at each joint. As the velocity of movement increases, greater mechanical power is required. An elderly person may carry out all the necessary tasks for daily life at a much slower pace than a younger person, yet the energy cost to both will be similar and the mechanical power required at each joint will be maintained within the capacity of each joint. Chapter 6 explores how mechanical work, energy, and power are derived from kinematic and kinetic measurements.

The precise forces produced by the muscles and transmitted through the tendons, ligaments, and bones can be directly measured only with indwelling force sensors or estimated by modeling the musculoskeletal system. The actual forces may be larger as a result of the person using an inefficient technique to move the joint or internal stabilizing forces from ligaments used to prevent collapse in the face of unexpected perturbations, but at least the minimal required forces are determined. Alternatively, the activation patterns of the muscles can be measured to better quantify the role of the muscles during performance of the tasks.

Muscle activation can be studied using electromyography (EMG). Typically, sensors are attached to the skin to record the muscles' electrical activity. The input from these sensors is amplified by other instruments and the output signal is digitally stored or viewed on an oscilloscope or other output device. EMG is frequently used in ergonomic evaluation to determine which muscles are under stress and at risk for injury and during sport performance to determine the phase of a motion during which a muscle group is the most active. Chapter 8 covers techniques for recording and interpreting EMG signals.

Why not rely on the information provided by the kinematic or kinetic analyses to predict which muscles are active? The techniques used to describe movement characteristics are not perfect predictors of the underlying activities in muscles. Part of the task of placing a briefcase on a table could be accomplished by allowing gravity to lower the briefcase rather than by activating specific muscles. Moreover, some muscles that are not directly involved in the task may be activated anyway. Walking up the stairs, particularly when carrying something, requires postural stabilization be performed to prevent a fall. Activating other muscle groups stabilizes posture.

Other analytical tools allow scientists to answer questions not directly amenable to measurement techniques. Although kinematic, kinetic, and EMG analyses are indispensable for studying actual movements, the question that remains is whether stairs can be climbed in a more efficient or effective manner. After all, if the goal is simply to accomplish the motor task, it could be done in several different ways, including hopping on one or both legs from stair to stair or crawling up on hands and knees. But what is the optimal way for a person to ascend stairs? Forward dynamics models address such questions by simulating a movement given a set of internally applied forces and torques. Once the model is customized for a specific individual, optimal control techniques are used to find the "best" set of forces and torques needed to accomplish the task. If *best* is defined as the movement pattern that minimizes overall muscular effort, the optimization model finds the patterns of kinetics and kinematics necessary to climb the stairs with the least muscular effort. A different pattern is found if *best* is defined as the movement pattern that presents the subject with the least likelihood of falling while on the stairs. When results from optimization models are compared with the actual movement produced by our human stair climber, ways in which she can improve her performance may come to light. Muscle models that mimic the force-generating capabilities of actual muscles can be used to provide values for the internal forces in these forward dynamics models. Use of these models is essential to biomechanics research because the technology for measuring individual muscle forces is highly invasive and unsuitable for use in most research situations. Chapter 9 reports on the modeling of muscles to better understand their function, and chapter 10 discusses the topics of computer simulation and forward dynamics.

Many different types of data are required to perform these analyses. For example, quantifying the motions of reflective markers placed over joint centers requires that the video data be digitized using high-speed computers to obtain the positions of the body segments during the activities under study. To determine velocities and accelerations, mathematical time derivatives are computed using algorithms that require special smoothing techniques to be applied. The researcher must know which technique is appropriate for use and then evaluate whether the technique was successful. These topics are discussed in detail throughout the text. Chapter 11, in particular, describes various data-smoothing and -processing techniques that result in reliable, noise-free data.

NUMERICAL ACCURACY AND SIGNIFICANT DIGITS

The following chapters have many example problems with numerical solutions, but with the use of modern calculators, the number of significant digits in a numerical answer needs addressing. After performing computations on a calculator, you often have more digits in your display than you care to report. A rule needs to be applied to keep answers reasonable. One convention used in engineering (Beer and

Johnston 1977) holds that to conserve an accuracy of 0.2% (historically based on the accuracy achieved by 10-inch slide rules) one needs three significant digits, unless the first significant digit is a one, in which case four significant digits are reported. For example, the accuracy of the number 456 is 456 ± 0.5 or in percentages $\frac{\pm 0.5}{456} \times 100\% = 0.1096\%$. On the other hand, the number 105, which represents the numbers between 104.5 and 105.5, has a percentage accuracy of $\frac{\pm 0.5}{105} \times 100\% = 0.476\%$. By adding a fourth decimal digit, however, the accuracy increases to $\pm 0.0476\%$, which is well under the 0.2% accuracy threshold. Here are some examples of reporting numbers in this way:

$$1/6 = 0.1667 \quad 4/5 = 0.200 \quad 1/8 = 0.1250$$
$$56\ 300 \quad 145.5 \quad 237 \quad 945 \quad 1.000 \quad 5580$$

Notice that with large or small decimal numbers spaces are added every third position from the decimal. This is the accepted SI format, however, it is permissible to leave out the space for numbers in the thousands, as shown in the last example. The following are incorrect forms:

76, 0.56, 6751, 25.05, 10.064, 932.0, and 22 456.56

Correctly written these numbers are

76.0, 0.560, 6750, 25.1, 10.06, 932, and 22 500

Note, it is good practice to include a zero before the decimal in fractional numbers as illustrated by the second number in the examples above. In complex problems it is often necessary to break up a problem into several steps. Intermediate results must therefore be recorded before reaching the final answer. In such a case, to maintain 0.2% accuracy in the final answer, retain an additional significant figure in all intermediate results and then report the final result rounded to the required number of significant digits (i.e., 3 or 4).

Of course the physical (versus numerical) accuracy of an answer to any problem is based upon the accuracy of the data used in the problem. If a measurement is only accurate to two significant digits then any quantity derived from this measurement is also only accurate to two significant digits. In this text, however, we will assume that all measurements reported in the problems are accurate to three or four digits and therefore all answers must follow the rule outlined above.

SUMMARY

Biomechanical analysis techniques allow us to solve many problems involving the interaction of humans and other animals with the physical environment. Studying the coordinated actions of limbed animals assists engineers in developing robots and vehicles such as those used to explore Earth's moon. Understanding the types of angular motions that might put a joint at risk for injury has allowed us to develop knee braces that limit potentially hazardous joint motions while minimally restricting movement. The analysis of muscle activity during functional movements has contributed to the design of artificial limbs. In the chapters that follow, readers will gain sufficient familiarity with these tools and techniques to begin applying them to real-world problems.

Part I

Kinematics

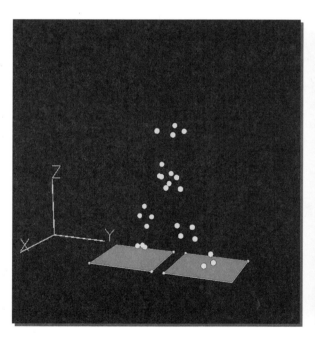
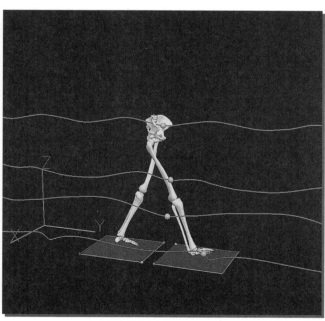

CHAPTER 1

Planar Kinematics

D. Gordon E. Robertson
and Graham E. Caldwell

Kinematics is the study of bodies in motion without regard to the causes of the motion. It is concerned with describing and quantifying both the linear and angular positions of bodies and their time derivatives. In this chapter and the next, we

- examine how to describe a body's position;
- define how to determine the number of independent quantities (called *degrees of freedom*) necessary to describe a point or a body in space;
- define how to measure and calculate changes in linear position *(displacement)* and the time derivatives *velocity* and *acceleration;*
- define how to measure and calculate changes in angular position *(angular displacement)* and the time derivatives *angular velocity* and *angular acceleration;*
- describe how to present the results of a kinematic analysis; and
- explain how to directly measure position, velocity, and acceleration by using motion-capture systems or transducers.

Examples showing how kinematic measurements are used in biomechanics research and, in particular, methods for processing kinematic variables for *planar* (two-dimensional; 2-D) analyses are presented in this chapter. In chapter 2, additional concepts for collecting and analyzing *spatial* (three-dimensional; 3-D) kinematics are introduced.

Kinematics is the preferred analytical tool for researchers interested in questions such as, Who is faster? What is the range of motion of a joint? and How do two motion patterns differ? An important application of kinematic data is their use as input values for inverse dynamics analyses performed to estimate the forces and moments acting across the joints of a linked system of rigid bodies (see chapters 5, 6, and 7). Thus, kinematic analysis may be an end in itself or an intermediate step that enables subsequent kinetic analysis. Whether kinematic variables are the primary goal of a research project or merely the first step in a series of analyses, they need to be quantified accurately.

DESCRIPTION OF POSITION

To quantitatively describe the position of a point or body, we must first identify the tools we will use. Our main tool is the *Cartesian coordinate system,* within which we establish one or more *frames of reference.* One that is desirable but not always necessary to define is an *inertial* or *Newtonian frame of reference,* which is also called an *absolute reference system,* a *global reference system,* or a *global coordinate system* (GCS). This type of reference system is constructed from stationary axes that are fixed in their orientation so that the X-axis is parallel to the floor. The coordinate system is defined by

- an *origin* defined by the 2-D coordinates (0,0) or the 3-D location (0,0,0) and
- two or three mutually orthogonal *axes* (at right angles to each other), each passing through the origin.

In this chapter, we adhere to the GCS axis convention adopted by the International Society of Biomechanics (ISB), shown in figure 1.1. In this definition, the X-axis direction corresponds to the principal horizontal direction of motion. The Y-axis is orthogonal to the X-axis, pointing upward vertically, whereas the Z-axis is a right-perpendicular to the X-Y plane (approximately medial/lateral to the subject). Note that an axis system can be *right-handed* or *left-handed*. The GCS described here is a right-handed system, with the third axis (Z) pointing to the right with respect to the plane formed by the X- and Y-axes (see figure 1.1). Right-handed axes are by convention the most common system.

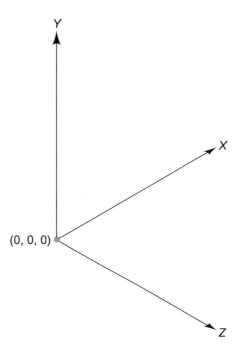

○ **Figure 1.1** A right-handed GCS using the convention adopted by the ISB. The subject's movement progresses along the X-axis.

In other texts, readers will encounter coordinate systems that do not follow the ISB convention. In principle, axis designations are arbitrary and easily understood as long as the researcher defines which one is being used. Some mathematics and engineering textbooks, most force platform manufacturers, and many 3-D biomechanics applications (such as those in chapters 2 and 7) often use GCSs that differ from the ISB specification. In the field of 3-D biomechanics, for example, often the Y-axis corresponds with the principal horizontal direction of movement, the X-axis is orthogonal to the Y-axis in the horizontal plane (approximately medial/lateral to the subject); the Z-axis is a right-perpendicular to the horizontal X-Y plane, pointing upward vertically. To familiarize

readers with some of these differences, we chose to use various conventions in this text. In each chapter, the convention adopted is clearly stated. As stated earlier, the system in this chapter adheres to the ISB convention, and we use the X-Y plane to discuss sagittal plane movement. This convention is the most commonly used 2-D system in the biomechanics literature.

The designation of the origin is the cornerstone for our ability to quantify position within the GCS. Any point in the GCS can be described by its position in relation to the origin, given by its coordinates in 2-D (X,Y) or 3-D (X,Y,Z), as shown in figure 1.2. Although the exact location of the origin is arbitrary, in biomechanics it is usually placed at ground level in a convenient location with respect to the motion studied. For example, when a force platform is used, the center or a corner of the force platform is a suitable location for the origin.

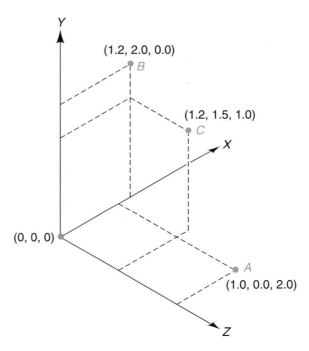

○ **Figure 1.2** Example of point locations and Cartesian coordinates in the GCS depicted in figure 1.1. Points are located using their (X,Y,Z) coordinate triplets. Point A is on the X-Z "floor" of the GCS, 1 unit along the X-axis and 2 units along the Z-axis. Point B is on the X-Y "wall" of the GCS, 1.2 units along the X-axis and 2 units along the Y-axis. Point C is located 1.2 units along the X-axis, 1.5 units along the Y-axis, and 1 unit along the Z-axis.

Now that we have established our reference system, we can use it to describe the location of any point of interest, such as the positions of markers affixed to the subject at palpable bony landmarks on or near joint centers. However, to describe the position of an object, or rigid body, rather than a specific point, we need to specify additional information. First, we must describe the location of a specific point on or within the object, such as the coordinates of its center of mass (see chapter 3) or its proximal and distal ends. In addition, because the object has a finite volume and shape, we must describe its orientation with respect to our established reference axes. To do this, we establish a second reference frame that has its origin and axes attached to, and therefore is able to move with, the body (figure 1.3). In general, this is known as a *relative* or *local coordinate system* (LCS); when applied to a human body segment, this system may also be called an *anatomical, cardinal,* or *segmental coordinate system*. Often, the LCS origin is placed at the segmental center of mass or at the proximal joint center, and the axes are aligned to roughly coincide with the GCS when the subject is in the anatomical position.

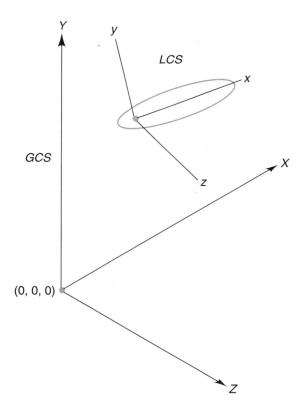

○ Figure 1.3 LCS attached to an object located within the GCS.

The relative positioning of the LCS axes with respect to the GCS defines the orientation of a rigid body or segment. At least three angles of rotation are needed to describe the LCS orientation in 3-D. Several different sets of angles can be used to specify the LCS orientation, but each of these sets contains three independent angles. For example, the system used by the aircraft industry employs the terms *yaw, pitch,* and *roll.* Yaw is a left-right rotation of a plane in flight, pitch is a nose up-nose down motion, and roll is rotation about the plane's long axis. These rotations correspond to rotations about the Y- (vertical), Z- (lateral), and X- (anteroposterior) axes defined in the ISB's GCS convention. More complete descriptions of the LCS and 3-D angles of rotation are given in chapter 2.

DEGREES OF FREEDOM

From the preceding discussion, we see that the location of a given point in space can be described by the three pieces of information contained in its coordinate location (X,Y,Z). The complete description of a rigid body, however, requires six pieces of information: the (X,Y,Z) location of its center of mass and the three angles that describe its orientation. The number of independent parameters (pieces of information) that uniquely define the location of a point or body is known as the object's degrees of freedom (DOF). Thus, a point has three DOF, whereas a rigid body has six DOF (figure 1.4a).

Although the complete description of motion involves spatial (3-D) movement, in many cases human motion can be described primarily in one specific plane. For example, walking and running involve relatively large excursions of segments within the sagittal plane defined by the X- and Y-axes of the GCS. Movements in the frontal and transverse planes exhibit less range of motion during walking or running. Therefore, many of the essential details of motion for walking and running can be determined from a sagittal plane analysis. This simplifies measurement, analysis, and interpretation for describing a movement and serves as an excellent starting place for understanding movements that are mainly planar in nature. The immediate advantage is a reduction in DOF from three (X,Y,Z) to two (X,Y) to describe the position of a point. For a rigid body in two dimensions, the DOF are reduced from six to three, with only two coordinates (X,Y) and one angle (θ) serving to locate the object (figure 1.4b) in the plane. The remainder of this chapter focuses on planar (2-D) kinematics. However, many of the concepts raised within the context of planar kinematics also apply to spatial (3-D) kinematics, as described in chapter 2.

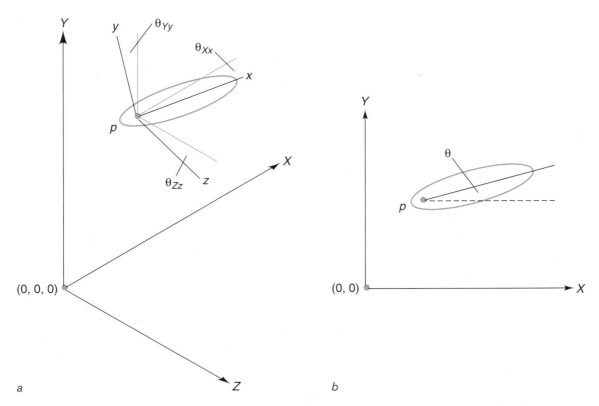

Figure 1.4 *(a)* DOF in a 3-D coordinate system. The segmental endpoint *p* can be described by its (X,Y,Z) coordinates and thus has three DOF. To describe the position of the segment itself, three angles (θ_{Xx}, θ_{Yy}, and θ_{Zz}) that describe the orientation of the LCS axes x, y, and z with respect to the GCS axes X, Y, and Z must be specified. Thus, in 3-D, a rigid body has six DOF (X, Y, Z, θ_{Xx}, θ_{Yy}, and θ_{Zz}). *(b)* DOF in a 2-D coordinate system. The segmental endpoint *p* can be described by its (X,Y) coordinates and thus has two DOF. To describe the position of the segment itself, an angle θ describing the orientation of the segment with respect to the GCS axes X and Y must be specified. Thus, in 2-D, a rigid body has three DOF (X, Y, and θ).

KINEMATIC DATA COLLECTION

The most common method for collecting kinematic data uses an imaging or motion-caption system to record the motion of markers affixed to a moving subject, followed by manual or automatic digitizing to obtain the coordinates of the markers. These coordinates are then processed to obtain the kinematic variables that describe segmental or joint movements. The most common imaging systems use video, digital video, or charge-coupled device (CCD) cameras (e.g., APAS, Elite, Motion Analysis, Peak Performance, Qualisys, and Simi). They record motion using ambient light or light reflected by markers affixed to the body (figure 1.5). In laboratory settings, the cameras have their own lights and the markers have reflective tape that amplifies the marker's brightness compared to the skin, clothing, and background. Other video systems use infrared light or infrared cameras to identify marker locations. Some systems use reflective infrared light (as does Vicon), whereas others (e.g., Optotrak) use active infrared light-emitting diodes (IREDs).

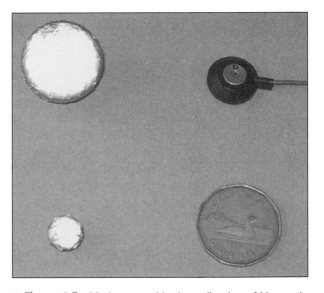

Figure 1.5 Markers used in the collection of kinematic data with imaging systems. Shown on the left are two typical passive reflective sphere markers used with video systems. On the right is an active IRED marker that pulses infrared light to signify its location to a base sensing receiver. A Canadian dollar (a "Loonie") is shown for size comparison.

Active marker systems require a control unit that pulses the individual IREDs in sequence for correct marker identification.

To study planar motion, a single camera placed with its optical axis perpendicular to the plane of motion is sufficient. However, many laboratories use multiple cameras to record 3-D coordinates from both sides of the body (figure 1.6). Locating 3-D coordinates actually requires only two cameras. However, because markers may be blocked by a body part or rotate out of the line of sight of either camera, multicamera systems grant a view of each marker by at least two cameras throughout the movement. Thus, a multicamera system is advantageous even for studying planar movements. In addition, with multicamera systems, the exact placement and orientation of each camera are not critical, as readers will see in the discussion of calibration that follows.

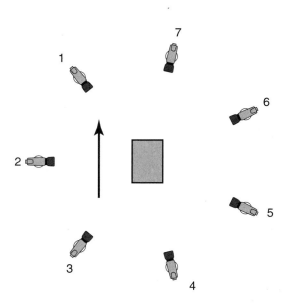

○ **Figure 1.6** Typical multicamera (seven) setup found in a laboratory for studying human motion. This view is from directly overhead, looking down at the laboratory floor with its imbedded force plate (gray rectangle). The subject movement direction is indicated by the arrow.

One of the advantages of modern imaging systems is that most of them have automated digitizing that quickly calculates and displays the coordinate position data from multiple markers throughout an entire movement sequence. Before the advent of such systems, 16 mm cinefilm was used to record human motion. Cinefilm has a number of advantages over video, including finer resolution and a wide range of shutter and camera speeds. Cinefilm analysis, however, requires time-consuming manual digitizing of coordinates, with a few seconds of film

taking hours to digitize. When coupled with the delays associated with film processing, long turnaround times are the norm. Furthermore, because there is no way to view the film during or immediately after recording, errors in photogrammetry (focus, lighting, shutter speed, and so on) or camera misalignment are discovered long after the subject has left the laboratory. Video systems permit real-time viewing of subjects and immediate replay to check the veracity of the recorded images.

PRINCIPLES OF PHOTOGRAMMETRY

According to the American Society for Photogrammetry and Remote Sensing (1980), *photogrammetry* is the "art, science, and technology of obtaining reliable information about physical objects and the environment through processes of recording, measuring, and interpreting images...." Biomechanists mostly concern themselves with the factors essential for obtaining clear images on photographic or videographic media. Few biomechanists today use photography or cinematography to collect data because of their high cost and long processing and manual digitizing times. Videography is a much more common method, but with all these media, many of the same principles are encountered in the quest to create usable photographic images. In this section, we discuss the factors most important to the biomechanics researcher. One issue, perspective error, is discussed in a later section, Calibration of Imaging Systems. More detailed information about photography, cinematography, and videography, such as the Additive Photographic Exposure System (APEX), must be obtained from specialized sources.

Field of view is defined as the rectangular area seen by the recording medium (film, video) after passing through the camera's optics (see figure 1.7). Care must be taken to ensure that the movement falls within the camera's field of view. Motion before and after the period of interest also must be recorded to prevent inaccuracy near the ends of the coordinate data record resulting from the smoothing process (see chapter 11, as well as Signal, Noise, and Data Smoothing later in this chapter). Unfortunately, this may reduce the size of the subject's image, which should be as large as possible to improve the signal-to-noise ratio. A less obvious problem is that the digitizing system used to quantify the motion may reduce the recorded field of view through a slight magnification that effectively "hides" markers as they approach the edges of the field of view. Furthermore, motion at the edges of the image may be distorted as a result of poor optics or the use of a wide-angle lens. It is generally a good idea to

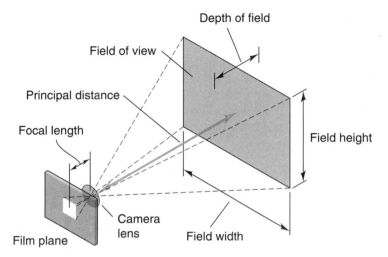

° **Figure 1.7** Photographic dimensions.

ensure that the trajectories of markers do not pass near the edges of the field of view of each camera in the system.

One of the most important concerns when recording the motion of markers is the exposure time, which in a photographic system is related to the camera speed and shutter speed. *Exposure time* is the duration of time that the recording medium is exposed to light passing through the camera's lens. *Camera speed,* also called the *frame rate,* is how fast the camera records images on the medium. Typical video cameras record 30 frames per second (fps) in NTSC format (North America) and 25 fps in SECAM or PAL format (Europe), depending on the line frequency of the electrical system (60 Hz in North America, 50 Hz in Europe). Standard cine cameras used for filming movies record at 24 fps, but cameras typically used by biomechanists record from 10 to 500 fps. Specially designed video cameras record at rates such as 60, 120, and 240 fps, and expensive optical systems reach 2000 fps. Doubling the line frequency or recording two or more images per frame achieves these rates.

Another technique used by most video digitizing systems to double the frame rate is interlacing, which means that only half of the lines on a video screen are refreshed each 1/60th (North America) or 1/50th (Europe) of a second (s). Separating the odd and even screen lines into two fields reduces the image quality, but doubles the frame rate, producing artificial frame rates of 60 or 50 fps. This also means that markers appear to shift up and down at the rate of 30 or 25 Hz. Low-pass filtering the data or digitizing a stationary marker eliminates this problem. This doubling technique is recommended for recording fast motions, but when slow motions are

to be analyzed, use the full video frame for better picture quality.

Clearly, the faster a camera records images, the shorter the exposure time is. If the camera is running at 30 fps, the exposure time must be less than 1/30 s. Most cameras use a shutter that can change the exposure time to as little as 1/2000 s. When markers are poorly lit, these brief exposure times may reduce visibility and prevent automatic digitizers from locating the marker's coordinates. On the other hand, a long shutter speed (e.g., 1/60 s) makes a fast-moving marker look like a streak across the screen instead of a single spot. In general, shutter speeds of 1/500 or 1/1000 s prevent streaking and make digitizing the coordinates of the marker more reliable. Sufficient light must be available to use these exposure times. Using a camera-mounted light and reflective markers will help to ensure visibility, as will larger markers, brighter lights, and moving the camera or, preferably, the lights closer to the markers. When recording new types of motion, run a pilot test using different shutter speeds, marker sizes, and lighting arrangements to determine the optimal setup.

The final major photographic consideration concerns *focus* and *depth of field,* which is the distance in front of and behind the subject that is considered to be "in focus." The camera's distance setting should be set equal to the distance between the lens and the object being filmed. This distance is called the principal distance. This number is marked on the camera lens and may be set manually or determined by autofocusing. Autofocusing generally is not useful in biomechanics because the camera may focus on the background or some other object if the subject is not in the field of view when recording starts. Then, as the subject enters the field of view, the optics refocus and temporarily blur the image until the camera can focus on the moving subject. Thus, when using autofocus, it is best to focus with the subject standing in the middle of the field of view and then turn off the autofocusing system. Alternatively, measure the distance from the camera to the plane of motion and then set this distance manually on the lens.

Although generally not a factor with video cameras, a camera's aperture setting has several important influences on picture quality and depth of field. Modern video cameras have automatic irises that open and close like eyes, permitting more or less light through the lens to produce a correctly exposed image. If too much light enters the lens during too long an exposure time, the image will be *overexposed*, making digitizing difficult or impossible. For example, if two markers are close together in an overexposed image, they may appear to be one marker. In contrast, if too little light is available because of poor lighting or brief exposure times, the markers will be *underexposed* and may be indistinguishable by the digitizing system. By changing the aperture of the lens, more or less light is permitted to reach the film. The *aperture* of a lens is the size of the opening of the lens's iris. The *iris* typically is a set of "leaves" that open and close to change the amount of light that passes through the lens. The *f-number* or *f-stop* of a lens is the ratio of the aperture of the lens divided by the *focal length*, which is the apparent distance between the front of the lens and the recording medium. As the iris closes, less light passes through the lens. Even with a completely open iris, however, some light is lost between the front of the lens and the back because of imperfections in the optical glass, filters, lens coatings, and the refraction and reflection of light within the lens barrel.

Each lens is marked with its maximum possible aperture. For example, a standard lens may be rated with an f-stop of 2, which means it will reduce the light by one-quarter with its iris fully open. Standard f-stops are based on powers of the root of two and rounded off to simplify their application. Table 1.1 shows standard f-stops and the amount of light each one restricts through the lens. At each higher f-stop, the amount of light permitted to reach the film is halved. Thus, increasing the f-stop from f/2 to f/2.8 (one f-stop) reduces the light by 1/2. Conversely, each decrease in f-stop, such as from f/11 to f/8, doubles the amount of light that reaches the film. Zoom lenses, as a result of their complex structure, reduce the amount of light that passes through them by between 1/4 (two f-stops) and 1/32 (five f-stops).

Changing the aperture also affects the depth of field. As the iris closes down (is changed to a higher f-stop), there is less curvature in the part of the lens that focuses the image. This keeps the image in focus better and widens the depth of field. Some lenses show the relationship between the depth of field and aperture, making it possible to determine what region on either side of the focal distance will be in focus in the image. This usually is not possible on zoom lenses, however, because the focal length of the lens also affects the depth of field. For example, lenses with long focal lengths—*telephoto lenses*—magnify the image and reduce the depth of field. Conversely, *wide-angle lenses* reduce the image size, allowing larger fields of view but increase the depth of field. However, wide-angle lenses should be avoided for biomechanics research because they distort the image and create artificial motions that must be corrected with complex transformations.

CALIBRATION OF IMAGING SYSTEMS

For any system of collecting kinematic data, a suitable means of calibration must be used to ensure that the image coordinates are correctly scaled to size. For

TABLE 1.1   *Standard Photographic Apertures, Exposure Times, and Film Speeds*

APEX value[1]	0	1	2	3	4	5	6	7	8	9	10
Aperture (f-stop)	1	1.4	2	2.8	4	5.6	8	11	16	22	32
Exposure time (s)	1	1/2	1/4	1/8	1/15	1/30	1/60	1/125	1/250	1/500	1/1,000
Film speed (ISO or ASA)[2]	3	6	12	25	50	100	200	400	800	1600	3200

[1] APEX stands for the Additive Photographic Exposure System. Each increase in APEX value indicates a decrease in the light level by one-half.

[2] Each faster film speed requires half the amount of light that the previous speed needed for proper exposure.

imaging systems, there are two basic methods. For 2-D systems with one camera, the simplest method is to use a calibrated ruler or surveyor's rod placed in the subject's plane of motion. The camera must be perpendicular to the plane of motion; otherwise, there will be linear distortions of the types shown in figure 1.8. By digitizing the length of the ruler, a scaling factor (s) is determined for scaling in both directions (X,Y), that is,

$$s = \frac{\text{actual length (meters)}}{\text{digitized length (arbitrary units)}} \qquad (1.1)$$

$$x = su \qquad (1.2)$$

$$y = sv \qquad (1.3)$$

where u and v are the digitized coordinates of a marker and x and y are the scaled coordinates.

The preferred method of calibration—essential for multicamera systems—is to establish a series of *control points*. Control points are markers, attached to a structure or affixed to the film site or laboratory, whose exact coordinates are known. For example, figure 1.9d shows a grid of 15 points on a board used to calibrate planar motion across a laboratory walkway. For 2-D analyses, at least four noncollinear points are required. At least six noncoplanar locations on a 3-D structure are needed for 3-D analyses.

Figure 1.9, b and c, shows several 3-D structures with control points for calibrating in three dimensions. Other types of control-point structures are also used, such as Woltring's (1980) method that uses several views of a plane of markers. Some commercial systems film a calibrated wand (figure 1.9a) that is wafted around the volume of the movement space (Dapena, Harman, and Miller 1982).

After filming the control points, equations are computed to scale the digitized coordinates into real metric units. In single-camera 2-D setups, image distortion resulting from misalignment of the camera's optical axis perpendicular to the plane of motion can be corrected. The common method of enabling the transformation of the data from digitized coordinates to real metric units is called *fractional linear transformation* (FLT) when applied to two dimensions or *direct linear transformation* (DLT) when applied to three dimensions (Abdel-Aziz and Karara 1971; Walton 1981; Woltring 1980). The digitized coordinates that result are said to be *refined* rather than *scaled* because the data are altered in a more complex way.

When a camera is tilted with respect to the plane of motion, distances are distorted as illustrated in figure 1.8b. FLT and DLT correct these types of errors. In 2-D analyses, however, *perspective errors* may occur in which objects appear to shorten as

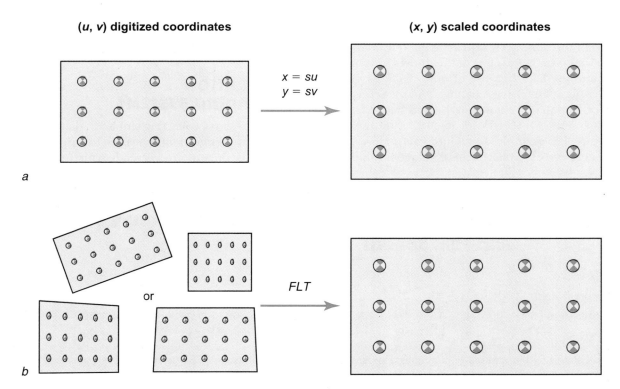

(u, v) digitized coordinates

(x, y) scaled coordinates

$x = su$
$y = sv$

a

FLT

or

b

○ **Figure 1.8** Comparison of scale factor method *(a)* and fractional linear transform method *(b)* for transforming digitized images to real-life measurements.

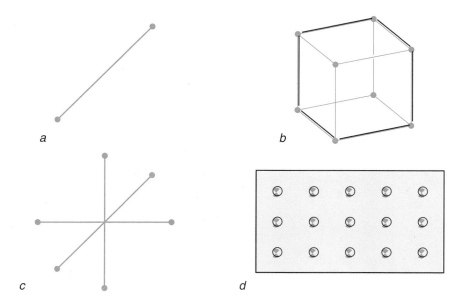

a

b

c

d

Figure 1.9 Types of calibration structures for 3-D motion calibration.

they move farther from the camera. One way to minimize this effect is to use a telephoto lens and to zoom in on the subject to make the motion fill as much as possible of the camera's field of vision. This technique flattens the subject, reducing perspective errors. Of course, one needs a large enough room and clear lines of view to use this method. The FLT equation takes the form

$$x = \frac{c_1 u + c_2 v + c_3}{1 + c_7 u + c_8 v} \qquad (1.4)$$

$$y = \frac{c_4 u + c_5 v + c_6}{1 + c_7 u + c_8 v} \qquad (1.5)$$

where c_1 to c_8 are the coefficients of the FLT, u and v are the digitized marker coordinates, and x and y are the marker's refined coordinates in metric units. Note that when the camera's optical axis is very close to perpendicular to the calibration plane, the coefficients c_2, c_4, c_7, and c_8 are nearly zero. The coefficients c_1 and c_5 become scaling factors, similar to those that would have been digitized if a ruler was used. The reason these numbers are not equal is because most digitizers have different scales for horizontal and vertical. The coefficients c_3 and c_6 correspond to the differences between the origin of the calibration system and the origin of the digitization system.

To check the accuracy of these systems, compare the actual coordinate locations of the control points with their refined (i.e., transformed) coordinates. The standard error or root mean square (RMS) of the differences between the true coordinate locations

and their digitized and refined locations reflects the accuracy of the system. Better still, film, digitize, and refine a second set of known coordinates to compute the system's accuracy.

TWO-DIMENSIONAL MARKER SELECTION

We have described how a Cartesian coordinate system locates a point in space and how imaging systems record the location of reflective markers. Logical questions for a biomechanist interested in studying human movement are as follows:

- Where do I place these reflective markers on my subject?
- How many markers should I use?
- Do I need other markers (not on the subject) in the camera's field of view?

The answers to these questions vary depending on the nature of the movement under study and the exact research questions being asked. An excellent starting point is to construct a model of the important anatomical parts involved in the motion. For planar motion, this model can be represented by a stick figure, as shown in figure 1.10. For example, the model for a person running might include body segments representing the foot, leg, and thigh of each lower extremity; the upper and lower portions of each upper extremity; and a trunk segment. For a seated cyclist, a segment representing the bicycle crank arm might be needed along with body segments representing the foot, lower leg, and thigh

○ **Figure 1.10**   *(a)* A simplified sketch (left) and stick-figure representation of a runner. *(b)* A simplified sketch (left) and stick-figure representation of a cyclist.

of only one lower extremity, plus the trunk. The simplified model for the cyclist is possible because the bicycle crank imposes symmetrical motion of the legs. Also, if the cyclist is instructed to keep his hands stationary on the handlebars, upper extremity motion is minimal and can be excluded from the model.

Once a suitable model has been constructed, it can guide the researcher in marker selection and placement. At least two points must be quantified for each segment in which planar motion is to be modeled. Often, markers are placed at the estimated centers of rotation at each end of the segment or over proximal and distal anatomical landmarks (defined in chapter 3). Two points are needed to define the planar angular orientation of the segment (see Angular Kinematics later in this chapter). Therefore, for our cycling example, roughly 10 markers would suffice. For running, however, we would need more markers to properly represent the body (figure 1.11). Additionally, if we were recording the runner's GRFs, we would need another marker affixed to the force plate to correctly locate the position of the force vector on the runner's foot (see chapters 4 and 5).

The exact number of markers needed also depends on the nature of the 2-D motion at the joints between adjacent segments. A simple *pin joint* or *hinge joint* (e.g., phalangeal joints of fingers and toes) has only one degree of freedom—the freedom to rotate about the pin or hinge. Note that here, *degrees of freedom* has a slightly different meaning than in our description of quantifying the position of a point or rigid body. The hip is also modeled as a hinge joint for planar analysis, despite its ball-and-socket con-

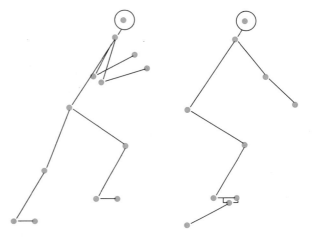

○ **Figure 1.11**   Markers needed for representing a runner (left) and a cyclist.

struction that allows three different spatial rotations. For hinge joints, the endpoint marker for adjacent segments can be placed directly over the point representing the hinge (figure 1.12). Other joints are more complex; the knee joint, for example, permits flexion-extension and some translation across the tibial plateau (called *shear*). This means that the knee has two degrees of freedom (one for rotation and one for translation). In such situations, it is impossible to place a marker that will always represent the position of the moving hinge, and separate endpoint markers must be used for the adjacent segments (figure 1.12). In some cases, researchers choose to ignore small translational motions and simply model the joint as a hinge. They must select an appropriate level of detail for the specific research question being studied. See chapter 3, tables 3.1 and 3.3, for

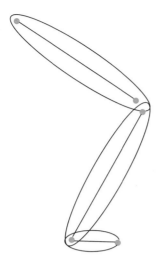

○ **Figure 1.12** A joint's possible motions influence the selection of marker locations. The ankle and hip joints can be modeled as hinges and designated with one marker. Two markers can be used to designate the distal end of the thigh and proximal end of the leg, because at the knee both rotation and translation are possible.

common marker locations used in 2-D studies and for measurements for computing the locations of "virtual" points without actual markers (e.g., segment centers of rotation and centers of mass).

MARKER-FREE KINEMATICS

In many situations, it is impractical, impossible, or undesirable to place markers on the subject performing the motion, such as athletes in competitions and patients in clinical situations who are unable to endure an extended experimental preparation period. In these situations, the researcher must digitize the movement record manually, using identifiable anatomical landmarks to locate the points necessary for the movement model. Recently, however, efforts have been made to develop automatic, marker-free motion-analysis systems (D'Apuzzo 2001; Trewartha, Yeadon, and Knight 2001). Such a system would be a major breakthrough in the analysis of human motion, because it would permit shorter data-acquisition sessions in clinical and research laboratories and extend data collection beyond the laboratory to more ecological environments.

Marker-free systems use computer graphics techniques to match the shape of a body segment with a predetermined shape (Trewartha, Yeadon, and Knight 2001). Initially, a model of the body that matches the general size and morphology of the subject is constructed. A video of the moving subject is recorded and converted to digital form.

Computer software then attempts to align the computer model with the image of the actual performer in each frame of video data. If the computer can find an acceptable position for the model in every frame, that effectively replicates the motion of the subject. The software extracts the positions of individual points and segments from the model, effectively *digitizing* the locations of desired points. Although such systems are now only in development, it is only a matter of time before they will be regularly used for human movement analyses.

LINEAR KINEMATICS

The Cartesian coordinate system permits quantification of the position of a point or rigid body in either 3-D space or on a 2-D plane. It is also the starting place for a complete kinematic description of any human motion. In this section, we introduce the kinematic variables of *displacement, velocity,* and *acceleration,* all of which describe the manner in which the position of a point changes over a period of time. The mathematical processes of differentiation and integration—the main concepts of calculus—relate these kinematic variables. Displacement is defined as the change in position. *Velocity* is the time derivative of displacement, defined as the rate of change of displacement with respect to time. *Acceleration* is the time derivative of velocity, defined as the rate of change of velocity with respect to time. Acceleration is therefore the second derivative of displacement with respect to time. These three kinematic variables can be used to understand the motion characteristics of a movement, to compare the motion of two different individuals, or to show how motion has been affected by some intervention. Sometimes, determining the time derivative of acceleration, called *jerk,* is desirable to assess the severity of head impacts during car crashes or collisions with surfaces or projectiles. Table 1.2 lists the commonly used kinematic measures and their associated symbols and units in the International System of Units (SI), including angular kinematic variables, which are discussed later.

In some situations, acceleration can be measured directly with a device aptly called an accelerometer. Integral calculus is then used to find the velocity and displacement data. Velocity is the first integral of acceleration with respect to time, whereas displacement is the integral of velocity with respect to time. Information on accelerometers is presented later in this chapter, and methods of integration are detailed in the section on impulse-momentum in chapter 4. The remainder of the present discussion concentrates on the process of differentiation.

TABLE 1.2   *Kinematic Variables and Their SI Units and Symbols*

Measure	Definition	Units	Symbols
Linear position, path length, or linear displacement		**m**, cm, km	*x, y, z, s* (arc), *d* (moment), *r* (radius)
Linear velocity	*ds/dt*	**m/s**, km/h	*v*
Linear acceleration	*dv/dt, d²s/dt²*	**m/s²**, *g* (= 9.81 m/s²)	*a*
Linear jerk (jolt)	*da/dt*	**m/s³**, *g*/s	*j*
Angular position, plane (2-D) angle, or angular displacement		**rad**, °, r (r = revolution)	θ, β, γ, φ
Angular velocity	*dθ/dt*	**rad/s**, deg/s, r/s	ω
Angular acceleration	*dω/dt, d²θ/dt²*	**rad/s²**	α
Solid (3-D) angle		**sr** (steradian)	Ω

Common unit is in **bold**.

COMPUTING TIME DERIVATIVES (DIFFERENTIATION)

There are several ways to compute the time derivatives of displacement once the position of a point has been established as a function of time for a particular movement. The starting point for the kinematic analysis is the coordinate data that were digitized at equal time increments throughout the movement sequence. The exact number of data points and the time increment depend on the duration of the movement and the sampling rate of the motion-capture system. For example, a 2 s movement captured at 200 fps yields an input data stream of 400 numbers, with each point separated by 5 milliseconds (ms; 0.005 s). For the differentiation process, the X and Y coordinates are treated separately, meaning that these 2 s of motion produce 400 X positions and 400 Y positions for each marker digitized in the motion sequence; the earlier example of a runner with 20 markers would therefore generate 40 separate data streams, each one 400 numbers in length. These data are often listed in row-by-column format, with the columns containing the X and Y coordinates of different markers and the rows showing the incremental steps in time throughout the movement. Table 1.3 is a portion of such a data table.

There are three categories of methods for computing derivatives. Analytical methods involve the differentiation of mathematical functions and are taught in high school and college differential calculus courses. Graphical methods use the concept of the instantaneous slope (rise over run) of a graphed function. Finally, numerical methods apply relatively simple computing formulae to a set of data points that represent any varying function. All three categories are used in biomechanics, as each have strengths and weaknesses. To use analytical techniques, the positional data collected at equal time increments must first be fitted to an appropriate mathematical function. Once the position data are in equation form, analytical techniques produce equations that represent the corresponding velocity and acceleration patterns. A strength of this procedure is that it generates velocity and acceleration data that are free of numerical errors. Graphical methods are slow and do not lend themselves to either numerical or analytical data forms, but the ability to graphically differentiate is of great value in checking the validity of results from the other two computation methods. The most common approach is numerical differentiation, primarily because of the manner in which experimental data are collected. The columns of position coordinates equally spaced in time are in the precise format needed for applying numerical techniques. However, if not used carefully numerical differentiation methods can result in computational errors. The following sections describe the steps necessary to minimize these computational errors.

There are a variety of different numerical derivative formulae. The equations presented here to obtain velocity and acceleration values are examples of *finite difference calculus*. The method used is the *central difference* method.

TABLE 1.3   *Sample Marker Data From a Walking Trial*

Frame #	Time	Right shoulder		Right hip		Right knee		Right ankle		Ball of right foot	
		X (cm)	Y (cm)	X (cm)	Y (cm)	X (cm)	Y (cm)	X (cm)	Y (cm)	X (cm)	Y (cm)
1	0.000	−7.3	154.3	−6.1	92.4	−3.8	50.2	−8.6	13.7	1.3	1.1
2	0.020	−7.2	154.3	−6.3	92.2	−3.0	50.1	−8.6	13.8	1.3	1.1
3	0.040	−7.0	154.2	−6.7	91.8	−2.0	49.9	−8.5	13.8	1.3	1.1
4	0.060	−6.7	153.9	−7.2	91.3	−0.8	49.7	−8.4	13.9	1.3	1.1
5	0.080	−6.4	153.2	−7.8	90.6	0.6	49.4	−8.2	14.0	1.3	1.1
6	0.100	−6.0	152.0	−8.5	89.6	2.1	48.9	−8.1	14.0	1.3	1.1
7	0.120	−5.6	150.3	−9.4	88.4	3.5	48.5	−7.9	14.1	1.3	1.1
8	0.140	−5.2	148.1	−10.3	86.9	4.9	47.9	−7.7	14.1	1.2	1.1
9	0.160	−4.8	145.4	−11.3	85.2	6.2	47.4	−7.5	14.2	1.2	1.1
10	0.180	−4.3	142.4	−12.3	83.3	7.3	46.8	−7.3	14.2	1.2	1.1
11	0.200	−3.9	139.0	−13.3	81.3	8.3	46.3	−7.2	14.2	1.2	1.1
12	0.220	−3.4	135.5	−14.3	79.2	9.1	45.8	−7.1	14.2	1.2	1.1
13	0.240	−2.9	131.8	−15.2	77.1	9.8	45.3	−7.0	14.2	1.2	1.1
14	0.260	−2.5	128.2	−16.0	75.0	10.4	44.8	−6.9	14.2	1.2	1.1
15	0.280	−2.1	124.6	−16.7	72.9	10.9	44.4	−6.8	14.2	1.2	1.1
16	0.300	−1.7	121.1	−17.4	71.0	11.4	44.1	−6.7	14.3	1.2	1.1
17	0.320	−1.3	117.7	−17.9	69.1	11.7	43.8	−6.7	14.3	1.2	1.1
18	0.340	−1.0	114.4	−18.4	67.3	12.0	43.5	−6.6	14.3	1.2	1.1
19	0.360	−0.6	111.3	−18.8	65.6	12.2	43.3	−6.6	14.4	1.2	1.1

$$v_i = \frac{s_{i+1} - s_{i-1}}{2\Delta t} \tag{1.6}$$

$$a_i = \frac{v_{i+1} - v_{i-1}}{2\Delta t} = \frac{s_{i+2} - 2s_i + s_{i-2}}{4(\Delta t)^2} \tag{1.7}$$

$$\text{or} \quad a_i = \frac{s_{i+1} - 2s_i + s_{i-1}}{\Delta t^2} \tag{1.8}$$

where v_i and a_i are the velocity and acceleration of the marker at time i, Δt is the sampling interval of the data (in seconds), and s is the linear position (*x* or *y* in meters; m). Notice that to obtain velocity and acceleration at a particular instant *(i)*, data from the time intervals before *(i − 1, i − 2)* and after *(i + 1, i + 2)* that instant are required. This is problematic at the start *(i = 1)* and end *(i = n)* of a data stream, because not all of these points are available. One way to obtain a velocity at time *1* or *n* is to use the following *forward* and *backward* difference equations (Miller and Nelson 1973).

$$v_1 = \frac{s_2 - s_1}{\Delta t} \tag{1.9}$$

$$v_n = \frac{s_n - s_{n-1}}{\Delta t} \tag{1.10}$$

$$a_1 = \frac{s_1 - 2s_2 + s_3}{\Delta t^2} \tag{1.11}$$

$$a_n = \frac{s_n - 2s_{n-1} + s_{n-2}}{\Delta t^2} \tag{1.12}$$

A better way to obtain these first and last derivatives is to collect extra data, both before and after the motion of interest. If this cannot be done, *pad* the ends of the data with extra data, take the derivatives, and then discard the padding points. Padding points can be derived by mirroring the data at each end (Smith 1989) or linearly or nonlinearly extrapolating from the endpoints.

SIGNAL, NOISE, AND DATA SMOOTHING

Unfortunately, as a result of errors introduced by the digitization process, numerical calculations of velocity and, in particular, acceleration yield results contaminated with high-frequency noise. Figure 1.13 shows the acceleration pattern of a toe marker that was computed from position data digitized from video. Notice that an irregular pattern occurs even during the period when the marker is supposed to be motionless on the ground (from 0.03 to 0.07 s). No amount of careful digitizing can eliminate this problem.

These noise spikes occur because small errors in the digitizing process represent large accelerations. Finding the second derivative of a pure sine wave analytically illustrates this phenomenon. Mathematically, this produces another sine wave that is phase shifted by 180°. However, if noise is present in the original sine wave, the second derivative is very different. Figure 1.14 compares the second derivatives of a pure sine wave and one to which random or white noise of amplitude ±0.01% (in other words, a signal-to-noise ratio of 10000:1) was added before double differentiation. Clearly, the noise dominates the second derivative signal. For kinematic time derivatives, noise in the position data must be removed to obtain a valid acceleration pattern. There are a number of acceptable *smoothing* methods for removing the high-frequency noise induced by the digitizing process, including low-pass digital filtering, piecewise quintic splines, and Fourier series reconstruction (see chapter 11).

ACCELEROMETERS

One commonly used technology for directly measuring the kinematic variable of acceleration is the accelerometer. There are three types of accelerometers: strain gauge, piezoresistive, and piezoelectric. Piezoelectric accelerometers use the piezoelectric effect to measure acceleration. The piezoelectric effect occurs when certain crystals, such as quartz, are mechanically stressed causing a voltage. Piezoelectric accelerometers typically have higher frequency responses than strain-gauge accelerometers, but they do not have a true static response, so they should not be used for recording slow motions or periods of inactivity. Figure 1.15 illustrates the inner workings of a strain-gauge accelerometer. Tiny strain gauges are attached to a cantilevered beam to measure the bending of the beam. If the beam is subjected to acceleration, the inertial mass at its free end causes the beam to bend in proportion to the imposed acceleration. Clearly, a sharp blow can easily damage such a fine element. Accelerometers are available as uniaxial units for measuring acceleration in one direction or in triaxial packages for measuring acceleration in three orthogonal directions. Note that the calibration of an accelerometer depends on the type of sensing element it has. Strain-gauge and piezoresistive units exhibit a static (DC) response and are calibrated by aligning their sensitive axis within the gravitational field. Because piezoelectric accelerometers lack this DC response, they must be calibrated at the factory using dynamic techniques. Padgaonkar, Krieger, and King (1975)

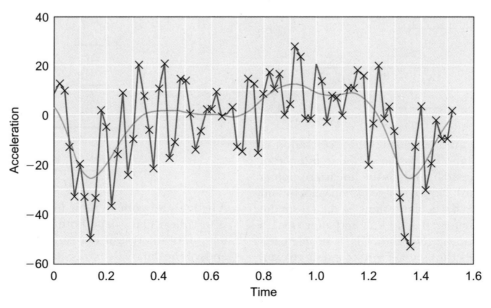

○ **Figure 1.13** Vertical acceleration history of filtered (blue line) and unfiltered (gray line marked with ✕s) toe marker during walking. Notice the unfiltered data varies irregularly around the filtered data.

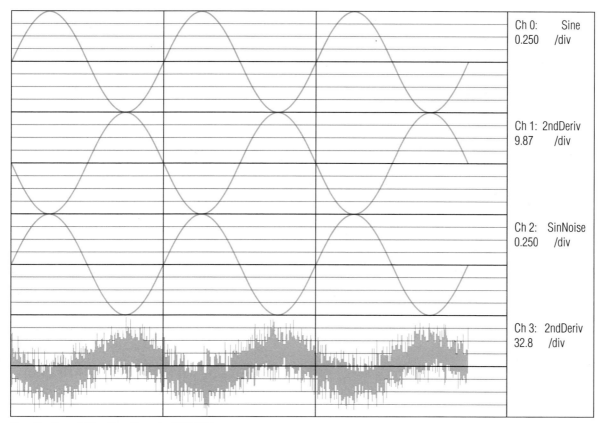

Ch 0:   Sine
0.250   /div

Ch 1: 2ndDeriv
9.87   /div

Ch 2:   SinNoise
0.250   /div

Ch 3: 2ndDeriv
32.8   /div

Sampling time: 3.00 s Sampling rate: 1000.00Hz

o Figure 1.14 Histories of sine wave and its second derivative followed by the same sine wave with white noise added and its second derivative. The top waveform is the original sine wave sampled at 1000 Hz over three seconds. The second waveform is the second derivative of the sine wave. The third waveform is the sine wave with noise added (random numbers of range ±0.0001). The fourth waveform is the second derivative of the sine wave with the added noise.

outline an accelerometric system that can quantify the triaxial linear and angular accelerations (six DOF) of a single segment but requires nine uniaxial accelerometers to achieve stable results.

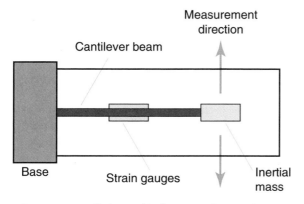

Measurement direction

Cantilever beam

Base

Strain gauges

Inertial mass

o Figure 1.15 Schematic diagram of a strain-gauge accelerometer.

GRAPHICAL PRESENTATION OF LINEAR KINEMATICS

There are several methods of graphically presenting kinematic measures. Plotting formats for linear kinematic variables include *trajectories* (X-Y or X-Y-Z plots), in which the path of the object is displayed, and time series or *histories*, in which variables are graphed as a function of time. Figure 1.16 shows a trajectory and a time series for two segmental markers. Time series representation is the most prevalent plot used in biomechanical studies because it permits comparison of different kinematic features that occur simultaneously. It is also useful for comparison with kinetic variables such as force and biophysical signals such as muscle activity (recorded with EMG). In this way, the motion can be linked with the underlying forces or physiological events. Trajectory plots are useful when studying motion that occurs in both directions within the plane of motion (e.g., the motion of the knee joint during running, which has significant

From the Scientific Literature

Mayagoitia, R.E., A.V. Nene, and P.H. Veltnik. 2002. Accelerometer and rate gyroscope measurement of kinematics: An inexpensive alternative to optical motion analysis systems. *Journal of Biomechanics* **35:537-42.**

These authors used four uniaxial seismic accelerometers and two rate gyroscopes to measure the kinematics of two lower-extremity segments during walking on a treadmill. They simultaneously quantified the motion with a well-calibrated optical system (Vicon). Although accelerometry usually is a real-time system, the complex mathematics required to derive the 2-D kinematics of the segments for comparison with the optical were done offline. The system performed favorably compared to a more expensive optical imaging system. The researchers noted that although their system was tested in a laboratory setting, it is capable, with an appropriate data logger, of collecting motion data in "almost any environment." Of course, it has the usual limitations of electronic systems, such as being unsuitable for measuring more than a few segments at a time rather than simultaneously measuring many segments during whole-body motions.

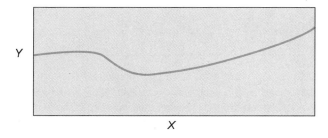

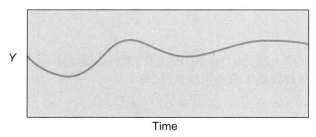

° Figure 1.16 Comparison of *(a)* path trajectory graph and *(b)* a position history.

vertical and horizontal components). They also illustrate how a movement in a secondary direction accomplishes a prescribed movement in another direction. For example, Bobbert and van Zandwijk (1999) demonstrated how the center of mass moves in the anterior/posterior direction during a vertical jump by plotting the X-Y trajectories of the jumper's center of mass.

Ensemble Averages

In biomechanics research articles, authors often want to present data that are pooled across several trials of a particular movement condition. This involves combining the trials for an individual subject or combining the data of several subjects to get a sense of the kinematic patterns produced by the group. In both cases, the patterns produced by the pooling of data are known as *ensemble averages*. The first step in creating these average curves is to standardize the movement time for all the trials, because each one has a different duration. For example, a subject might complete the first trial in 0.8 s and the second in 0.9 s. This *time normalization* is accomplished by identifying a given portion of the motion (performed in all trials) as 100% and rescaling the sampling rate as a percentage of the duration. For cyclic motions such as gait, one complete cycle (right-foot toe-off to right-foot toe-off, for example) is normally the complete scaling period. Alternatively, the stance phase time alone could be used.

To accomplish time normalization, the original data for a given kinematic variable must be *interpolated* to find the values of that variable at specific times. This produces data equally spaced throughout the rescaled movement time (often each 1 or 2% of the motion). For example, if data were collected at a sampling rate of 200 Hz during a movement that lasted 0.9 s (900 ms), there would be 180 data points separated by intervals of 5 ms. If the 900 ms were rescaled to 100%, to identify data values at 1% intervals we would need to find 101 (0 to 100%) data points separated by intervals of 9 ms. To do this, we interpolate between data that are 5 ms apart to estimate the values for data 9 ms apart. This can be done in several ways, with the two most common being *linear interpolation* and *splining* (with *cubic* or *quintic* splines).

For linear interpolation, it is assumed that the collected data were separated by such a small time increment that motion between any two consecutive times was linear in nature. For example, if a marker moved in a given direction from position 1.185 m at time 0.310 s to position 1.190 m at time 0.315 s, linear interpolation computes the following position-time data:

0.310 s	1.185 m
0.311 s	*1.186 m*
0.312 s	*1.187 m*
0.313 s	*1.188 m*
0.314 s	*1.189 m*
0.315 s	1.190 m

For data collected at sufficiently high sampling rates, this method works well. In cases where movements occur rapidly and the sampling rate is not excessively high, a splining method can be used. The concept behind splines is the same as that with linear interpolation, although there is no assumption that the data within sampling intervals are linear. Cubic splining fits the entire data set (100% in rescaled time) to a series of cubic (third-order) polynomial equations, allowing the user to evaluate the equations at any chosen time, not just the time of the original data. For this reason, these equations are called *interpolating splines*. The quintic spline is similar to the cubic, except that fifth-order polynomials are used (Wood 1982). The advantage of quintic splines is that their second derivatives are cubic splines, whereas the second derivatives of cubic splines are lines. Thus, using quintic splines is preferred if the data are to be double differentiated. The resulting "acceleration" curves will therefore be continuous curves rather than a series of connected line segments.

After rescaling each trial's data, an ensemble average is created. From every trial to be included in the ensemble, the data representing the first time point (e.g., 0%) are summed and the mean and standard deviation (SD) are computed. This process is repeated for each time point of the movement, yield-

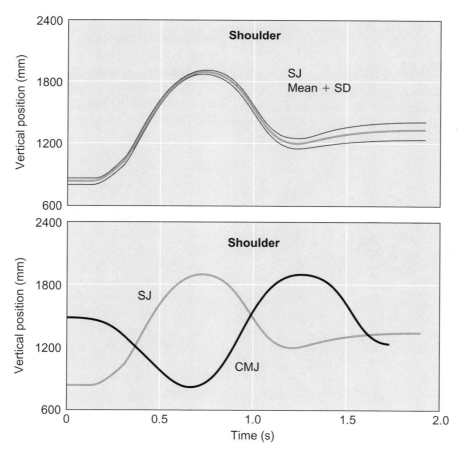

○ **Figure 1.17**   Ensemble average plots of the vertical position of a marker attached to the shoulder of a subject while performing vertical jumps. The top panel shows the mean ±1 standard deviation of 4 squat jumping (SJ) trials. The bottom panel compares the ensemble from the squat jumps with the ensemble from 3 countermovement jumps (CMJ).

ing a 101-point series for both the ensemble average and the SD about that average. The ensemble average data for different movement conditions within a study may be graphed and compared to illustrate variations in the motion. Alternatively, the ensemble average could be plotted along with ±1 SD bounds or the 95th confidence interval (±1.96 × SD) to illustrate the degree of variation within the data (figure 1.17). This process is not unique to linear kinematics and can be applied to any type of signal, including angular kinematics, kinetics, and muscle activity variables (e.g., angular velocity, forces, moments of force, and EMGs).

ANGULAR KINEMATICS

Angular position measures can be divided into two classes. The first class concerns the angular position or orientation of single bodies. These are called *segment* or *absolute angles,* because they are usually referenced to an absolute or Newtonian frame of reference. The second class concerns the angle between two, usually adjacent segments of the body. These are called *relative, joint,* or *cardinal angles,* because they measure the angular position of one segment relative to another.

SEGMENT ANGLES

As stated earlier in the section on markers, at least two points must be quantified to describe the angular position of a human body segment in a 2-D plane. These absolute angles follow a consistent rule called

the *right-hand rule* (figure 1.18), which specifies that positive rotations are counterclockwise and negative rotations are clockwise. Curving the fingers of the right hand in the direction of the angle or rotation and then comparing the direction of the thumb to the reference axes determine the sign of an angle or rotation about a particular axis. If the thumb points in the direction of a positive axis, then the angle or rotation is positive. For planar analyses, segment angles are often defined as the angle of the segment with respect to a right-horizontal line originating from the proximal end of the segment. Other conventions are possible, and researchers must identify the one they use.

Angular Conventions

Two conventions are used to quantify segment angles. The first measures angles like a compass, with angles ranging from 0 to 360°, whereas the second allows a range from +180 to −180° (figure 1.19). These conventions yield the same values for angles between 0 and +180°, but different ones for angles between 180 and 360°. With the second convention (see figure 1.20), these angles range from 0 to −180°, making them easier to visualize.

Discontinuity Problems

With both conventions, a problem arises when a segment crosses the 0/360° line or the ±180° line. For example, when a segment moves clockwise from an angle of 10 to 350°, the change is recorded as (350° − 10° =) +340° instead of the correct value of

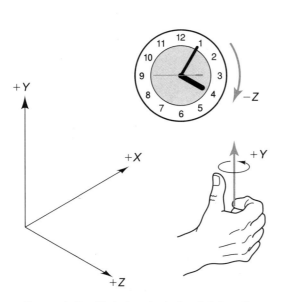

○ **Figure 1.18**   Right-hand rule for defining directions of rotations.

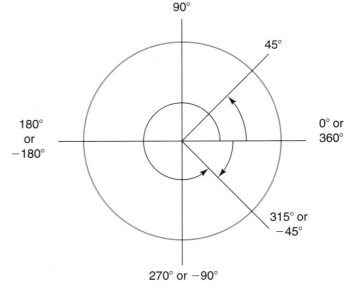

○ **Figure 1.19**   Two conventions for defining the absolute angles of segments. One convention always measures angles in the range 0 to 360° while the other uses a range from 180 to −180°.

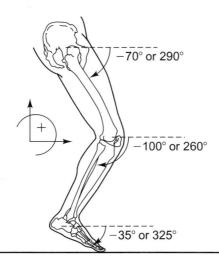

−70° or 290°

−100° or 260°

−35° or 325°

○ **Figure 1.20** Examples of absolute angles of the lower extremity. All angles are taken from a right horizontal from segment's proximal end.

−20°. Similarly, for the ±180° convention, whenever a segment moves counterclockwise from +170° to −160°, the difference is said to be (−160° − 170° =) −330° instead of the correct change of +30°. In both cases, it is possible that the segments rotated in the opposite directions through the larger angle changes, although this is unlikely if the data-sampling rate was correct. To solve this dilemma, it is assumed that no angle changes by greater than 180° from one frame to the next. When graphing angular histories, discontinuities also arise when a segment crosses 0/360° or ±180°. This can be resolved by adding or subtracting 180° whenever an absolute angular change of more than 180° occurs. Of course, this correction can create angles that exceed ±180°.

JOINT ANGLES

The human body is a series of segments linked by joints, so measurement and description of relative or joint angles are often useful. Quantification of a joint angle requires a minimum of three coordinates or two absolute angles, as shown in figure 1.21.

When defining joint motions, adjacent joints may have different directions for the same type of motion. For example, if knee flexion is a positive rotation according to the GCS, then flexion of the hip is a negative rotation (see figure 1.22). Notice that a *biomechanical* system that respects the right-hand rule is presented. Also shown is a *medical* system that is used by physiotherapists, anatomists, and the medical community to define the anatomical positions of joints. With the latter system, negative angles are avoided by specifying the type of joint motion (such as flexion, extension, hyperextension, and so on).

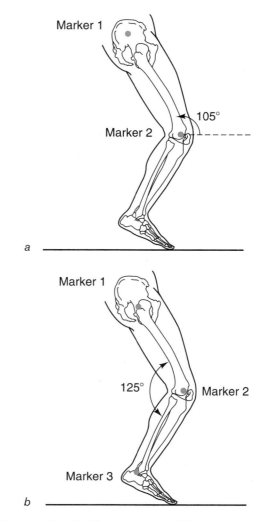

Marker 1

Marker 2

105°

a

Marker 1

125° Marker 2

Marker 3

b

○ **Figure 1.21** (a) Absolute versus (b) relative angles.

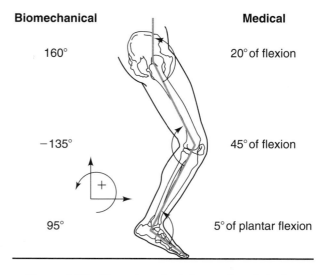

Biomechanical

160°

−135°

95°

Medical

20° of flexion

45° of flexion

5° of plantar flexion

○ **Figure 1.22** Examples of relative angles of the lower extremity.

POLAR COORDINATES

Angular motion of a rigid body also produces linear motion of individual points (such as markers) attached to the body. Furthermore, the amount of linear displacement of a point is dependent on its location with respect to the axis about which the body rotates. Consider the motion of the minute hand of a clock (figure 1.23). A marker placed at the center of the clock will undergo no linear displacement as the minutes tick by, but a marker at the end of the hand will sweep out a circular path.

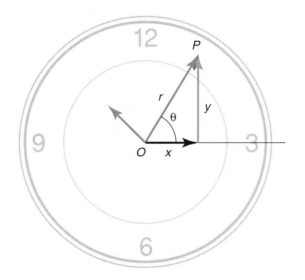

○ **Figure 1.23**   The polar (r, θ) and Cartesian (x, y) coordinates for the minute hand of a clock.

The mathematical description of such angular-to-linear motion is reflected in the use of *polar coordinates,* an alternative to Cartesian coordinates. In the polar system, as with the Cartesian system, two degrees of freedom describe the planar position of a point on a plane. A line is constructed from the origin of the axis system to the point (figure 1.24). The length *(r)* of the line represents one DOF, whereas the angle (θ) between the line and one of the reference axes (usually the axis right-horizontal from the clock center) describes the second DOF. These are called the polar coordinates of the point and are written (r, θ). Note in figure 1.24 that drawing a line from the point to the reference axis forms a right triangle. Using simple trigonometry, conversion from polar to Cartesian coordinates is accomplished as follows:

$$x = r \cos \theta \tag{1.13}$$

$$y = r \sin \theta \tag{1.14}$$

If we consider the center of the clock to be the origin, a marker farther from the center will have a

greater length, *r*. Thus, for one complete rotation, a point at the end of the minute hand sweeps out a longer path than a point at the end of the shorter hour hand. Although Cartesian coordinates are used most often in biomechanics research, for some applications it is more convenient to use polar coordinates. It is always possible, given the Cartesian coordinates *(x, y)* of a point, to compute its polar coordinates (r, θ) using the following trigonometric relationships:

$$r = \sqrt{\quad\quad} \tag{1.15}$$

$$\theta = \tan^{-1}(\ /x) \tag{1.16}$$

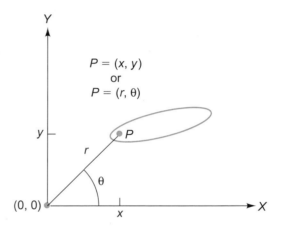

○ **Figure 1.24**   Polar coordinates can represent the position of a point as readily as Cartesian coordinates. Polar coordinates use the length of a line segment joining the point and the origin *(r)* as one coordinate, and the angle between the line segment and a fixed axis (θ) as the second coordinate. The location of point *P* can be expressed either as *(x, y)* in the Cartesian system or as (r, θ) in polar coordinates. Note that in either system the point has 2 DOF.

ANGULAR TIME DERIVATIVES

Just as with linear kinematic variables, differential and integral calculus are also used to determine these angular variables. As stated earlier in this chapter, *angular displacement* is defined as the change in angular position. *Angular velocity* is defined as the rate of change of angular displacement with respect to time, and *angular acceleration* is the rate of change of angular velocity with respect to time. Conversely, angular velocity is the integral of angular acceleration with respect to time and angular displacement is the time integral of angular velocity. These three kinematic variables are used to describe the angular motion of rigid bodies during a motion sequence. The symbols representing units of measure for these angular variables were presented in table 1.2.

Once the continuity of the angular histories has been ensured, the angular velocities and accelerations can be computed using finite difference equations that are similar in form to those used for linear kinematics. The central finite difference equations for computing angular velocity (ω) and acceleration (α) are

$$\omega_i = \frac{\theta_{i+1} - \theta_{i-1}}{2\Delta t} \qquad (1.17)$$

$$\alpha_i = \frac{\omega_{i+1} - \omega_{i-1}}{2\Delta t} \qquad (1.18)$$

$$\alpha_i = \frac{\theta_{i+1} - 2\theta_i + \theta_{i-1}}{\Delta t^2} \qquad (1.19)$$

where θ represents the angular position and Δt represents the time duration between adjacent samples (see Miller and Nelson 1973). The character i represents the particular instant in time that is being analyzed. As is the case with linear data, before applying these equations the raw angular positions must be smoothed to remove high-frequency noise (Pezzack, Norman, and Winter 1977). If the angular positions were derived from marker coordinates that were already filtered, no further smoothing is necessary.

From the Scientific Literature

Pezzack, J.C., R.W. Norman, and D.A. Winter. 1977. An assessment of derivative determining techniques used for motion analysis. *Journal of Biomechanics* 10:377-82.

This paper uses both direct kinematic measures of angle and angular acceleration and indirect kinematics of markers to assess two different methods of smoothing digitized marker kinematics, particularly angular acceleration. First, data from an aluminum arm that could be rotated only in the horizontal plane (one DOF) were collected by filming the arm's motion from above while simultaneously measuring its angular position with a potentiometer (see appendix C) and its linear acceleration with a uniaxial accelerometer. The aluminum arm was manually moved in several different ways. The accelerometer was mounted so that it measured transverse acceleration (a_t), which was then converted to angular acceleration by dividing by the distance from the arm's center of rotation to the accelerometer ($\alpha = a_t/r$). Next, the film data were digitized and the angular positions of the arm were computed. Then, the angular acceleration of the arm was computed in three different ways—without data smoothing, after smoothing with Chebyshev least-squares polynomials of orders 10 to 20, and after digital filtering with a fourth-order, zero-lag Butterworth filter.

The results (figure 1.25) showed very good agreement between the two angular displacement measures (film vs. potentiometer), but significant differences existed among the three methods of computing angular acceleration. First, the unsmoothed accelerations were very noisy, although the waveform, on average, did follow the signal derived from the direct measure of acceleration. Second, the least-squares polynomial that accurately fit the displacement signal did not follow the accelerometer signal after two derivatives were taken. Clearly, fitting a single polynomial to even a relatively simple human motion was not suitable. Third, after two time derivatives, the digitally filtered signal did closely match the true acceleration as measured by the accelerometer. Some attenuation of the signal occurred at the peaks, but these could have been corrected by increasing the cutoff frequency. See the following section and chapter 11 for more details on the use and implementation of digital filtering.

The importance of this paper cannot be overestimated. In addition to successfully reducing the effects of high-frequency noise in digitized film, the authors also provided the data from one of their trials so that other researchers could evaluate their data-smoothing techniques (Lanshammer 1982b). Subsequently, two other techniques were also shown to have acceptable smoothing capabilities. Wood (1982) used piecewise quintic splines and Hatze (1981) used optimally regularized Fourier series to appropriately smooth human motion data.

(continued)

(continued)

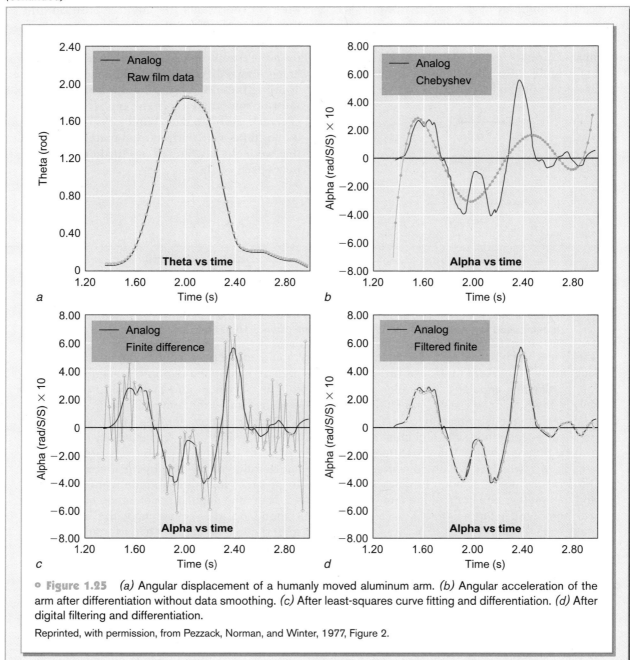

Figure 1.25 *(a)* Angular displacement of a humanly moved aluminum arm. *(b)* Angular acceleration of the arm after differentiation without data smoothing. *(c)* After least-squares curve fitting and differentiation. *(d)* After digital filtering and differentiation.

Reprinted, with permission, from Pezzack, Norman, and Winter, 1977, Figure 2.

ANGULAR TO LINEAR CONVERSION

In the earlier section on polar coordinates, we showed how linear and angular motion are related. When a rigid body undergoes angular rotation, it is possible to calculate linear velocity and acceleration from the known angular velocity and acceleration. Consider the example of a body rotating about a fixed axis of rotation at point Q in figure 1.26. Designate a marker attached at the far end of the body

as point P so that points Q and P are separated by length *r*. As the body rotates, point P describes an arc (or circle) on the underlying surface. Now, attach to the body a 2-D reference frame with its origin at point P. One axis, called the normal axis, is at right angles to the curvature of the path, whereas the tangential axis is tangential to the path. Affixed to the body, this reference frame will rotate with the body so that the normal and tangential axes change their orientations within the GCS. The angular motion

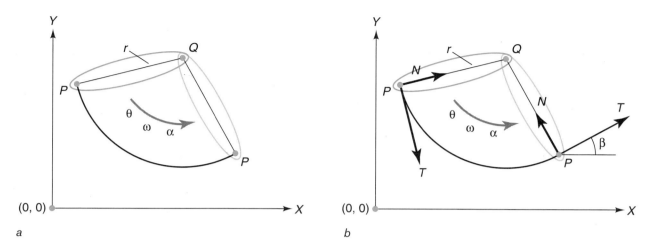

° **Figure 1.26** *(a)* Angular motion (θ, ω, and α) of a rigid body produces linear motion of points attached to it. Point *P* is located at a distance *r* from fixed point *Q*. As the body rotates through the angle θ, *P* scribes an arc such that its (*x,y*) coordinates are constantly changing. *(b)* LCS attached at *P* allows conversion from angular motion of the body to linear motion of the point *P*. The *tangential axis* (*T*) of the LCS describes a tangent to the arc scribed by *P*, while the orthogonal axis *N* is directed toward the axis of rotation point *Q*. Angle β is the angle between the tangential axis and the right-horizontal, and can be used to convert the tangential velocity ($v_t = r\omega$) into its x and y components.

of the body is described by its angular velocity (ω) and angular acceleration (α) within the GCS. Note that such a system could represent a human body segment such as the thigh, where Q represents the stationary hip joint, P the moving knee joint, and *r* the length of the thigh.

The linear velocity of point P within the rotating reference system can be calculated from the equation $v_t = r\omega$, where v_t is the *tangential velocity* (i.e., in the direction of the tangential axis). Note that the *normal velocity* (v_n) is zero, because the length *r* is a constant (because points Q and P are fixed relative to each other). Point P may also undergo tangential acceleration, computed from the equation $a_t = r\alpha$. This acceleration will be nonzero if the body increases or decreases its angular velocity. If the body rotates at a constant angular speed, the angular and tangential accelerations will be zero.

Curiously, although the normal velocity is zero for circular motions, there must always be a nonzero *normal acceleration* (a_n) if the body is traveling in a circular path. The normal acceleration is calculated as $a_n = r\omega^2$. This acceleration, also called *centripetal acceleration,* is caused by the continual change of direction of the point P within the GCS. Readers should consult an engineering mechanics text such as Beer and Johnston (1977) for more detailed consideration of these accelerations.

The normal and tangential velocities and accelerations are defined within the reference system attached to the rotating body, but can be converted to the GCS using knowledge of the body's angular orientation and simple trigonometric identities. In figure 1.26, the angle β is the angle formed by the tangential velocity and the right-horizontal, which is parallel to the X-axis in the GCS. The vector can be resolved into the components v_x and v_y, corresponding to the principal directions of the GCS, using the equations $v_x = v_t \cos \beta$ and $v_y = v_t \sin \beta$. Similar transformations can be applied to the accelerations a_n and a_t to express them in the GCS, as well.

ELECTROGONIOMETERS

A goniometer is a manual device for measuring joint angles (figure 1.27a). It is essentially a protractor with two arms—one affixed to the protractor and one that rotates to measure angles. To measure joint angles electronically during motion, electrogoniometers are used. Typically, they are much cheaper than imaging systems and data can be collected and viewed immediately. Unfortunately, these devices do encumber movement because various electronics must be attached to the subject and cables must be run to a data-collection system.

The most common type of electrogoniometer uses a potentiometer as the sensing element. A potentiometer is essentially a variable resistor (see appendix C). A constant voltage is applied across its terminals and a *wiper* that turns with the potentiometer taps off voltage in an amount proportional to the amount of the turn (see figure 1.27c). One part of the potentiometer is attached to one segment of the joint and the other to the adjacent segment.

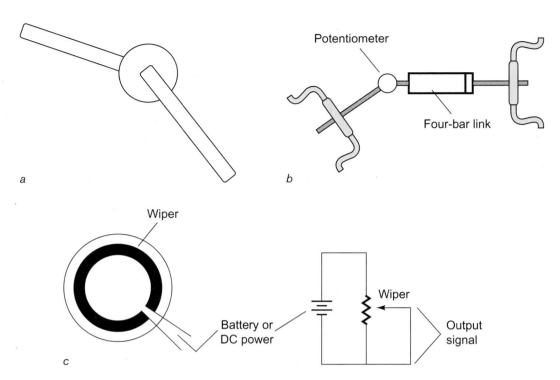

Figure 1.27   *(a)* Manual goniometer, *(b)* electrogoniometer, and *(c)* schematic of a potentiometer and electrogoniometer circuit.

Any angular motion of the joint causes the potentiometer to rotate and therefore change its output voltage. Other types of sensors include digital encoders, polarized-light photography, strain gauges, and fiber-optic cables.

One problem with goniometers is that not all joints act as pure hinges; any translational motion of the joint creates an erroneous angular rotation of the electrogoniometer. In principle, designing self-aligning mechanisms such as the four-bar linkage illustrated in figure 1.27b solves this problem. A similar self-aligning element was an integral component of the CARS-UBC (or MERU) electrogoniometer system developed by Hannah, Cousins, and Foort (1978). This device simultaneously measured triaxial angular motion of the ankle, knee, and hip of both legs in real time—a total of 18 signals. Although this system was clinically valuable for some patients, it is not as useful for examining people with severe disabilities because of the device's encumbrance. Another significant problem with all electrogoniometers is that they only measure joint angles. This prevents them from recording absolute motion of the segments with respect to a Newtonian frame of reference (in other words, a GCS), which is necessary for performing inverse dynamics analyses (see chapter 5). However, this type of system is useful in

clinical environments where immediate joint kinematic information is required.

When selecting or building an electrogoniometer, use a sensing element that is continually differentiable. For example, as the wiper of most inexpensive potentiometers moves, it jumps from one loop of a coil of wire to another to vary its resistance. This contaminates the output position data with *discontinuity spikes* that disrupt the calculation of derivatives for obtaining angular velocity or acceleration. To eliminate these discontinuous steps, more expensive potentiometers have a continuous strip of conductive material. There is usually a break at one spot along the strip, and this portion must be placed outside the joint's range of motion (see figure 1.27c).

Another solution is to select a different type of sensor. One company, Measurand, devised a device, called ShapeSensor, which uses a fine fiber-optic cable that transmits less light as the bend in the cable increases, allowing continuous measurement of the amount of joint rotation. Other systems (such as one made by Biometrics Ltd., formerly Penny and Giles) use strain gauges to quantify the degree of bending in a steel wire—and therefore angular position—about two axes (two DOF). For example, a transducer that crosses the elbow can simultaneously measure elbow flexion and forearm supination. Yet another solution is to use a polarized-light goniometer, which uses two sensors that are sensitive to polarized light (Chapman, Caldwell, and Selbie 1985; Mitchelson 1975). One sensor is placed on the rotating segment and the other is affixed to a stationary object. The relative positions of the sensors are determined by the orientation of each sensor within the polarized light plane.

Calibration of electrogoniometers is relatively straightforward. If the electrogoniometer design allows, it can be attached directly to a manual goniometer. Moving the manual goniometer from one known position to another while recording data from the electrogoniometer will provide a voltage measure equivalent to an actual angular displacement. From these measurements a calibration coefficient is computed. If the electrogoniometer cannot be directly attached to a manual goniometer, a similar procedure can be used while the electrogoniometer is attached to the joint of interest.

ANGULAR KINEMATIC DATA PRESENTATION

The presentation of angular kinematic data is similar to that for linear kinematics, with the most common format being the graphing of θ, ω, or α as a function of time throughout the movement (a history or time series). This format is useful for comparison with other kinematic and kinetic signals that occur simultaneously, particularly when trying to relate the observed kinematics to the underlying forces and torques (see chapters 4 and 5). Another display format unique to angular kinematics is the *angle-angle diagram,* which illustrates the coordinated motion of two segments or joints by plotting one vs. the other. The graphic must be carefully described and annotated because the sense of time during the movement is lost, but this weakness can be overcome by marking specific movement events on the angle-angle plot and placing small arrows that indicate the relative timing sequence.

Another unique format for presenting angular results is known as a *phase plot, phase portrait,* or *phase diagram.* Here, the relationship between θ and ω for a specific segment or joint is depicted, with θ on the horizontal axis and ω on the vertical axis. This presentation has become popular among

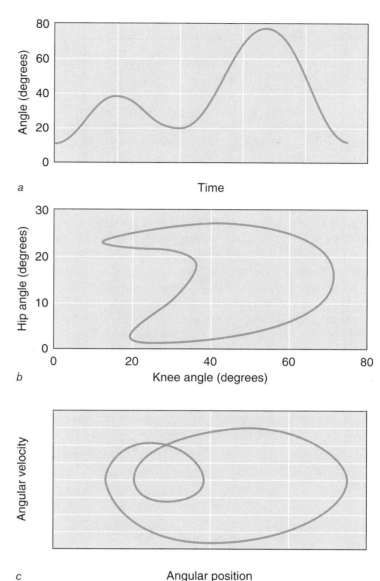

a Time

b Knee angle (degrees)

c Angular position

° **Figure 1.28**   *(a)* Time series of knee angle; *(b)* angle-angle plot of hip versus knee angle; and *(c)* phase plot of knee angular velocity versus angular position.

biomechanists studying movement from a *dynamical systems* perspective, in which the emphasis is on meaningfully representing the kinematic state of a system as a window to the underlying control of that system. The phase plot therefore is seen as an expression of the control of the segment or joint. Figure 1.28 shows examples of an angular kinematic time series, angle-angle diagram, and phase portrait. Note that the presentation of angular kinematic data for groups of trials or subjects can make use of the ensemble averaging techniques that were discussed in the earlier section on the presentation of linear kinematic variables.

SUMMARY

This chapter outlined the major tools used by biomechanists to collect, process, and present data about the motion patterns of planar human movements. Data that describe motions are called kinematics. The next chapter will expand on these concepts to handle motions in 3-D. Although there are many different tools available, most kinematic studies begin with acquiring data from some sort of imaging or motion-capture system, usually video. Equations for differentiating these data to obtain both linear and angular velocities and accelerations were presented. Additional technical information about processing and removing invalid information from these data may be found in chapter 11.

Although describing motion may be an end in itself, the most important reason for collecting kinematic data is to derive various kinetic quantities. Kinetics concerns the causes of motion and often can be indirectly computed from kinematic data and the inertial properties of bodies (see Chapter 3). Chapters 5, 6, and 7 outline how to derive such kinetic measures as mechanical work, energy, power, and momentum from kinematics.

SUGGESTED READINGS

Beer, F.P., and E.R. Johnston, Jr. 1977. *Vector Mechanics for Engineers: Statics and Dynamics.* 3rd ed. Montreal: McGraw-Hill.

Hamill, J., and K.M. Knutzen. 1995. *Biomechanical Basis of Human Movement.* Baltimore: Williams and Wilkins.

Nigg, B.M., and W. Herzog. 1994. *Biomechanics of the Musculo-skeletal System.* Toronto: John Wiley & Sons.

Robertson, D.G.E. 1997. *Introduction to Biomechanics for Human Motion Analysis.* Waterloo, Ontario: Waterloo Biomechanics.

Winter, D.A. 1990. *Biomechanics and Motor Control of Human Movement.* 2nd ed. Toronto: John Wiley & Sons.

Zatsiorsky, V.M. 1998. *Kinematics of Human Motion.* Champaign, IL: Human Kinetics.

Three-Dimensional Kinematics

Joseph Hamill and W. Scott Selbie

The term *kinematics* has been defined in the context of motion in two-dimensional (2-D) space. That definition can be extended to three-dimensional (3-D) kinematics, which is simply the description of motion in 3-D space without regard to the forces that cause the motion. Many of the same concepts apply to 2-D and 3-D analysis, and 3-D should not be considered substantially more difficult than 2-D.

The question often arises if computing 2-D angles in the sagittal, frontal, and transverse planes constitutes a 3-D analysis. It should be understood that multiple planar views of 2-D angles do not result in a true 3-D angle. In fact, reporting multiple planar views of angles is incorrect. The gold standard for joint angles is the calculation of these angles in 3-D space.

This chapter is not meant to be the definitive work on the theory of 3-D kinematics, but rather a more practical, how-to-get-started chapter. The basic principles of 3-D kinematics are presented without reference to the theoretical background. Although many conventions for calculating angles in 3-D space have been used in the biomechanics literature, we briefly present three methods of calculating 3-D lower-extremity angles. For more detailed and theoretical views of 3-D kinematics, readers should refer to works by Zatsiorsky (1998), Nigg and Herzog (1994), and Allard et al. (1998). Thus, the purpose of this chapter is

- to present computational tools necessary for the calculation of 3-D kinematics,

- to discuss how 3-D data are collected,
- to determine how these data are used to evaluate positions and orientations of the body or body segments in space, and
- to present three methods of calculating segment and joint angles.

SCALARS, VECTORS, AND MATRICES

In 2-D kinematics, the mathematical operations that were used consisted mainly of algebra and trigonometry. However, in a 3-D analysis, we must extend the mathematics to include vector operations and matrix algebra. Although this sounds foreboding, it should not be considered so. Knowledge of these concepts actually makes the computations much easier and helps us take care of the "bookkeeping" in our calculations.

VECTORS AND SCALARS

A quantity that is completely specified by its magnitude is referred to as a *scalar*. Examples of scalars are mass, density, work, energy, and volume. They are treated mathematically as real numbers and, as such, obey all of the usual rules of algebra.

For this chapter, however, the use of a physical quantity that expresses direction is important. The location of any vector $\vec{A}$ in 3-D space relative to an origin is referred to as a position vector. A vector is a quantity that requires both direction and magnitude for its complete specification and must add according to the parallelogram law. In this chapter, a vector is designated by an arrow over the vector name (e.g., $\vec{A}$).

The method most often used for defining a position in 3-D space is the Cartesian coordinate system, in which a position vector has three coordinates that uniquely distinguish the point in space. These coordinates are mutually orthogonal. Therefore, any location in this 3-D space can be defined using Cartesian coordinates or the projection of each component onto the reference frame. For example, vector $\vec{A}$ in figure 2.1 is located at distances A_x from the Y-Z plane, A_y from the X-Z plane, and A_z from the X-Y plane. The location of this position is defined by the coordinates (A_x, A_y, A_z). Two vectors are considered equal if all of the respective components of the vectors are equal. That is, vectors $\vec{A}$ and $\vec{B}$ are equal if $A_x = B_x$, $A_y = B_y$, and $A_z = B_z$.

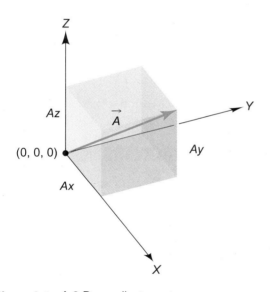

○ **Figure 2.1**   A 3-D coordinate system.

Another method of representing the components of point A is by using *unit vectors*. Unit vectors are defined as vectors of unit length along each of the axes of the coordinate system and are specified as $\vec{i}$, $\vec{j}$, and $\vec{k}$ along the X, Y, and Z axes, respectively (figure 2.2). Specifically, a unit vector, $\vec{e}_A$ is defined as

$$\vec{e}_A = \frac{\vec{A}}{\left\|\vec{A}\right\|} \quad \text{where: } \left\|\vec{e}_A\right\| = 1 \qquad (2.1)$$

or a vector with unit length and in the direction of $\vec{A}$. The double bars mean the norm or magnitude

of the vector; this operation is defined in appendix D. The unit vectors associated with each of the axes of the coordinate system are defined as

$$\text{in the x-direction, } \vec{i} = (1,0,0) \qquad (2.2)$$

$$\text{in the y-direction, } \vec{j} = (0,1,0) \text{ and} \qquad (2.3)$$

$$\text{in the z-direction, } \vec{k} = (0,0,1) \qquad (2.4)$$

Any vector can then be represented using its Cartesian coordinates or as the vector sum of its unit vectors. For example, the vector $\vec{A}$ can be represented as

$$\vec{A} = (A_x, A_y, A_z) \qquad (2.5)$$

or in unit vector form by

$$\vec{A} = A_x \vec{i} + A_y \vec{j} + A_z \vec{k} \qquad (2.6)$$

A vector can be changed to a unit vector by dividing each component by the norm or magnitude of the vector. That is, for a vector $\vec{A}$, the unit vector in the direction of $\vec{A}$ is

$$\vec{e}_A = \frac{A_x}{\left\|\vec{A}\right\|}\vec{i} + \frac{A_y}{\left\|\vec{A}\right\|}\vec{j} + \frac{A_z}{\left\|\vec{A}\right\|}\vec{k} \qquad (2.7)$$

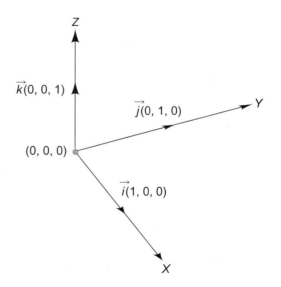

○ **Figure 2.2**   Orientation of the unit vector coordinates in a coordinate system.

VECTOR OPERATIONS

In a 3-D analysis, several operations and concepts concerning vectors are critically important to gaining an understanding of motion in 3-D space. However, they are presented only briefly in appendix D of this book. Readers are urged to refer to more detailed presentations of the concepts in other sources.

MATRICES

Many of the computations done in 3-D kinematics can be accomplished relatively efficiently using *matrix algebra*. A *matrix* is any rectangular array of numbers. Each number in the array is an *element*. If the array has m rows and n columns, the matrix is referred to as an $m \times n$ matrix. Each element of the matrix can then be identified by its row and column positions. The number of rows and columns in a matrix denotes the order of the matrix. For example, the 3×3 matrix [A] can be written as

$$[A] = \begin{bmatrix} a_{11} & a_{12} & a_{13} \\ a_{21} & a_{22} & a_{23} \\ a_{31} & a_{32} & a_{33} \end{bmatrix} \qquad (2.8)$$

The element a_{23} is the element in the second row and the third column, or the element in the (2,3) place. Generally, each element in matrix [A] can be referred to as a_{ij}, where i is the row number and j is the column number. A matrix having only one row with several columns is a *row matrix*. For example,

$$[A] = \begin{bmatrix} a_{11} & a_{12} & a_{13} \end{bmatrix} \qquad (2.9)$$

Conversely, a matrix can have several rows but only one column. This type of matrix is referred to as a *column matrix*. For example,

$$[A] = \begin{bmatrix} a_{11} \\ a_{21} \\ a_{31} \end{bmatrix} \qquad (2.10)$$

We will encounter a number of special matrices in this chapter. A *square matrix* has the same number of rows and columns. A **diagonal matrix** is a square matrix in which the elements on the diagonal $(a_{11}, a_{22}, \ldots, a_{nn})$ are nonzero and the others are zero. This matrix [A] is a square, diagonal matrix:

$$[A] = \begin{bmatrix} a_{11} & 0 & 0 \\ 0 & a_{22} & 0 \\ 0 & 0 & a_{33} \end{bmatrix} \qquad (2.11)$$

An *identity matrix* is a square matrix whose diagonal elements equal 1. Designated as [I] matrices, they are written like this:

$$[I] = \begin{bmatrix} 1 & 0 & 0 \\ 0 & 1 & 0 \\ 0 & 0 & 1 \end{bmatrix} \qquad (2.12)$$

The *transpose* of matrix [A] is designated as $[A]^T$ and is one in which the rows and columns are interchanged.

$$\text{If } [A] = \begin{bmatrix} a_{11} & a_{12} & a_{13} \\ a_{21} & a_{22} & a_{23} \\ a_{31} & a_{32} & a_{33} \end{bmatrix},$$

$$(2.13)$$

$$\text{then } [A]^T = \begin{bmatrix} a_{11} & a_{21} & a_{31} \\ a_{12} & a_{22} & a_{32} \\ a_{13} & a_{23} & a_{33} \end{bmatrix}$$

MATRIX OPERATIONS

A 3-D analysis involves many operations concerning matrices with which you must become familiar. These operations are presented in appendix E of this book. As with vector operations, readers are urged to refer to more detailed presentations of the concepts in other sources.

COLLECTION OF THREE-DIMENSIONAL DATA

All current multicamera motion-analysis systems can collect 3-D data. The orientation and arrangement of the cameras are not as critical in a 3-D setup as in a 2-D setup (figure 2.3).

Each of the cameras only has to provide a set of 2-D coordinates, with each digitized landmark appearing in at least two cameras. From these sets of 2-D coordinates, 3-D spatial coordinates are generated. In most cases, the *direct linear transformation* (DLT) method (Abdel-Aziz and Karara 1971) is used to calculate the 3-D coordinates. The DLT method establishes a linear relationship between the 2-D camera coordinates of each body landmark and the representation of this landmark in 3-D space.

The technique to do this is relatively straightforward and, in this chapter, we only outline the methodology. More detailed explanations of DLT methods are presented in papers by Miller, Shapiro, and McLaughlin (1980); Shapiro (1978); and Marzan and Karara (1975). A set of n points, called *control points*, defines an *object space* or a coordinate system in the laboratory. The control points have known locations in real units in 3-D space. The set

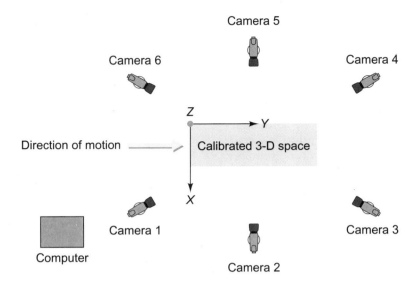

○ Figure 2.3   Typical multicamera experimental setup for a 3-D kinematic analysis of a walking stride.

of control points must contain at least six noncoplanar points $(n \geq 6)$, but usually many more than that are used. In fact, up to 20 control points are often used. The images of these control points are then captured by all cameras in a multicamera setup and digitized to produce 2-D coordinates of each point in each camera view. From the 2-D coordinates of the n noncoplanar control points, a set of $2n$ equations is developed for each camera view. This set of equations is then solved for 11 DLT parameters (L_1 through L_{11}) relative to each camera. The DLT parameters reflect the relationship between 3-D space and the 2-D camera image. The equations for one camera are

$$x_i + L_1 X_i + L_2 Y_i + L_3 Z_i + L_4 + \\ L_9 x_i X_i + L_{10} x_i Y_i + L_{11} x_i Z_i = 0 \quad (2.14)$$

and

$$y_i + L_5 X_i + L_6 Y_i + L_7 Z_i + L_8 + \\ L_9 y_i X_i + L_{10} y_i Y_i + L_{11} y_i Z_i = 0 \quad (2.15)$$

where i is the number of the control point; x_i and y_i are the digitized 2-D coordinates for the ith control point; X_i, Y_i, and Z_i are the known object space coordinates of the ith control point; and L_1 to L_{11} are the DLT parameters.

The $2n$ set of equations with 11 unknown parameters forms an overdetermined solution (i.e., more equations than unknowns) and is solved using a least-squares technique (Miller, Shapiro, and McLaughlin 1980). The DLT parameters contain implicit information relating to the camera orientations and locations. Once these DLT parameters are known, the control points can be removed from the laboratory. The markers that are placed on a subject can then be

captured by the same cameras as the subject moves through the calibrated object space. When these markers have been digitized in 2-D space in at least two cameras, the 2-D coordinates of the landmarks and the DLT parameters can be entered into the equations to solve for the 3-D coordinates using the same equation. In this case, however, we solve for X, Y, and Z, the 3-D coordinates of the points.

The DLT method is not the only method used to determine 3-D coordinates from a set of 2-D coordinates. Other methods include nonlinear transformations. The techniques of all are similar in that known locations of markers in 3-D space are combined with their digitized images in 2-D space to determine a set of nonlinear equations from which the locations of unknown markers can be determined.

COORDINATE SYSTEMS

Two coordinate systems must be defined when conducting a 3-D analysis. These are the *global* or *laboratory coordinate system* (GCS) and the *local coordinate system* (LCS).

GLOBAL OR LABORATORY COORDINATE SYSTEM

The GCS is also referred to as the *inertial reference system*, and it is determined when the object space is defined during the 3-D data capture. This coordinate system—generally a right-handed, orthogonal system with an arbitrary origin—defines the fixed coordinate system in the laboratory from which all positions are ultimately derived. In this chapter, the GCS is designated using uppercase letters with the arbitrary designation of X, Y, and Z. In addition, the X-axis is pointed nominally in the mediolateral direction, the Y-axis anteroposteriorly, and the Z-axis vertically. The unit vectors for the GCS are $\vec{i}$, $\vec{j}$, and $\vec{k}$, respectively.

LOCAL COORDINATE SYSTEM

The LCS is a reference system that is fixed within a body or a segment and moves with the body or segment. Like the GCS, the LCS is also a right-handed, orthogonal coordinate system with the origin generally placed at the center of mass of the body or segment. In this chapter, the LCS is designated in lowercase letters x, y, and z and the unit vectors are $\vec{i'}$, $\vec{j'}$, and $\vec{k'}$, respectively. The LCS is oriented such that the x-axis of the LCS points mediolaterally, the y-axis points anteroposteriorally, and the z-axis axially. The orientation of the LCS to the GCS defines the orientation of the body or segment in space, and it changes as the body or segment moves through the 3-D space. Calculating the orientation of the LCS in the GCS is discussed in a later section of this chapter. How the LCS is defined, however, depends on how the markers to be digitized or tracked are placed on the body or segment in question. The GCS and LCS and their respective unit vectors are depicted in figure 2.4.

MARKER SYSTEMS

Many of the same assumptions that we used for 2-D analysis apply to 3-D. The most important of these is that the body consists of *rigid* segments. Segments are considered rigid if the length of the segment does not change. This, of course, is not true; the skeletal structures we are dealing with are not rigid structures. However, by making this assumption we avoid "messy" mathematical situations for which we cannot find a solution. We must use a set of at least three noncollinear points on each segment to define a rigid body in 3-D space. *Noncollinear* means that the points are not in a straight line.

It should also be understood that when we measure the kinematics of a body or segment, we are attempting to determine the actions of the skeletal structures. This is not a simple proposition, because we are limited in how we represent the individual skeletal structures. Although at least three noncollinear markers must be placed on each segment, these markers can be placed and oriented on a segment in multiple configurations. Examples of

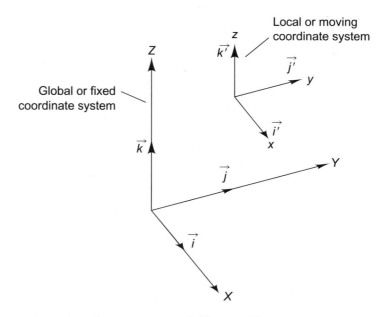

° **Figure 2.4**   Orientation of the GCS and LCS reference frames.

three such configurations are presented in figures 2.5 and 2.6.

Broadly speaking, there are four categories of configuration: (1) markers mounted on bone pins, (2) skin-mounted markers on specific anatomical landmarks on the segment, (3) arrays of markers on a rigid surface that is attached to the body, and (4) a combination of markers on anatomical landmarks and arrays of markers. Great care must be taken to ensure that the weight of the marker or the marker-attachment devices moving relative to the bones do not produce movement artifacts (Karlsson and Tranberg 1999). The marker configuration that we use in this chapter is a combination of arrays of markers placed on a rigid surface and markers placed on anatomical landmarks (figure 2.6).

An array of three noncollinear markers on a pin that is inserted directly into the bone is the gold standard against which other marker systems often are held. Directly attaching the markers to the bone gives the most accurate measurement of the motion of that bone (Fuller et al. 1997; Reinschmidt et al. 1997). The least accurate marker set uses three individual, noncollinear markers placed directly on the skin (Fuller et al. 1997; Karlsson and Tranberg 1999; Reinschmidt et al. 1997). Because the markers move independently of each other, a great deal of error is introduced into subsequent calculations. Fuller et al. (1997) reported displacements of the individual markers relative to the bone of up to 20 mm, whereas other studies reported values of up to 40 mm (Reinschmidt et al. 1997). Reinschmidt et al. (1997) reported that there was good agreement

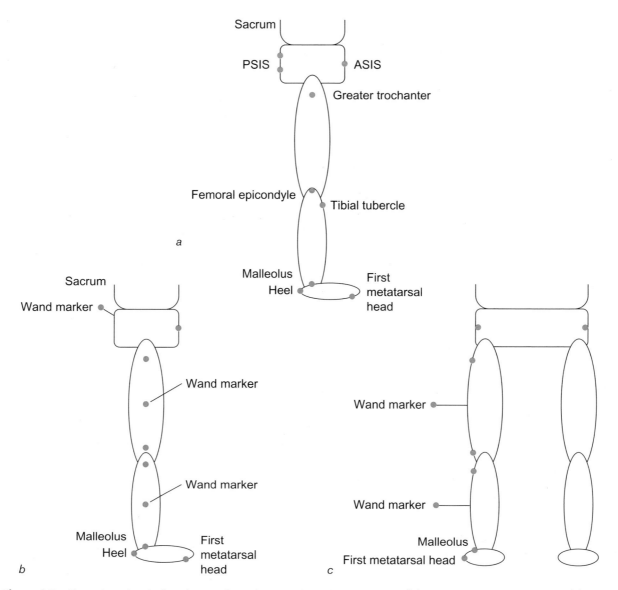

° **Figure 2.5**   Examples of typical marker configurations used in gait laboratories: *(a)* using anatomical landmarks; *(b)* marker set using a combination of anatomical markers and markers on wands; and *(c)* frontal view of (b).

between skin and bone markers in knee flexion/extension, but a great deal of disagreement in abduction/adduction and internal/external rotation when compared to the respective ranges of motion. The average error relative to the range of motion was 21% for flexion/extension, 63% for abduction/adduction, and 70% for internal/external rotation. It should be noted that the motion of skin markers is also task dependent. For motions in which there is an impact transient, for example, the motion of skin-mounted markers was greater than that of bone markers.

In many instances, researchers have placed the noncollinear markers on a rigid structure (figure 2.6) and attached the structure to the segment (Manal et al. 2000; McClay and Manal 1999). Because all of the markers are affixed to the rigid structure, the markers all move relative to each other and are not independent of each other. Angeloni et al. (1993) reported that significantly greater displacements occurred with skin-mounted markers than with markers mounted on rigid plates and suggested that the latter markers were preferable to skin-mounted markers for both practicality and accuracy. Manal et al. (2000) evaluated several configurations of arrays of markers for the tibia and concluded that a set of four markers attached to a rigid shell is optimal. The authors suggested that

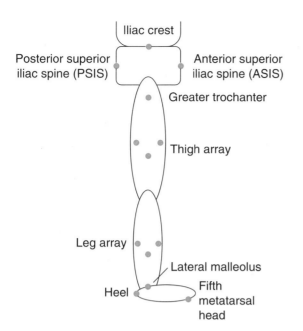

Iliac crest

Posterior superior
iliac spine (PSIS)

Anterior superior
iliac spine (ASIS)

Greater trochanter

Thigh array

Leg array

Lateral malleolus

Heel

Fifth
metatarsal
head

○ **Figure 2.6** Example of a marker configuration using rigid arrays on the thigh and leg and anatomical markers on the pelvis and foot. This marker system is used in this chapter to illustrate the calculation of joint angles.

such a shell should be placed more distally to avoid the soft tissue of the more proximal portion of the segment. However, rotational deviations of ±2° about the mediolateral and anteroposterior axes and ±4°

about the longitudinal axis occurred even with the optimal marker set.

In several marker sets, the triad of markers on a segment may consist of two markers on anatomical locations and a third marker placed on a rod that projects laterally from the segment (figure 2.5b). This marker is used to better measure 3-D rotation of the segment about its longitudinal axis. Karlsson and Tranberg (1999) suggested that the inertia of markers placed on rods might be greater than skin movement when measuring faster motions.

Therefore, when deciding on the marker configuration to be used, researchers must be aware of the limitations that the marker set can impose on the measurement of 3-D coordinates. Optimally, markers should be placed on a pin inserted into the bone itself. However, this is not always feasible because of the pain and risk associated with the procedure. In addition, the procedure may change the motion you wish to study. Consequently, researchers must decide which of the externally placed marker configurations is optimal for the task under investigation. Furthermore, the accuracy of measurement from segment to segment and axis to axis is different, although relatively reliable measures can be made for all segments in flexion/extension movements. However, secondary axis motions are less reliable and valid, particularly internal/external rotation about the long axis of the segments.

From the Scientific Literature

Fuller, J., L.-J. Liu, M.C. Murphy, and R.W. Mann. 1997. A comparison of lower-extremity skeletal kinematics measured using skin- and pin-mounted markers. *Human Movement Science* **16**:219-42.

The purpose of this study was to evaluate the validity of skin-mounted markers in measuring the 3-D kinematics of the underlying bone. Kinematic data for marker arrays mounted on skeletal pins screwed directly into the bone were compared to data for markers and marker arrays placed directly on the skin. The key findings of the study as they pertain to this chapter were that

- soft-tissue motion relative to the underlying bone was up to 20 mm, depending on the task;

- the power spectra of the skin- and pin-mounted markers covered similar frequency bands and there was no evidence of a distinct, frequency-dependent soft-tissue artifact;

- joint angles calculated from the skin-mounted arrays were significantly different from the expected values; and

- skin-mounted marker data exhibited a transient response to heel strike in gait. For low-mass markers, however, this could be removed with optimal smoothing techniques.

DETERMINATION OF THE LOCAL COORDINATE SYSTEM

A number of methods can be used to determine the LCS from the external markers. In this chapter, we describe only one of these techniques. In many instances, researchers collect data from a standing calibration trial with extra calibration markers placed on the subject to determine the joint centers. They then use the external markers in conjunction with the calibration markers to develop a segment coordinate system that is internal and aligned with the long axis of the segment. This approach is discussed in chapter 7, on 3-D kinetics. In this chapter, however, we use the external markers only to determine the orientation of a segment. This technique requires an extra step that is discussed in the next section.

CALCULATION PROCEDURE

For this chapter, we use triads of markers placed on the lateral side of the limb (figure 2.6). We also assume that the triads are oriented parallel to the sagittal plane. The anterior superior illiac spine (ASIS), posterior superior iliac spine (PSIS), and greater trochanter markers represent the pelvis; triads on rigid plates represent the thigh and leg; and the lateral malleolus, heel, and fifth metatarsal markers represent the foot. This marker configuration has two basic orientations of the triad, both of which are shown in figure 2.7. The individual markers on the rigid plate are numbered as indicated in figure 2.7.

Note that the individual markers on the rigid plate have coordinates based on the GCS. However, the

unit vectors of the LCS have coordinates relative to the origin of the LCS. That is, for the marker arrangement shown in figure 2.7a, the origin of the LCS is marker $\vec{p}_1$. The first step is to subtract the position of marker $\vec{p}_1$ from marker $\vec{p}_2$ and divide by the norm of the vector created by $(\vec{p}_2 - \vec{p}_1)$. This gives us a unit vector in the $\vec{j}^i$ direction and describes the antero-posterior axis of the segment. When two vectors are subtracted, in this case $(\vec{p}_2 - \vec{p}_1)$, the resulting vector begins at marker $\vec{p}_1$ and points toward marker $\vec{p}_2$. For each triad of markers, then,

$$\vec{j}^i = \frac{\left(\vec{p}_2 - \vec{p}_1\right)}{\left\|\left(\vec{p}_2 - \vec{p}_1\right)\right\|} \qquad (2.16)$$

Note that by dividing by the norm of the vector $(\vec{p}_2 - \vec{p}_1)$, the coordinates of the $\vec{j}^i$ vector are the coordinates of a unit vector.

In this marker arrangement, the $\vec{i}^i$ axis—the medial/lateral axis of the segment—is the second LCS axis to determine. This calculation is done in two steps. We again subtract $(\vec{p}_2 - \vec{p}_1)$, which creates a vector originating at $\vec{p}_1$ and pointing toward $\vec{p}_2$. Next, we subtract $(\vec{p}_3 - \vec{p}_1)$, leaving us with a vector starting at $\vec{p}_1$ and pointing toward $\vec{p}_3$. Last, we do the following cross-product operation:

$$\vec{i}^i = \frac{\left(\vec{p}_3 - \vec{p}_1\right) \times \left(\vec{p}_2 - \vec{p}_1\right)}{\left\|\left(\vec{p}_3 - \vec{p}_1\right) \times \left(\vec{p}_2 - \vec{p}_1\right)\right\|} \qquad (2.17)$$

Remember that the result of this operation is a vector and dividing by the norm or magnitude of

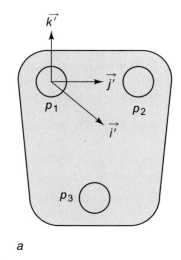
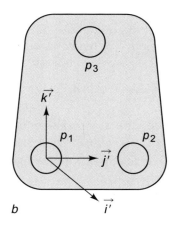

a   b

○ **Figure 2.7** Two configurations of triads and the LCS generated from each triad. The markers in each triad are designated as p_1, p_2, and p_3. Each marker has X, Y, and Z coordinates in the GCS.

$(\vec{p}_3 - \vec{p}_1) \times (\vec{p}_2 - \vec{p}_1)$ leaves us with the unit vector $\vec{i'}$. The order in which the vectors $(\vec{p}_3 - \vec{p}_1)$ and $(\vec{p}_2 - \vec{p}_1)$ are crossed is determined by the right-hand rule. That is, we want to represent the $\vec{i'}$ vector so that the only order in which we can cross the two vectors is the order in which they are presented in the preceding formula. The cross-product operation is not commutative, and if we calculate $(\vec{p}_2 - \vec{p}_1) \times (\vec{p}_3 - \vec{p}_1)$, the resulting vector is the $\vec{i'}$ vector.

The third axis of the LCS, $\vec{k'}$, is determined by the cross product of the two unit vectors—$\vec{i'}$ and $\vec{j'}$—that we already calculated. The order of vectors in the cross product is once again determined by the right-hand rule. To have a vector point in the vertical direction, the vectors $\vec{i'}$ and $\vec{j'}$ must be crossed as follows:

$$\vec{k'} = \vec{i'} \times \vec{j'} \tag{2.18}$$

Because $\vec{i'}$ and $\vec{j'}$ are unit vectors, the resulting vector is also a unit vector. The $\vec{k'}$ axis represents the axial or longitudinal axis of the segment.

If the marker set is oriented as shown in figure 2.7b, only slight changes to the equations just shown are required to calculate the LCS. The equations take the form:

$$\vec{j'} = \frac{\left(\vec{p}_2 - \vec{p}_1\right)}{\left\|\left(\vec{p}_2 - \vec{p}_1\right)\right\|} \tag{2.19}$$

$$\vec{i'} = \frac{\left(\vec{p}_2 - \vec{p}_1\right) \times \left(\vec{p}_3 - \vec{p}_1\right)}{\left\|\left(\vec{p}_2 - \vec{p}_1\right) \times \left(\vec{p}_3 - \vec{p}_1\right)\right\|} \tag{2.20}$$

$$\vec{k'} = \vec{i'} \times \vec{j'} \tag{2.21}$$

CALIBRATION PROCEDURE

When employing the type of marker system being used in this chapter, instead of defining an internal coordinate system, a calibration routine must be defined to align the LCS with the GCS. The following calibration procedure, based on the technique presented by Areblad et al. (1990), defines the orientations of the LCS and GCS independently of the positions of the markers. Generally, this type of calibration is accomplished by orienting the segment so that the axes of the segment are aligned with the GCS. For example, for the lower extremity, the thigh, leg, and foot would be oriented so that the long axes of the thigh and leg aligned with the Z-axis of the GCS and the long axis of the foot aligned with the Y-axis of the GCS (figure 2.8).

Thus, theoretically, the x-, y-, and z-axes of the segment are aligned with the X-, Y-, and Z-axes of the GCS. The orientation of the marker system is then captured as a calibration picture and the positions of the markers are converted to 3-D coordinates. From these static 3-D coordinates a transformation matrix is calculated and later applied to the actual movement trials.

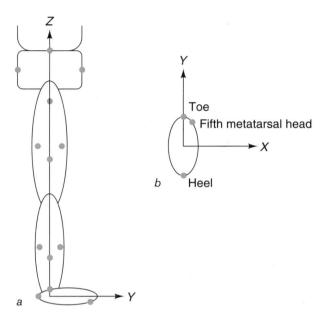

○ **Figure 2.8** Orientation of the lower-extremity segments to the GCS for the calibration picture: *(a)* sagittal view oriented to the Z-Y plane of the GCS, and *(b)* transverse view of the foot in the X-Y plane.

From the 3-D coordinates of the calibration picture, the LCS of the segment is computed as described in the previous section. The computed values are then placed in a *calibration coordinate system* (CalCS) matrix. This system can be expressed as a 3×3 matrix in the form

$$[\text{CalCS}] = \begin{bmatrix} i'_x & j'_x & k'_x \\ i'_y & j'_y & k'_y \\ i'_z & j'_z & k'_z \end{bmatrix} \tag{2.22}$$

where the columns represent the 3-D coordinates of the $\vec{i'}$, $\vec{j'}$, and $\vec{k'}$ unit vectors of the LCS of the calibration frame.

The next step is to calculate the inverse matrix of the [CalCS]. The method is described in appendix E of this text. A rotational transformation matrix ([RTM]) of the segment can then be formed from the product of the inverse of [CalCS] and the identity matrix of the GCS as follows:

$$[\text{RTM}] = [\text{CalCS}]^{-1} \begin{bmatrix} 1 & 0 & 0 \\ 0 & 1 & 0 \\ 0 & 0 & 1 \end{bmatrix} \qquad (2.23)$$

It should be noted that

$$[\text{CalCS}]^{\text{T}} = [\text{CalCS}]^{-1} \qquad (2.24)$$

The [RTM] is used to define the zero orientation of the segment relative to the GCS and then is used in each movement trial. That is, during movement, when the markers achieve the same orientation they had in the calibration frame, the angle is 0°. During each ith instant in time (i.e., each frame) of the movement being studied, the LCS for the segment is calculated and designated the *provisional coordinate system* at the ith frame, or $[\text{PCS}]_i$. The [RTM] is then multiplied by the $[\text{PCS}]_i$ in every frame to generate a *segment coordinate system* [SCS] as follows:

$$[\text{SCS}]_i = [\text{PCS}]_i [\text{RTM}] \qquad (2.25)$$

The [SCS] at each instant in time then becomes the new LCS for the segment and is used in any subsequent calculations.

When studying multiple segments, a [PCS] for each segment is determined from the calibration frame and an [RTM] for each segment is calculated. In turn, an LCS matrix for each segment at each instant in time is determined. This expresses the orientation of each segment relative to the calibration orientation.

TRANSFORMATIONS BETWEEN REFERENCE SYSTEMS

We can look at the same movement in different ways based on the reference system we are using. The descriptions of a body or segment moving in space can be transformed into different reference systems. *Transformation* converts coordinates expressed in one reference system to those expressed in another coordinate system. We generally refer to transformations as being either linear or rotational.

LINEAR TRANSFORMATION

If the vector describing the relative positions of origin of two coordinate systems is described by a vector $\vec{V}$, then the components of $\vec{V}$ are V_x, V_y, and V_z (figure 2.9). Let us call one reference system LCS' and the other LCS". The translation transformation between these coordinate systems is then defined by the vector $\vec{V}$, whose components are written as a column matrix in the form

$$\vec{V} = \begin{bmatrix} V_x \\ V_y \\ V_z \end{bmatrix} \qquad (2.26)$$

If we assume that there is no rotation of LCS" relative to LCS', converting the coordinates of a point P in LCS"—$\vec{P'}$—to coordinates in LCS'—$\vec{P'}$—can be accomplished by

$$\vec{P'} = \vec{V} + \vec{P''} \qquad (2.27)$$

or

$$\begin{bmatrix} x' \\ y' \\ z' \end{bmatrix} = \begin{bmatrix} V_x \\ V_y \\ V_z \end{bmatrix} + \begin{bmatrix} x'' \\ y'' \\ z'' \end{bmatrix} \qquad (2.28)$$

Conversely, converting the coordinates of a point in LCS'—$\vec{P'}$—to coordinates in LCS"—$\vec{P''}$—can be accomplished by

$$\vec{P''} = \vec{P'} - \vec{V} \qquad (2.29)$$

or

$$\begin{bmatrix} x'' \\ y'' \\ z'' \end{bmatrix} = \begin{bmatrix} x' \\ y' \\ z' \end{bmatrix} - \begin{bmatrix} V_x \\ V_y \\ V_z \end{bmatrix} \qquad (2.30)$$

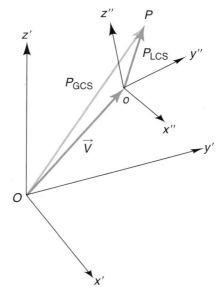

Figure 2.9 Translation of the LCS" from LCS' is defined by the coordinates V_x, V_y, and V_z of $\vec{V}$.

ROTATIONAL TRANSFORMATION

In this section, we present one method for calculating an [RTM]. The vector components of one coordinate system, T_P (e.g., the proximal segment LCS), are ($\vec{i'}$, $\vec{j'}$, and $\vec{k'}$) and those for another coordinate system, T_D (e.g., the distal segment LCS), are ($\vec{i''}$, $\vec{j''}$, and $\vec{k''}$). First, we construct unit vector matrices for each coordinate system (equations 2.31 and 2.32). To develop the [RTM], we take the dot product of a unit vector matrix (equation 2.33) from one coordinate system and the unit vector matrix of another coordinate system.

$$[T_P] = \begin{bmatrix} \vec{i'}_x & \vec{i'}_y & \vec{i'}_z \\ \vec{j'}_x & \vec{j'}_y & \vec{j'}_z \\ \vec{k'}_x & \vec{k'}_y & \vec{k'}_z \end{bmatrix} \tag{2.31}$$

$$[T_D] = \begin{bmatrix} \vec{i''}_x & \vec{i''}_y & \vec{i''}_z \\ \vec{j''}_x & \vec{j''}_y & \vec{j''}_z \\ \vec{k''}_x & \vec{k''}_y & \vec{k''}_z \end{bmatrix} \tag{2.32}$$

$$[T_R] = [T_D][T_P]^T \tag{2.33}$$

This results in

$$[T_R] = \begin{bmatrix} \vec{i''} \cdot \vec{i'} & \vec{i''} \cdot \vec{j'} & \vec{i''} \cdot \vec{k'} \\ \vec{j''} \cdot \vec{i'} & \vec{j''} \cdot \vec{j'} & \vec{j''} \cdot \vec{k'} \\ \vec{k''} \cdot \vec{i'} & \vec{k''} \cdot \vec{j'} & \vec{k''} \cdot \vec{k'} \end{bmatrix} \tag{2.34}$$

The matrix $[T_R]$ is a rotation transformation matrix. The calculation of the rotation matrix using the method described in this chapter assumes that the markers are placed on a rigid body and that there is zero noise in the marker coordinates. To compensate for these limitations, researchers have used other, more sophisticated methods of calculating rotation matrices. Two such methods are linear algebra with matrix perturbation theory (Laub and Schiflett 1982) and singular value decomposition (Arun, Huang, and Blostein 1987). These methods attempt to take into account the noise in the marker coordinates. Readers are referred to the works cited for more details on the methods.

JOINT ANGLES

Several different methods can be used to determine the relative orientation of the two coordinate systems or joint angles (e.g., Chao 1980; Grood and Suntay 1983; Spoor and Veldpaus 1980; Woltring 1991). Three of the most commonly used methods, the Cardan/Euler (e.g., Davis et al. 1991; Engsberg, Grimston, and Wackwitz 1988), joint coordinate system (e.g., Grood and Suntay 1983; Soutas-Little et al. 1987), and helical axis (e.g., Woltring 1991) methods, are presented in this chapter. These are probably the most commonly used approaches in biomechanics literature.

CARDAN/EULER ANGLES

Projection angles are formed by the projections of the vectors of one coordinate system on the orthogonal planes of another coordinate system. For example, in figure 2.10, the vector y of a coordinate system LCS" will have three projection angles on each of the three planes of the other coordinate system, LCS'.

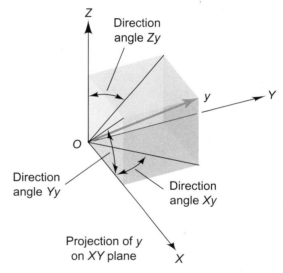

○ **Figure 2.10** The three projection angles of y on the ZX, ZY, and XY planes.

For the three vectors (x", y", and z") of LCS", nine projection angles can be determined. However, only three of these are independent of each other, and the others are redundant. The orientation of LCS" in space can thus be determined using three independent projection angles that correspond to three rotational DOF. However, these angles or rotations must be performed in a specific order because the rotations are not commutative. In total, there are 12 sequences of rotations that can be used. However, the sequences generally take the same order. That is, the first rotation is about an axis in LCS' (x', y', or z'), the second is about a floating axis (an axis that changes according to the orientations of the first and third axes), and the third is about an axis fixed in LCS" (x", y", or z"). Six of the 12 sequences have a terminal rotation axis identical to the first rotation axis. These six sequences are *Euler rotations* (e.g., x' y" x", z' y" z", and so on) and the angles that define these rotations are *Euler angles*. However, six other rotations have a different terminal rotation axis (e.g., x' y" z", z' y" x", and so on) and the angles based on this rotation sequence are often referred to as *Cardan angles*. Both Euler and Cardan angles describe the orientation of one coordinate system relative to another coordinate system as a sequence of ordered rotations from the initial position of one coordinate system. In biomechanics, a Cardan sequence of rotations appears to be used more often than a Euler sequence.

To illustrate the notion of a rotating coordinate system, we can look at 2-D rotation about a single axis (in this case, the X-axis). In figure 2.11, we see the case of one LCS rotated from the right horizontal of another LCS by an angle φ.

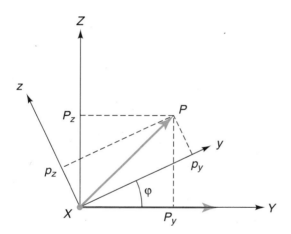

○ Figure 2.11 Vector $\vec{P}$ can be located in the LCS (p_y,p_z) and in another LCS coordinate system (P_y,P_z). A rotation matrix can be determined knowing the angle that the LCS (y,z) is rotated from LCS (Y,Z) about the X-axis (pointing out from the page). The vector $\vec{P}$ can be located in either coordinate system using this rotation matrix.

The rotation matrix $[R_x]$ describing this transformation is

$$[R_x] = \begin{bmatrix} \cos\varphi & \sin\varphi \\ -\sin\varphi & \cos\varphi \end{bmatrix} \qquad (2.35)$$

A vector $\vec{P}$ can be represented in either one LCS ($\vec{P'} = \vec{P'_y}, \vec{P'_z}$) or the other LCS ($\vec{P''} = \vec{P''_y}, \vec{P''_z}$). We can use this rotation matrix to describe the position of the vector in the GCS in terms of the rotation matrix and the coordinates in the LCS as follows:

$$\vec{P''} = [R_x]\vec{P'} \qquad (2.36)$$

or

$$\begin{bmatrix} y'' \\ z'' \end{bmatrix} = [R_x]\begin{bmatrix} y' \\ z' \end{bmatrix} \qquad (2.37)$$

Similarly, we can present the vector in the LCS in terms of the transpose of the rotation matrix and the coordinates in the GCS as follows:

$$\vec{P'} = [R_x]^T \vec{P''} \qquad (2.38)$$

or

$$\begin{bmatrix} y' \\ z' \end{bmatrix} = [R_x]^T\begin{bmatrix} y'' \\ z'' \end{bmatrix} \qquad (2.39)$$

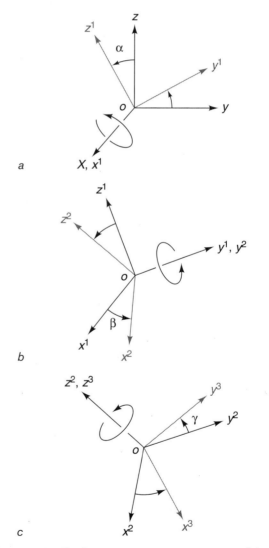

○ Figure 2.12 Cardan sequence of rotations about *(a)* the X-axis of the GCS (α), *(b)* the y-axis (β), and *(c)* the z-axis of the LCS (γ).

We can perform analogous operations to determine the orientation of a segment in 3-D space. However, we now have to perform rotations about all three axes. A common Cardan rotation sequence often used in biomechanics is an *Xyz sequence* (Cole et al. 1993). This sequence involves, first, rotation about the medially directed axis (X); second, rotation about the anteriorly directed axis (y); and third, about the vertical axis (z). These rotations are diagrammed in figure 2.12.

Using figure 2.12, we see that the first rotation takes place about the X-axis and leads to new orientations of the y- and z-axes (y^1 and z^1), with the X-axis remaining in the same orientation and now labeled x^1. The second rotation about the y^1 axis leads to new positions of the x^1 and z^1 axes (x^2 and z^2). For the third rotation about the z^2 axis, the x^2 and y^2 axes assume the new orientation of x^3 and y^3.

For illustrative purposes, the angles for the Xyz sequence are designated α (alpha) for the first rotation, β (beta) for the second rotation, and γ (gamma) for the third rotation. The rotation matrix [R] for an Xyz rotation sequence then can be described as

$$[R] = [R_z][R_y][R_x] \qquad (2.40)$$

where

$$[R_x] = \begin{bmatrix} 1 & 0 & 0 \\ 0 & \cos\alpha & \sin\alpha \\ 0 & -\sin\alpha & \cos\alpha \end{bmatrix} \qquad (2.41)$$

$$[R_y] = \begin{bmatrix} \cos\beta & 0 & -\sin\beta \\ 0 & 1 & 0 \\ \sin\beta & 0 & \cos\beta \end{bmatrix} \qquad (2.42)$$

$$[R_z] = \begin{bmatrix} \cos\gamma & \sin\gamma & 0 \\ -\sin\gamma & \cos\gamma & 0 \\ 0 & 0 & 1 \end{bmatrix} \qquad (2.43)$$

The rotations can then be expressed as the successive rotations of these matrices to produce the combined rotation matrix

The elements in the combined matrix represent the relative orientation of one LCS about a second LCS. This matrix is often called the *decomposition matrix*, where the Cardan angles are decomposed into their projections of one LCS onto the axes of a second LCS.

The three Cardan/Euler angles can be used to represent the anatomical joint actions of flexion/extension, adduction/abduction, and axial (external/internal) rotation. This sequence is equivalent to an Xyz order of rotations generally chosen to represent those joint actions and has been recommended as a standard for biomechanists (Cole et al. 1993). The axes about which the rotations take place are (1) flexion/extension—the laterally directed proximal segment axis, (2) abduction/adduction—a nodal axis (that is, an axis that is perpendicular to both the flexion/extension and axial rotation axes), and (3) axial rotation—the longitudinal axis of the distal segment (An and Chao 1991). We can use this rotation sequence to demonstrate how to calculate the three Cardan angles. We assume the same marker set and calculation of the LCS that was used in prior sections of this chapter. The appropriate angles are designated as α, flexion/extension; β, abduction/adduction; and γ, external/internal rotation.

We assume that the LCS of one segment is oriented relative to the LCS of a second segment and the joint angles are those between the two segments. The movement at a joint is generally defined as motion of the distal segment relative to the proximal segment (Woltring 1991). For example, to determine the knee joint angles, we can develop a matrix based on the LCS of the thigh or proximal segment ($\vec{i}'$, $\vec{j}'$, $\vec{k}'$) and one based on the LCS of the leg or distal segment ($\vec{i}''$, $\vec{j}''$, $\vec{k}''$; see equations 2.31 and 2.32). The rotational transformation matrix $[T_R]$ is then developed as it was in equation 2.34.

The Cardan angle calculations are derived directly from the matrix [R] in equation 2.44. For example, the angle for the second rotation (β) is contained within the element R_{31} of this matrix. Note that element R_{31} of the [R] matrix is equivalent to element T_{R31} of the transformation matrix $[T_R]$. The angle, β, can then be calculated as

about the y-axis (abduction/adduction)

$$\beta = \sin^{-1}(R_{31}) = \sin^{-1}(T_{R31}) \qquad (2.45)$$

$$[R] = \begin{bmatrix} \cos\beta\cos\gamma & \cos\gamma\sin\beta\sin\alpha + \sin\gamma\cos\alpha & \sin\gamma\sin\alpha - \cos\gamma\sin\beta\cos\alpha \\ -\sin\gamma\cos\beta & \cos\alpha\cos\gamma - \sin\alpha\sin\beta\sin\gamma & \sin\gamma\sin\beta\cos\alpha + \cos\gamma\sin\alpha \\ \sin\beta & -\cos\beta\sin\alpha & \cos\alpha\cos\beta \end{bmatrix} \qquad (2.44)$$

The first rotation, α, is calculated from the entry in the third row and second column of the [R] matrix ($-R_{32}$) and the cosine of the rotation about the y-axis ($\cos \beta$). Thus,

about the x-axis (flexion/extension)

$$\alpha = \sin^{-1} \frac{-R_{32}}{\cos \beta} = \sin^{-1} \frac{-T_{R32}}{\cos \beta}$$

if $\dfrac{-T_{R32}}{\cos \beta} > 0$, then $\alpha = \pi - \varphi$ (2.46)

and if $\dfrac{-T_{R32}}{\cos \beta} > \pi$ then $\alpha = \alpha - 2\pi$

For the third angle, γ, we use the entry in the second row and first column of the [R] matrix (R_{21}) and the cosine of β. Thus,

about the z-axis (axial rotation)

$$\gamma = \sin^{-1} \frac{-R_{21}}{\cos \beta} = \sin^{-1} \frac{-T_{R21}}{\cos \beta}$$

if $\dfrac{-T_{R21}}{\cos \theta} > 0$, then $\gamma = \pi - \gamma$ (2.47)

and if $\dfrac{-T_{R21}}{\cos \theta} > \pi$ then $\gamma = \gamma - 2\pi$

Figure 2.13 represents a time series of the 3-D hip angles during walking that was calculated using this approach. From these graphs, we can see that the flexion/extension graph is very similar to a comparable 2-D graph. It is also clear that the range of motion of the abduction/adduction and axial rotation angles is not very large relative to the flexion angle. Davis et al. (1991) stated that abduction/adduction and axial rotation angles are not used in their clinical laboratory because of the poor signal-to-noise ratio associated with these data. However, while these angles are small, it is thought that they may contain information that is pertinent to understanding human locomotion.

Calculate the hip angles with the same formulae that were used for the knee angles, but use the pelvis LCS and thigh LCS. The same formulae are also solved for the ankle angles, using the leg LCS and foot LCS, but the terminology concerning the actions is different. For the ankle, the flexion/extension angle, α, refers to ankle dorsiflexion/plantar flexion; the second angle, β, is more appropriately referred to as ankle inversion/eversion; and the axial rotation angle, γ, is referred to as abduction/adduction.

For the left lower extremity, the LCS of each segment is computed in exactly the same manner as for the right limb. However, the i' direction is medially

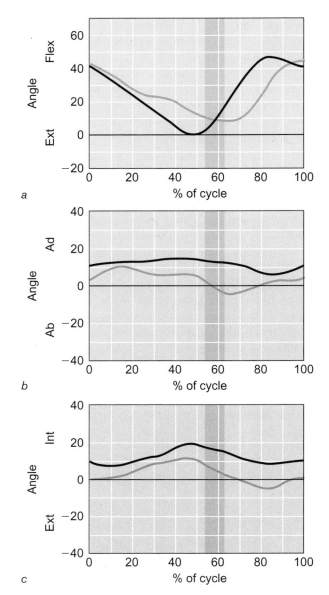

a

b

c

○ **Figure 2.13** Walking 3-D hip angles of a normal subject (blue line) and a cerebral palsy subject (black line) over a stride cycle: *(a)* flexion/extension, *(b)* adduction/abduction, and *(c)* internal/external axial rotation.

2.13a Adapted, by permission, from D.H. Sutherland, K.R. Kaufman, and J.R. Moitoza, 1994, "Kinematics of normal human walking. In *Human Walking*, edited by J. Rose and J.G. Gamble (Baltimore: Williams and Wilkins), 23-44.

directed, whereas it was laterally directed in the right limb. Maintaining the orientation LCS of each segment results in only a slight adjustment to the formulae for abduction/adduction and axial rotation angles and no adjustment to the flexion/extension angle. To maintain the same polarity between right and left limbs, the left limb abduction/adduction and axial rotation angles simply need to be multiplied by –1.

From the Scientific Literature

Davis, R.B., S. Ounpuu, D. Tyburski, and J.R. Gage. 1991. A gait analysis data collection and reduction technique. *Human Movement Science* 10:575-87.

Human motion can be characterized by means of gait analysis. Gait analysis is the systematic measurement, description, and assessment of quantities that are representative of human motion. In this paper, Davis et al. describe the development of the Newington Children's Hospital Gait Laboratory, with emphasis on the equipment used in the laboratory, the marker set, and the computations used to quantify human locomotion. In this laboratory, for each segment LCS, Y is medially directed, X is anteriorly directed, and Z is vertically directed. Joint angle determination is based on Euler angles with a Y-X-Z rotation sequence, corresponding to a flexion/extension, adduction/abduction, and internal/external sequence (see figure 2.14).

The authors stated that a number of refinements of the data-collection procedure were under way. These included improving the hip-joint centering algorithm and developing a better foot model.

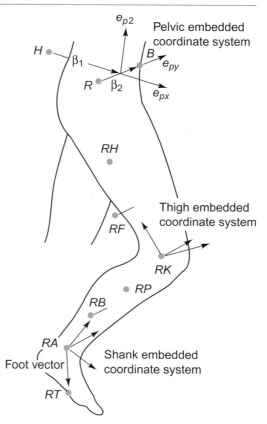

° **Figure 2.14**   Lower-extremity embedded coordinate systems used to compute joint angles.

Reprinted, by permission, from R.B. Davis et al., 1991, "A gait analysis data collection and reduction technique," *Human Movement Science* 10: 575-587.

JOINT COORDINATE SYSTEM

The joint coordinate system (JCS) was first proposed to describe the motion of the knee joint (Grood and Suntay 1983) and has since been applied to the other lower-extremity joints. The method was developed so that all three rotations between body segments had a functional, anatomical meaning. The JCS approach uses one coordinate axis from each LCS of the two segments that constitute the joint. In the example of the knee, the longitudinal axis of the JCS is the $\vec{k}'$ axis of the LCS of the distal segment and the laterally directed axis is the $\vec{i}'$ axis from the LCS of the proximal segment. The third axis is a floating axis that is the cross product of the longitudinal and lateral axes and thus is perpendicular to the plane formed by these directed axes ($\vec{k}'_{distal} \times \vec{i}'_{distal}$). It should be clear that the vertical and lateral axes of this system are not necessarily perpendicular. The JCS is, therefore, not an orthogonal system. In addition, the coordinate system may not have a common origin. The JCS for an instant in time for the knee joint is schematically presented in figure 2.15.

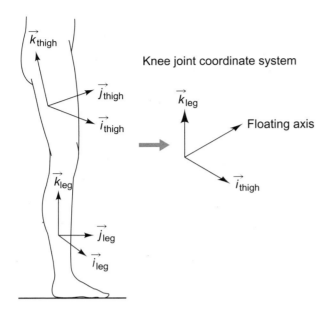

Knee joint coordinate system

Floating axis

Figure 2.15 A diagrammatic representation of the JCS of the knee. The vertical axis is the k-axis of the leg LCS, the medial axis is the i-axis of the thigh LCS, and the floating axis is the cross product of the other two axes.

The angles for the JCS are once again designated as α for flexion/extension, β for abduction/adduction, and γ for external/internal rotation. Flexion/extension is assumed to be a rotation about the laterally directed axis of the proximal segment ($i'_{proximal}$), abduction/adduction is a rotation about the floating axis, and axial rotation is a rotation about the vertical axis of the distal segment ($\vec{k}'_{distal}$). The floating axis (FA) is defined as

$$\vec{FA} = \frac{\left(\vec{k}'_{distal} \times \vec{i}'_{proximal} \right)}{\left| \vec{k}'_{distal} \times \vec{i}'_{proximal} \right|} \qquad (2.48)$$

Based on the marker set that has been used throughout this chapter, the right limb angle calculations are as follows:

flexion/extension

$$\alpha = 90 - \cos^{-1}\left(\vec{k}'_{proximal} \bullet \vec{FA} \right) \qquad (2.49)$$

abduction/adduction

$$\beta = -90 - \cos^{-1}\left(\vec{k}'_{distal} \bullet \vec{i}'_{proximal} \right) \qquad (2.50)$$

axial rotation

$$\gamma = 90 - \cos^{-1}\left(\vec{i}'_{distal} \bullet FA \right) \qquad (2.51)$$

The left limb joint angles are calculated in exactly the same manner. However, to maintain the same polarity between left and right limbs, the abduction/adduction and axial rotation angles must be multiplied by –1.

HELICAL ANGLES

Another method of representing the orientation of one reference system to another is based on the *finite helical axis* or *screw axis* (Woltring et al. 1985; Woltring 1991). In this technique, a position vector and an orientation vector are defined. Any finite movement from a reference position can be described in terms of a rotation about and translation along a single directed line or axis (i.e., the helical or screw axis) in space with unit direction $\vec{n}$ (figure 2.16). Note that in many cases this axis will not coincide with any of the defined axes of the LCS of either segment. Thus, the instantaneous position and orientation of one LCS are defined and described with respect to the other LCS (Spoor and Veldpaus 1980).

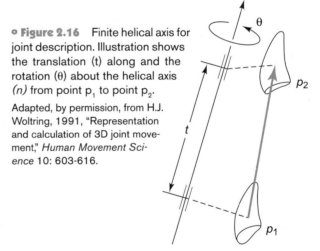

Figure 2.16 Finite helical axis for joint description. Illustration shows the translation (t) along and the rotation (θ) about the helical axis *(n)* from point p₁ to point p₂.

Adapted, by permission, from H.J. Woltring, 1991, "Representation and calculation of 3D joint movement," *Human Movement Science* 10: 603-616.

The orientation vector is defined from the transformation matrix (T_R) as calculated previously (see equation 2.34). The orientation vector components are calculated using the relationships outlined by Spoor and Veldpaus (1980). The orientation components can be determined as:

$$\sin\theta\vec{n} = \frac{1}{2}\begin{bmatrix} T_{R32} - T_{R23} \\ T_{R13} - T_{R31} \\ T_{R21} - T_{R12} \end{bmatrix} \qquad (2.52)$$

If $\vec{n}^T \vec{n} = 1$ and $\sin \theta \leq \frac{1}{2}\sqrt{}$, we can use the following equation to solve for $\sin \theta$:

From the Scientific Literature

McClay, I., and K. Manal. 1997. Coupling parameters in runners with normal and excessive pronation. *Journal of Applied Biomechanics* 13:109-24.

The purpose of this study was to investigate differences in the coupling behavior of foot and knee motions during the support phase of running in subjects classified as normal pronators and in those with excessive pronation. Three-dimensional data were collected and ankle and knee angles were computed using the JCS analysis described in the previous section (figure 2.17).

Excursion ratios between rear-foot eversion and tibial internal rotation were compared between the two groups and found to be significantly lower in the pronator subjects. Timing between peak eversion, knee flexion, and knee internal rotation was also assessed. The timing between peak knee and rear-foot angles was not significantly different between the groups, although times were more closely matched in the normal subjects. Results of this study suggested that increased motion of the rear foot can lead to excessive movement at the knee.

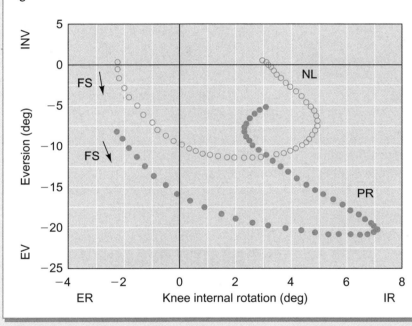

○ **Figure 2.17** Ensemble angle-angle diagram of ankle eversion and knee internal rotation for normal and excessive pronation subjects.

Reprinted, by permission, from I. McClay and K. Manal, 1997, "Coupling parameters in runners with normal and excessive pronation," *Journal of Applied Biomechanics* 13: 109-124.

$$\sin\theta = \left(\frac{1}{2}\sqrt{\left(T_{R32} - T_{R23}\right)^2 + \left(T_{R13} - T_{R31}\right)^2 + \left(T_{R21} - T_{R12}\right)^2}\right) \tag{2.53}$$

However, if $\sin\theta > \frac{1}{2}\sqrt{}$, we can use the following equation to solve for $\cos\theta$:

$$\cos\theta = \frac{1}{2}\left(T_{R11} + T_{R22} + T_{R33} - 1\right) \tag{2.54}$$

We can then calculate the unit vector, $\vec{n}$, along the helical axis as

$$\vec{n} = \frac{\frac{1}{2}\begin{bmatrix} T_{R32} - T_{R23} \\ T_{R13} - T_{R31} \\ T_{R21} - T_{R12} \end{bmatrix}}{\sin\theta} \tag{2.55}$$

Using these components, we can decompose the helical angle into three orthogonal angles:

$$\theta = \begin{bmatrix} \theta_x \\ \theta_y \\ \theta_z \end{bmatrix} = \begin{bmatrix} n_x \\ n_y \\ n_z \end{bmatrix} = \frac{\theta}{2\sin\theta}\begin{bmatrix} T_{R32} - T_{R23} \\ T_{R13} - T_{R31} \\ T_{R21} - T_{R12} \end{bmatrix} \tag{2.56}$$

The three angles computed using the helical axis procedure represent the orientation of the segment in the same manner as the three Cardan or the three JCS angles do.

CARDAN/EULER SYSTEM VS. JOINT COORDINATE SYSTEM VS. HELICAL ANGLES

Of the methods presented here, the Cardan/Euler and JCS techniques are the most widely used for calculating 3-D joint angles. Neither of these two approaches appears to have any obvious advantages or disadvantages over each other. In fact, the principle of the two approaches is the same and the angles calculated by each should give exactly the same results. There are advantages and drawbacks, however, to using all of these techniques.

The major advantage for Cardan/Euler angles is that they are widely used in biomechanics and provide a well-understood anatomical representation of joint movement. However, the sequence dependence of the angles is thought to be a disadvantage. More important is the problem of *gimbal lock*. This occurs when the second rotation equals ±π divided by 2, resulting in a mathematical singularity. However, this is generally a much greater problem in the upper extremity than in the lower extremity.

Helical angles are especially good when the rotation is very small, and they eliminate the problem of gimbal lock. However, the representation of joint motion provided by helical angles does not correspond with an anatomical representation that is clinically meaningful. In addition, helical angles are very sensitive to noisy coordinate data. Therefore, the coordinate data must be substantially smoothed before the helical angles are calculated.

SEGMENT ANGLES

Segment angles represent the orientation of the individual segment to the GCS. They can be calculated in much the same way that joint angles are. However, instead of using an LCS matrix for each segment (see equations 2.31 and 2.32), the LCS matrix of the segment in question and the identity matrix representing the GCS can be used. From these two matrices, a transformation matrix can be computed (see equation 2.34). When this is done, the segment angles can be computed. Usually, either a Cardan/Euler or helical approach can be used for this purpose.

SUMMARY

The definition of the term *kinematics* applies equally in both 2-D and 3-D space. It should be stressed that 3-D joint angles are not equivalent to planar-view 2-D angles. In order to calculate 3-D joint angles, a multicamera setup is necessary. The 2-D views of each camera are then converted to a 3-D view of the movement, generally via a direct linear transformation method. When establishing the 3-D view of the movement, a GCS must be defined.

Various marker setups, all of which have pros and cons, can be used. Minimally, however, three noncollinear markers should define each segment. These markers are used to define an LCS for each segment that moves with the segment as the segment moves through space. We can use the LCSs of two segments to determine a rotational transformation matrix. The matrix forms the basis for the calculation of both joint and segment angles.

Several methods are used to calculate joint angles, including (1) the Cardan/Euler approach, (2) the JCS approach, and (3) the finite helical axis approach. The most widely used in the biomechanics literature appear to be the Cardan/Euler and JCS approaches.

SUGGESTED READINGS

Allard, P., A. Capozzo, A. Lundberg, and C. Vaughan. 1998. *Three-Dimensional Analysis of Human Locomotion.* Chichester, U.K.: John Wiley & Sons.

Nigg, B.M., and W. Herzog. 1994. *Biomechanics of the Musculo-skeletal System.* New York: John Wiley & Sons.

Vaughan, C.L., B.L. Davis, and J.C. O'Connor. 1992. *Dynamics of Human Gait.* Champaign, IL: Human Kinetics.

Zatsiorsky, V.M. 1998. *Kinematics of Human Motion.* Champaign, IL: Human Kinetics.

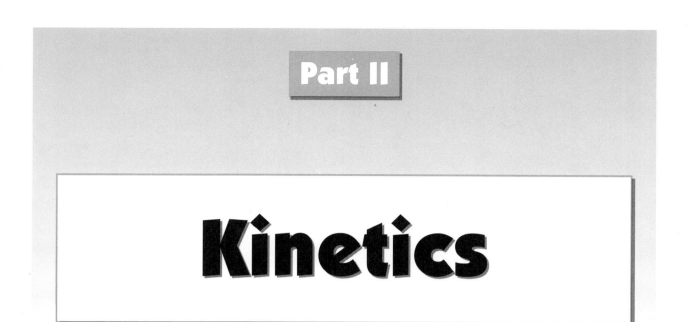

Kinetics

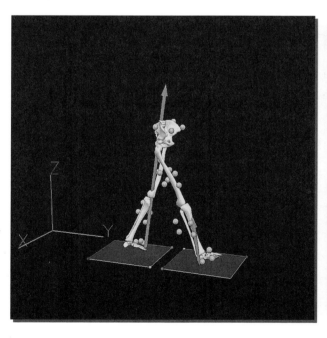
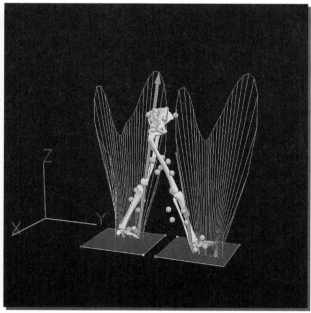

CHAPTER 3

Body Segment Parameters

D. Gordon E. Robertson

Anthropometry is the discipline concerned with the measurement of the physical characteristics of humans. Biomechanists are mainly interested in the inertial properties of the body and its segments, a subfield of anthropometry called body segment parameters. Before a kinetic analysis of human movement is possible, each segment's physical characteristics and inertial properties must be determined. The relevant characteristics are the segmental mass, locations of the segmental centers of gravity, and segmental mass moments of inertia.

This chapter is not meant to be a detailed account of the study of anthropometry or body segment parameters; they are described elsewhere (e.g., Contini 1972; Drillis, Contini, and Bluestein 1964; Nigg and Herzog 1994; Zatsiorsky 2002). The purpose of this chapter is

- to present background material on measuring and estimating body segment parameters,
- to outline computational methods for quantifying total body and segmental inertial characteristics for planar (2-D) motion analyses (essential for understanding chapters 5 and 6), and
- to outline computational methods for quantifying total body and segmental inertial characteristics for spatial (3-D) motion analyses (essential for understanding chapter 7).

To learn planar motion analysis only, the section titled Three-Dimensional (Spatial) Computational Methods can be omitted from the course of study. Equivalently, if only spatial (i.e., 3-D) analysis is undertaken, the moment of inertia part of the Two-Dimensional (Planar) Computational Methods section can be omitted.

METHODS FOR MEASURING AND ESTIMATING BODY SEGMENT PARAMETERS

A major concern for biomechanists is the assumption that body segments behave as rigid bodies during movements. This assumption obviously is not valid because bones bend, blood flows, ligaments stretch, and muscles contract. It is also common to model some body parts as single rigid bodies despite the fact that they consist of several segments. For instance, the foot is commonly considered to be one segment even though it clearly can bend at the metatarsal-phalangeal joints. Similarly, the trunk is often treated as one single rigid body, but sometimes as two or three. In reality, however, it is a series of interconnected rigid bodies including the many interconnected vertebrae, pelvis, and scapulae. These assumptions simplify the otherwise complex musculoskeletal system and eliminate the necessity of quantifying the changes in mass distribution caused by tissue deformation and movement of bodily fluids. By also assuming that segmental mass distribution is similar among members of a particular population, an individual's segmental parameters can be estimated by applying equations based on the averages obtained from samples taken from the population. Several sources offer these averaged parameters, but it is best to select from a population that closely matches the subject (Hay 1973).

Attempts to quantify body segment parameters generally fall into four categories: cadaver studies (e.g., Braune and Fischer 1889; Dempster 1955; Fischer 1906; Harless 1860), mathematical modeling (e.g., Hanavan 1964; Hatze 1980; Yeadon 1990a, 1990b; Yeadon and Morlock 1989), scanning and imaging techniques (e.g., Durkin and Dowling 2003; Durkin, Dowling, and Andrews 2002; Mungiole and Martin 1990; Zatsiorsky and Seluyanov 1983), and kinematic measurements (e.g., Hatze 1975; Dainis 1980; Vaughan, Davis, and O'Connor 1992). Each of these techniques has both advantages and disadvantages. We next review some of the literature that utilizes one or more of these technologies.

CADAVER STUDIES

Inertial properties (mass, center of mass, moment of inertia) are difficult to determine for a particular living person. If you were to quantify these properties for a robot, each segment would be separated and analyzed individually by performing specific tests. Because this is not possible for living persons, indirect methods must be used. For example, the *coefficient method* uses tables of proportions that predict the body segment parameters from simple, noninvasive measures such as total body mass, height, and segment lengths. The earliest attempts at this procedure date to the works of Harless (1860), Braune and Fischer (1889), and Fischer (1906), but the most significant advancement was the work done by W.T. Dempster, published in 1955 as the monograph *Space Requirements of the Seated Operator*. This document, which Dempster compiled while working for the U.S. Air Force, not only outlined the procedures for measuring body segment parameters from cadaveric materials, but also included tables for proportionally determining the body segment parameters needed to biomechanically analyze human motion.

Dempster collected data from living persons, anatomical specimens, and, most importantly (for biomechanists), from eight complete cadavers. First, he used the method of Reuleaux (1876) to determine the average center of rotation at each joint (see figure 3.1). At some joints, particularly the shoulder, the center of rotation was difficult to identify, and these locations became the endpoints for the various body segments. Table 3.1 describes these endpoints.

The cadavers were then segmented according to Dempster's own techniques and their lengths, masses, and volumes were carefully recorded. Dempster then calculated the location of the center of gravity (using a balancing technique) and the moment of inertia (using a pendulum technique) for each seg-

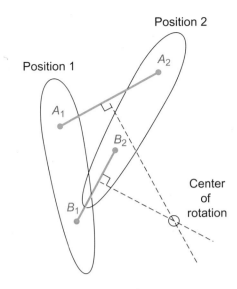

○ Figure 3.1   Reuleaux's (1876) method for calculating the center of rotation of a body. Points A and B are fixed on the rigid body. After movement from Position 1 to Position 2, dotted lines are drawn as right bisectors of the lines A_1-A_2 and B_1-B_2. The center of rotation is at the intersection of the dotted lines.

ment. Finally, Dempster created tables showing the segmental masses as proportions of the total body mass and the locations of the centers of gravity and lengths of the radii of gyration as proportions of the segments' lengths. (The radius of gyration was used as an indirect means of calculating rotational inertia. Its meaning and relationship with moment of inertia is presented in the next section.) These data in modified form (Miller and Nelson 1973; Plagenhoef 1971; Winter 1990) appear in table 3.2. Later, Barter (1957), working with Dempster's data, performed stepwise regression analysis to derive regression equations that more accurately compute segmental masses.

Many other body segment parameter studies have been conducted since Dempster's groundbreaking work. Two in particular, by Clauser, McConville, and Young (1969) and Chandler et al. (1975), are noteworthy because they defined body segments using palpable bony landmarks instead of estimated, averaged joint centers of rotation. They also segmented an additional 13 cadavers (only 6 were used by Chandler et al.) and presented the results in a table similar to Dempster's. The segment definitions appear in table 3.3 and the values are presented in table 3.4.

For their 3-D kinematic and kinetic investigation of the lower extremity during human gait, Vaughan, Davis, and O'Connor (1992) enhanced the methods used by Chandler et al. (1975). They created regression equations for the masses of the lower extremities that included various segmental anthropometrics, such as calf and midthigh circumference, in addition

Table 3.1 Dempster's Segment Endpoint Definitions

Segment	Proximal end	Distal end
Clavicular link	Sternoclavicular joint center: midpoint of palpable junction between proximal end of clavicle and upper border of sternum	Claviscapular joint center: midpoint of line between coracoid tuberosity of clavicle and acromioclavicular articulation at lateral end of clavicle
Scapular link	Claviscapular joint center: see Clavicular link, Distal end	Glenohumeral joint center: midpoint of palpable bony mass of head and tuberosities of humerus
Humeral link	Glenohumeral joint center: see Scapular link, Distal end	Elbow joint center: midpoint between lowest palpable point of medial epicondyle of humerus and a point 8 mm above the radiale (radiohumeral junction)
Radial link	Elbow joint center: see Humeral link, Distal end	Wrist joint center: distal wrist crease at palmaris longus tendon or palpable groove between lunate and capitate bone in line with metacarpal bone III
Hand link	Wrist joint center: see Radial link, Distal end	Center of mass of hand: midpoint between proximal transverse palmar crease and radial longitudinal crease in line with third digit
Femoral link	Hip joint center: tip of femoral trochanter 1 cm anterior to most laterally projecting part of greater trochanter	Knee joint center: midpoint between centers of posterior convexities of femoral condyles
Leg link	Knee joint center: see Femoral link, Distal end	Ankle joint center: level of a line between tip of lateral malleolus of fibula and a point 5 mm distal of the tibial malleolus
Foot link	Ankle joint center: see Leg link, Distal end	Center of gravity of foot: midway between ankle joint center and ball of foot at head of metatarsal II

Table 3.2 Dempster's Body Segment Parameters

Segment	Endpoints[a] (proximal to distal)	Segmental mass/ total mass (P)[b]	Center of mass/ segment length (R_proximal)[c]	(R_distal)[c]	Radius of gyration/ segment length (K_cg)[d]	(K_proximal)[d]	(K_distal)[d]
Hand	Wrist center to knuckle II of third finger	0.0060	0.506	0.494	0.298	0.587	0.577
Forearm	Elbow to wrist center	0.0160	0.430	0.570	0.303	0.526	0.647
Upper arm	Glenohumeral joint to elbow center	0.0280	0.436	0.564	0.322	0.542	0.645
Forearm and hand	Elbow to wrist center	0.0220	0.682	0.318	0.468	0.827	0.565
Upper extremity	Glenohumeral joint to wrist center	0.0500	0.530	0.470	0.368	0.645	0.596
Foot	Ankle to ball of foot	0.0145	0.500	0.500	0.475	0.690	0.690
Leg	Knee to ankle center	0.0465	0.433	0.567	0.302	0.528	0.643
Thigh	Hip to knee center	0.1000	0.433	0.567	0.323	0.540	0.653
Lower extremity	Hip to ankle center	0.1610	0.447	0.553	0.326	0.560	0.650

(continued)

Table 3.2   (continued)

Segment	Endpoints[a] (proximal to distal)	Segmental mass/ total mass	Center of mass/ segment length		Radius of gyration/ segment length		
		(P)[b]	(R$_{proximal}$)[c]	(R$_{distal}$)[c]	(K$_{cg}$)[d]	(K$_{proximal}$)[d]	(K$_{distal}$)[d]
Head	C7-T1 to ear canal	0.0810	1.000	0.000	0.495	1.116	0.495
Shoulder	Sternoclavicular joint to glenohumeral joint center	0.0158	0.712	0.288			
Thorax	C7-T1 to T12-L1	0.2160	0.820	0.180			
Abdomen	T12-L1 to L4-L5	0.1390	0.440	0.560			
Pelvis	L4-L5 to trochanter	0.1420	0.105	0.895			
Thorax and abdomen	C7-T1 to L4-L5	0.3550	0.630	0.370			
Abdomen and pelvis	T12-L1 to greater trochanter	0.2810	0.270	0.730			
Trunk	Greater trochanter to glenohumeral joint	0.4970	0.495	0.505	0.406	0.640	0.648
Head, arms, and trunk	Greater trochanter to glenohumeral joint	0.6780	0.626	0.374	0.496	0.798	0.621
Head, arms, and trunk	Greater trochanter to mid-rib	0.6780	1.142	−0.142	0.903	1.456	0.914

[a] Endpoints are defined in table 3.1

[b] A segment's mass as a proportion of the total body mass

[c] The distances from the proximal and distal ends of the segment to the segment's center of gravity as proportions of the segment's length

[d] The radii of gyration about the center of gravity, proximal and distal ends of the segment to the segment's center of gravity as proportions of the segment's length

Adapted, by permission, from D.A. Winter, 1990, *Biomechanics and motor control of human movement*, 2nd ed. (Toronto: John Wiley and Sons, Inc.).

Table 3.3   Clauser et al.'s Definitions of Endpoints of Body Segment Parameters

acromion	Point at the superior and external border of the acromion process of the scapula. This point is easier to find if the subject bends laterally at the trunk to relax the deltoid muscle.
glabella	Bony ridge under the eyebrows.
metacarpale III	Distal head of the third metacarpal (proximal knuckle of the middle finger).
occiput	Occipital process.
radiale	Point at the proximal and lateral border of the head of the radius. Palpate downward in the lateral dimple at the elbow. Have the subject pronate and supinate the forearm slowly so that the radius may be felt rotating beneath the skin.
sphyrion	Most distal point of the medial malleolus of the tibia. Place the marker on the fibula at the level of the sphyrion. Do not use the sphyrion fibulare, as it is more distal than the sphyrion.
stylion	Styloid process of the radius. The styloid process of the ulna may be used.

| tibiale | Most proximal point of the medial border of the head of the tibia. Use the tibiale externum, which is equivalent. Palpate the tendon of the quadriceps at the distal end of the patella and follow the border of the tibia laterally to the frontal border of the lateral collateral ligament. The point is easier to find if the subject unloads the leg and slightly flexes the knee. |
| trochanter | Most superior point on the greater trochanter of the femur, which is difficult to locate in women and people with excessive subcutaneous fat. Have the subject abduct and move the leg forward and backward. Palpate the femur with the index finger, and move upward gradually until the most superior palpable point on the trochanter is found. |

Table 3.4 Clauser, McConville, and Young's Body Segment Parameters

Segment	Endpoints[a] (proximal to distal)	Segmental mass/ total mass (P)[b]	Center of mass/ segment length		Radius of gyration/ segment length	
			$(R_{proximal})^c$	$(R_{distal})^c$	$(K_{cg})^{d, e}$	$(K_{proximal})^d$
Hand	Stylion to metacarpale III	0.0065	0.1802	0.8198	0.6019	0.6283
Forearm	Radiale to stylion	0.0161	0.3896	0.6104	0.3182	0.5030
Upper arm	Acromion to radiale	0.0263	0.5130	0.4870	0.3012	0.5949
Forearm and hand	Radiale to stylion	0.0227	0.6258	0.3742		
Upper extremity	Regression equation[f]	0.0490	0.4126	0.5874		
Foot	Heel to toe II	0.0147	0.4485	0.5515	0.4265	0.6189
Foot	Sphyrion to floor	0.0147	0.4622	0.5378		
Leg	Tibiale to sphyrion	0.0435	0.3705	0.6295	0.3567	0.5143
Thigh	Trochanter to tibiale	0.1027	0.3719	0.6281	0.3475	0.5090
Leg and foot	Tibiale to floor	0.0582	0.4747	0.5253		
Lower extremity	Trochanter to floor (sole)	0.1610	0.3821	0.6179		
Trunk	Chin-neck intersect. to trochanter[g]	0.5070	0.3803	0.6197	0.4297	0.5738
Head	Top of head to chin-neck intersect.	0.0728	0.4642	0.5358	0.6330	0.7850
Trunk and head	Chin-neck intersect. to trochanter	0.5801	0.5921	0.4079		
Total body		1.0000	0.4119	0.5881	0.7430	0.8495

[a] Endpoints are defined in table 3.3

[b] A segment's mass as a proportion of the total body mass

[c] The distances from the proximal and distal ends of the segment to the segment's center of gravity as proportions of the segment's length

[d] The radii of gyration about the center of gravity and the proximal end of the segment to the segment's center of gravity as proportions of the segment's length

[e] Computed from $K_{cg} = \sqrt{K_{proximal}^2 - R_{proximal}^2}$

[f] Regression equation for arm length = 1.126 (acromion to radiale distance) + 1.057 (radiale to stylion distance) + 12.52 (all lengths in cm)

[g] Chin-neck intersection. The point superior to the cartilage, at the level of the hyoid bone. Marker should be placed level with the intersection, but at the lateral aspect of the neck.

From Clauser et al., 1969, "Weight, volume and center of mass of segments of the human body," AMRL-TR-69-70 and Chandler et al., 1975, "Investigation of inertial properties of the human body," AMRL-TR-74-137, both from Wright-Patterson Air Force Base.

to segment length and body mass (table 3.5). Hinrichs (1985) also extended Chandler's work by applying regression equations for the calculation of segmental moments of inertia. For in-depth reviews of the historical and scientific development of body segment parameters, refer to Contini (1972), Drillis, Contini, and Bluestein (1964), and Hay (1973, 1974).

MATHEMATICAL MODELING

Mathematical modeling of the inertial properties of human body segments was pioneered by the work of Hanavan (1964) when it became necessary to model the body for 3-D analyses. Hanavan made the assumption that mass was uniformly distributed within each segment and that the segments were rigid bodies that could be represented by geometrical shapes. Most of the segments were modeled as frusta of right-circular cones, as shown in figure 3.10. (A *frustum* of a cone—*frusta* in the plural—is the base of a cone that has had its vertex removed.) The hands were modeled as spheres, the head as an ellipsoid, and the trunk segments as elliptical cylinders. Table 3.4 contains equations used for the computation of the principal mass moments of inertia of various uniform solids. By taking additional anthropometric measures, such as midthigh circumference, malleolus height, knee diameter, and biacromial breadth, equations were developed to compute the three principal moments of inertia. Hanavan's methods have since been enhanced to include more segments and more directly measured anthropometric measures. For example, Hatze (1980) developed a system that uses 242 direct anthropometric measurements to define the segmental properties of a 42-DOF, 17-segment model of the body.

Table 3.5 Body Segment Parameters for Equations Developed by Vaughan, Davis, and O'Connor

Segment	Axis	Regression equation[a]
Foot	Flexion/extension	$I_{f/e} = 0.00023\,m_{total}\left[4\left(h_m^2\right) + 3\left(l_{foot}^2\right)\right] + 0.00022$ $h_m = malleolus\ height\ (m)$
	Abduction/adduction	$I_{a/a} = 0.00021\,m_{total}\left[4\left(h_m^2\right) + 3\left(l_{foot}^2\right)\right] + 0.00067$
	Internal/external rotation	$I_{i/e} = 0.00041\,m_{total}\left[h_m^2 + w_{foot}^2\right] - 0.00008$ $w_{foot} = \text{foot width }(m)$
Leg	Flexion/extension	$I_{f/e} = 0.00347\,m_{total}\left[l_{leg}^2 + 0.076\,c_{leg}^2\right] + 0.00511$ $c_{leg} = \text{leg circumference }(m)$
	Abduction/adduction	$I_{a/a} = 0.00387\,m_{total}\left[l_{leg}^2 + 0.076\,c_{leg}^2\right] + 0.00138$
	Internal/external rotation	$I_{i/e} = 0.00041\,m_{total}\,c_{leg}^2 + 0.00012$
Thigh	Flexion/extension	$I_{f/e} = 0.00762\,m_{total}\left[l_{thigh}^2 + 0.076\,c_{thigh}^2\right] + 0.01153$ $c_{thigh} = \text{thigh circumference }(m)$
	Abduction/adduction	$I_{a/a} = 0.00762\,m_{total}\left[l_{thigh}^2 + 0.076\,c_{thigh}^2\right] + 0.01186$
	Internal/external rotation	$I_{i/e} = 0.00151\,m_{total}\,c_{thigh}^2 + 0.00305$

[a] Moments of inertia are in kilogram meters squared (kg·m2), m_{total} is the total body mass in kilograms, and I_{thigh}, I_{leg}, and I_{foot} are the segment lengths in meters.

Reprinted, by permission, from C.L. Vaughan, B.L. Davis, and J.C. O'Connor, 1992, *Dynamics of Human Gait* (Champaign, IL: Human Kinetics).

SCANNING AND IMAGING TECHNIQUES

Another approach to determining inertial properties and then estimating body segment parameters involves scanning the living body with various radiation techniques. For instance, Zatsiorsky and Seluyanov (1983, 1985) presented data from an extensive study on the body segment parameters of both male (n = 100) and female (n = 15) living subjects. To compute mass distribution, gamma mass scanning was used to quantify the density of incremental slices of each segment. This method enabled estimations to be made of the mass, center of mass, and principal moments of inertia in 3-D for a total of 15 segments. Their subjects included individuals younger than the ages of the cadavers used in previous studies; furthermore, they applied regression equations to customize body segment parameters. These values and methods were recently documented and republished by Zatsiorsky (2002, 583-616).

In addition to gamma mass scanning, other techniques that have been developed to quantify human body segment parameters include photogrammetry (Jensen 1976, 1978), magnetic resonance imaging (Mungiole and Martin 1990), and dual-energy X-ray absorptiometry (DEXA; Durkin and Dowling 2003; Durkin, Dowling, and Andrews 2002).

Furthermore, some researchers have provided data from populations that are not well represented by the previously mentioned studies, which were based primarily on middle-aged male or young adult Caucasians. For example, Jensen and colleagues provided data for children (1986, 1989) and pregnant women (1996), Schneider and Zernicke (1992) quantified regression equations for the body segment parameters of infants, and Pavol, Owings, and Grabiner (2002) estimated inertial properties for the elderly.

KINEMATIC TECHNIQUES

Kinematic techniques are methods that measure kinematic characteristics to indirectly determine the inertial properties of segments. Hatze (1975) developed an *oscillation technique* that derives the mass, center of mass, and moment of inertia of the segments of the extremities and the damping coefficients of joints. The technique, which cannot be used for the trunk segments, requires that a body part be set into oscillation with an instrumented spring. The muscles must

From the Scientific Literature

Schneider, K., and R.F. Zernicke. 1992. Mass, center of mass and moment of inertia estimates for infant limb segments. *Journal of Biomechanics* 25:145-8.

The purpose of this study was to develop linear regression equations that could quantify the body segment parameters for infants (0.4 to 1.5 years). The authors recognized that before the mechanics of infant motion could be analyzed, they needed to develop a new set of anthropometric proportions applicable to infants. Clearly, the mass distribution of infants is different from that of adults. This is the reason that adult seat belts cannot be used by infants or young children, whose total body center of gravity is proportionally higher than that of an adult.

The authors collected data from the upper limbs of 44 infants and the lower limbs of 70 infants. They adapted a 17-segment mathematical model developed by Hatze (1979, 1980) to data from 18 infants. Hatze's model requires 242 anthropometric measurements to compute the volumes of the 17 segments. These data were adjusted so that there was a close agreement between the model's estimates of total body mass and the infants' measured masses. Regression analysis was then used to derive mass, center of mass, and moment of inertia for the three segments of the upper limb and three segments of the lower limb. The researchers found that infants' segment masses and moments of inertia changed remarkably during the first 1 1/2 years of growth, but that the center of mass was not greatly influenced by age. They also concluded that, because of the high correlations achieved by their linear regression equations (64 to 98% variance accounted), nonlinear regression was unnecessary.

be relaxed so that they do not influence the damped oscillations of the spring-limb system. Mathematically derived equations based on small oscillation theory are then used to estimate the properties of the joint and segment based on a passively damped reduction of the oscillations of the spring-limb system.

Another technique for estimating mass moment of inertia is called the *quick-release method* (Drillis, Contini, and Bluestein 1964). This technique also assumes that the muscles are relaxed and that the acceleration of a rapidly accelerated segment is affected only by the segment's rotational inertia. Thus, by measuring the angular acceleration of a segment after release of a known force (F) or moment (M), the moment of inertia (I) of the segment can be derived from the following relationship:

$$I = M / \alpha = (Fd) / \alpha \qquad (3.1)$$

where M (or Fd) is the moment of force immediately before release and α is the angular acceleration after release. Obviously, this method can be used only for terminal segments because applying it to other body parts would be difficult.

TWO-DIMENSIONAL (PLANAR) COMPUTATIONAL METHODS

In the following sections, we present methods for computing body segment parameters for 2-D analyses based on the traditional proportional methods developed by Dempster (1955). These are the simplest types of equations that yield reasonable results for people of average dimensions. Researchers who need more accurate measures should consult the research literature for methods that are suitable for their population of subjects and utilize procedures and equipment that are within the financial and practical constraints of their project. Many excellent reviews of such literature are presented in the Suggested Readings section of this chapter.

SEGMENT MASS

The standard method for computing segment mass is to weigh the subject and then multiply the total body mass by the proportion that each segment contributes to the total. These proportions $(P\text{-values})$ may be taken from tables 3.2 or 3.4, but these values were derived from middle-aged male cadavers and may not be appropriate for all subjects, particularly children and women. For young adult subjects, the values presented by Zatsiorsky (2002) can be used. Segment mass (m_i) is defined as

$$m_s = P_s \, m_{total}, \qquad (3.2)$$

where m_{total} is the total body mass and P_s is the segment's mass proportion. Note that the sum of all the P-values should equal 1; otherwise, the calculated body weight will be incorrect. That is,

$$\sum_{s=1}^{S} P_s = 1.000 \qquad (3.3)$$

where S is the number of body segments and s is the segment number. The values in table 3.2 were derived from the work of Dempster (1955), but they have been adjusted so that they total unity (i.e., 100%). Dempster's original proportions totaled less than 1 as a result of fluid loss during dissection and other sources of error. Miller and Nelson (1973) worked out adjustments to compensate for these losses.

EXAMPLE 3.1

Use the proportions in table 3.2 to calculate the thigh mass of a person who weighs 180.0 kg.

➤   *See answer 3.1 on page 271.*

CENTER OF GRAVITY

Center of gravity and center of mass are in essentially the same locations. In biomechanics, the terms are used interchangeably. A difference occurs only when a body is far from the Earth's surface or otherwise affected by a large gravitational source. The center of gravity is the point where a motionless body, if supported at that point, will remain balanced if suspended—the balance point of a body. It is the point on a rigid body where the mass of the body can be considered to be concentrated for translational motion analyses. In other words, the translational motion of a rigid body concerns only the motion of the body's center of gravity. This lessens the amount of information that needs to be recorded about the body. The body's shape and structure can be ignored and only its center of gravity needs to be quantified.

Segment Center of Gravity

To simplify and generalize the process of calculating segment centers of gravity, Dempster (1955) developed the technique of representing the distances from each endpoint of a segment to that segment's center of gravity as proportions $(R\text{-values})$ of the segment's length (l). Given that $r_{proximal}$ and r_{distal} are the distances from the proximal and distal ends to the segment's center of gravity, respectively, these proportions are defined as

$$R_{proximal} = r_{proximal} / l \qquad (3.4)$$

$$R_{distal} = r_{distal} / l \qquad (3.5)$$

To compute a segment's center of gravity, the coordinates of the segment's endpoint must be quantified. Next, the *R*-value for the particular segment must be selected from a table of proportions suitable for the subject. In general, table 3.2 can be used for adult males because these proportions are averages derived from eight male cadavers. Other tables may be used for other populations. One can select either $R_{proximal}$ or R_{distal}, but by consensus, segment centers are usually defined from their proximal ends. The center is then computed as a proportion of the distance from the proximal end toward the distal end, as depicted in figure 3.2. These equations can then be used to determine the X-Y coordinates of the segment's center (x_{cg}, y_{cg}):

$$x_{cg} = x_{proximal} + R_{proximal} (x_{distal} - x_{proximal}) \quad (3.6)$$

$$y_{cg} = y_{proximal} + R_{proximal} (y_{distal} - y_{proximal}) \quad (3.7)$$

where $(x_{proximal}, y_{proximal})$ and (x_{distal}, y_{distal}) are the coordinates of the proximal and distal ends, respectively. Note that $R_{proximal} + R_{distal} = 1.000$, because they represent the total length (l) of the segment. Also, the actual distance from the proximal end to the center of gravity $(r_{proximal})$ can be computed from

$$r_{proximal} = R_{proximal} \, l \quad (3.8)$$

EXAMPLE 3.2

Calculate the center of gravity of the thigh using the proportions from table 3.2, given that the proximal end (the hip) of the thigh has the coordinates (−12.80, 83.3) cm and the distal end (the knee) has the coordinates (7.30, 46.8) cm.

➤   *See answer 3.2 on page 271.*

Limb and Total Body Center of Gravity

To compute the center of gravity of a limb or combination of segments, a "weighted" average of the segments that make up the limb is computed. These equations define this process:

$$x_{limb} = \frac{\sum_{s=1}^{L} P_s x_{cg_s}}{\sum_{s=1}^{L} P_s} \quad (3.9)$$

$$y_{limb} = \frac{\sum_{s=1}^{L} P_s y_{cg_s}}{\sum_{s=1}^{L} P_s} \quad (3.10)$$

where *L* is the number of segments in the limb, (x_{limb}, y_{limb}) represents the limb's center of gravity, (x_{cg}, y_{cg}) represents each segment's center of gravity, and the P_i are each segment's mass proportion. The total

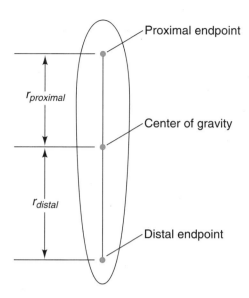

Figure 3.2   Center of gravity in relation to segment endpoints.

body's center of gravity is computed in a similar fashion. It is the weighted average of all the segments of the body. That is,

$$x_{total} = \sum_{s=1}^{L} P_s x_{cg_s} \quad (3.11)$$

$$y_{total} = \sum_{s=1}^{L} P_s y_{cg_s} \quad (3.12)$$

where (x_{total}, y_{total}) is the total body's center of gravity coordinates, (x_{cg}, y_{cg}) are the coordinates of the segments' centers of gravity, and *S* is the number of body segments. Each segment is weighted according to its mass proportion, P_s. In other words, each segment's center of gravity contributes proportionally to its *P*-value, which is its mass as a proportion of the total body mass (figure 3.3). The heavier the segment, the more it affects the location of the center of gravity. Lighter segments have less influence on the center of gravity. Notice that there is no divisor in these equations because, as shown previously, the sum of all the *P*-values is 1, namely, $\Sigma P_s = 1.0$.

MASS MOMENT OF INERTIA

The mass moment of inertia, also called *rotational inertia,* is the resistance of a body to change in its rotational motion. It is the angular or rotational equivalent of mass. In the following pages, the term *moment of inertia* is used instead of *mass moment of inertia.* There is another moment of inertia, one rarely used in biomechanics, that is called the area moment of inertia. It is a geometrical property that does not concern itself with how mass is distributed within a body. Because it is not used in this textbook,

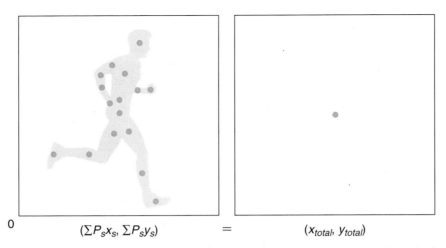

0    $(\Sigma P_s x_s, \Sigma P_s y_s)$ = (x_{total}, y_{total})

○ **Figure 3.3**   The total body center of gravity is a weighted average of the segment centers of gravity (x_s, y_s).

any reference to the term *moment of inertia* is to the mass moment of inertia.

Moments of inertia are needed whenever the rotational motion of a body is investigated. It is not as important a measure as mass because, relatively speaking, it has less influence on the motion of human bodies than does mass. This is because humans are more often concerned with translational motions and because human movements rarely rotate at high rates. Of course, such measures become important whenever acrobatic motions are examined, such as diving, trampolining, figure skating, gymnastics, skateboarding, and aerial skiing. In these sporting activities, 3-D measurements of moments of inertia are required. A simple method for computing the moment of inertia for planar (2-D) analyses of motion is presented next.

Segment Moment of Inertia

Computing the moment of inertia of a segment is not as straightforward as computing the center of gravity. This is because the moment of inertia is not linearly related to segment length. Classically, the moment of inertia is defined as the "second moment of mass"; it is the sum (integral) of mass particles times their squared distances (moments) from an axis. That is,

$$I_{axis} = \int r^2 dm \qquad (3.13)$$

where r is the distance of each mass particle *(dm)* from a point or axis of rotation.

Because moments of inertia differ nonlinearly with the axis they are computed about, they cannot be computed directly using proportions, as the other body segment parameters are. To simplify the computation of moment of inertia, an indirect method that involves calculating the radius of gyration is used. The radius of gyration is the distance that represents how far the mass of a rigid body would be from an axis of rotation if its mass was concentrated at a point. This concept is illustrated in figure 3.4. Thus, a rigid body can be considered

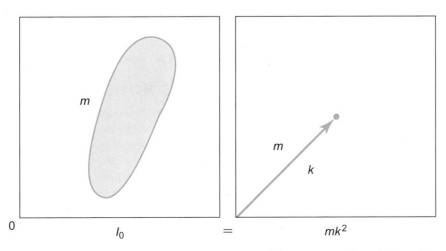

0   I_0 = mk^2

○ **Figure 3.4**   Relationship between the centroidal moment of inertia (I_0) of a rigid body and the radius of gyration (k) of a rotationally equivalent point mass *(m)*.

rotationally equivalent to a point mass that is located at this certain distance—called the radius of gyration, *k*—from the axis of rotation. This distance differs from the center of mass, which is the point at which a rigid body's mass can be considered to be concentrated for linear motions. Note that, whereas the center of gravity of a rigid body is always at the same location on the body, the radius of gyration varies depending on the axis about which the body is rotating. The value of determining radii of gyration is that they can be made proportional to a segment's length, thereby simplifying the task of determining segmental moments of inertia for any axis.

Thus, to compute a segment's moment of inertia using proportions of segment length, first compute the radius of gyration. The radius of gyration (k_{axis}) of a rigid body or segment is defined as the length (radius) that satisfies this relationship:

$$k_{axis} = \sqrt{\frac{I_{axis}}{m}} \qquad (3.14)$$

where I_{axis} is the moment of inertia about the axis *(axis)* and *m* is the mass of the rigid body or segment.

Radius of gyration changes depending on the axis of rotation selected. The minimum radius of a rigid body occurs when the body rotates about its own center of gravity, called its centroidal moment of inertia. Using table 3.2 or 3.3, the radius of gyration of a segment rotating about its center of gravity can be computed from

$$k_{cg} = K_{cg}\, l, \qquad (3.15)$$

where k_{cg} is the radius of gyration in meters and K_{cg} is the length of the radius of gyration as a proportion of the segment's length, *l*.

The segment's centroidal moment of inertia can then be computed from

$$I_{cg} = mk_{cg}^2 \qquad (3.16)$$

where I_{cg} is the segment's moment of inertia for rotations about the segment's center of gravity and *m* is the segment's mass in kilograms. Note that the moment of inertia of a segment about any arbitrary axis is equal to the segment's centroidal moment of inertia, I_{cg}, plus a term equal to the segment's mass times the square of the distance between the segment's center of gravity and the arbitrary axis. This is called the *parallel axis theorem*. That is,

$$I_{axis} = I_{cg} + mr^2 \qquad (3.17)$$

where I_{axis} is the moment of inertia about an arbitrary axis and *r* is the distance from the axis to the center of gravity. Because segments tend to rotate about either

their proximal or distal ends, it is often desirable to compute the segment's moment of inertia about them. Using the proportions $K_{proximal}$ and K_{distal} and knowing the moment of inertia about the segment's center of gravity (I_{cg}), applying the parallel axis theorem yields:

$$I_{proximal} = I_{cg} + m(K_{proximal} \times l)^2 \qquad (3.18)$$

$$I_{distal} = I_{cg} + m(K_{distal} \times l)^2 \qquad (3.19)$$

where $I_{proximal}$ and I_{distal} are the moments of inertia at the proximal and distal ends, respectively; *m* is the segment mass; and *l* is the segment length.

EXAMPLE 3.3

Calculate the thigh's moment of inertia about its center of gravity for a 180 kg person using the proportions from table 3.2 given that the proximal end (the hip) of the thigh has the coordinates (–12.80, 83.3) cm and the distal end (the knee) has the coordinates (7.30, 46.8) cm (the same as in example 3.2).

➤ *See answer 3.3 on page 271.*

A direct method (Plagenhoef 1971:43-4) for determining the moment of inertia of a rigid body involves measuring the period of oscillation (the duration of a cycle) of the object as it oscillates as a pendulum. Typically, 10 to 20 periods are timed and then the duration is computed for a single cycle. The equation for the moment of inertia of the object about the suspension point is computed from

$$I_{axis} = \frac{mgrt^2}{4\pi^2} \qquad (3.20)$$

where I_{axis} is the moment of inertia of the object about an axis through the pivot point of the pendulum in kilograms meters squared (kg·m²), *m* is the mass of the object in kilograms, *g* is 9.81 m/s², *r* is the distance in meters from the suspension point to the center of gravity of the object, and *t* is a single period of oscillation in seconds. Note that for this equation to be valid, the pendulum should not swing more than 5 to 10° to either side of the vertical.

Total Body Moment of Inertia

Although it is rarely used in biomechanics, occasionally a researcher needs to compute the total body moment of inertia. It is tempting to simply add all the segments' centroidal moments of inertia (I_{cg}), but this ignores the fact that each segment has a different location for its center of gravity. Therefore, the parallel axis theorem must be applied to each segment where the distance between each segment's center of gravity and the total body's center of gravity needs to be calculated (r_s in equation 3.21).

Consequently, the total body's centroidal moment of inertia (I_{total}) is the sum of each segment's centroidal moment of inertia (I_{cg}) plus transfer terms based on the parallel axis theorem. That is,

$$I_{total} = \sum_{s=1}^{S} I_{cg_s} + \sum_{s=1}^{S} m_s r_s^2 \qquad (3.21)$$

where S is the number of segments and r_s is the distance between the total body's center of gravity and each segment's center of gravity.

If it then becomes necessary to compute the total body's moment of inertia about a secondary axis (I_{axis}), the parallel axis theorem can be applied once again. Thus,

$$I_{axis} = I_{total} + mr^2 \qquad (3.22)$$

where r is the distance between the total body's center of gravity and the secondary axis.

Note that this moment of inertia can also be calculated more directly as

$$I_{axis} = \sum_{s=1}^{S} I_{cg_s} + \sum_{s=1}^{S} mr_s^2 \qquad (3.23)$$

where r_s is the distances between the segment centers and the secondary axis *(axis)*.

CENTER OF PERCUSSION

The center of percussion is not strictly speaking a body segment parameter. It is usually associated with sporting implements, such as baseball bats, rackets, and golf clubs. It is the point on an implement, sometimes called the *sweet spot,* where the implement when struck experiences no pressure at its grip or point of suspension. Thus, a bat suspended about its grip end and struck at its center of percussion will rotate only about the suspension point.

In contrast, if the bat is struck at its center of gravity, it moves in pure translation with no rotational motion. Any other striking point produces some combination of translation and rotation (figure 3.5). Note that striking at the center of percussion does not produce pure rotation of the body, only pure rotation of the body about its suspension point. Most implements are not held at their center of gravity, so contacting at the center of percussion produces minimal forces to the hands holding the implement.

The center of percussion can be determined either empirically by striking the implement in various places and observing which location causes pure rotation about the suspension point or computationally (Plagenhoef 1971):

$$q_{axis} = \frac{k_{axis}^2}{r_{axis}} \qquad (3.24)$$

where q_{axis} is the center of percussion, k_{axis} is the radius of gyration, and r_{axis} is the distance to the center of gravity from the axis of rotation at the point of suspension.

Note that there is a sequential relationship between r, k, and q. The radius of gyration (k) is always farther than the center of gravity (r) and the center of percussion (q) is always farther than the radius of gyration (k) from the axis of rotation (figure 3.6). The one exception is for a uniformly dense spherical body in which all three locations are coincident.

The center of percussion is not the only sweet spot. There are other locations on an implement that produce special effects. For example, all rackets, depending on their construction, shape, and how they are strung, have a location that provides maximum rebound. This location can be determined experimentally by dropping balls onto the motionless racket and observing which location produces the highest bounce (figure 3.7). The square root of the ratio of the bounce height to the drop height is called the **coefficient of restitution**:

$$c_{restitution} = \sqrt{h_{bounce} / h_{drop}} \qquad (3.25)$$

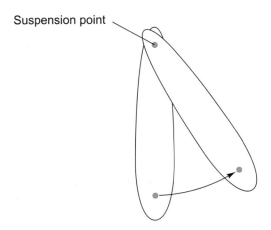
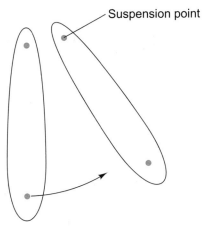

o Figure 3.5   Effects of striking a bat at the center of percussion and away from the center of percussion.

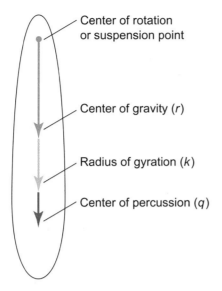

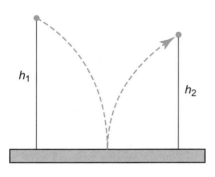

○ **Figure 3.6** Relative locations of the center of gravity *(r)*, radius of gyration *(k)*, and center of percussion *(q)*.

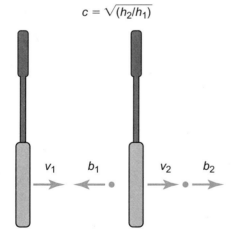

$$c = \sqrt{(h_2/h_1)}$$

$$c = -(v_2 - b_2)/(v_1 - b_1)$$

○ **Figure 3.7** Methods of calculating the coefficient of restitution *(c)*.

The larger the coefficient of restitution, the greater the ball's velocity after impact. The coefficient of restitution generally is calculated by measuring the velocities of the racket (or other implement) and the ball before and after an impact. Note that the ball's velocity should be measured only after the ball is undeformed after the impact. The coefficient is then defined by the ratio of the relative velocities after and before an impact (see Hatze 1993; Plagenhoef 1971):

$$c_{restitution} = -\left(\frac{v_{bat\ after} - v_{ball\ after}}{v_{bat\ before} - v_{ball\ before}} \right) \tag{3.26}$$

THREE-DIMENSIONAL (SPATIAL) COMPUTATIONAL METHODS

In the following sections, methods for computationally determining body segment parameters for 3-D analyses are presented. The main differences between 2- and 3-D analyses lie in the computation of mass moments of inertia. The methods outlined for calculating mass and center of mass are not necessarily different from the equations used for two dimensions that were presented previously, but for the purposes of this text and to use a consistent system for 3-D analysis, the methods employed by Vaughan, Davis, and O'Connor (1992) are emphasized.

SEGMENT MASS AND CENTER OF GRAVITY

There are few differences between the equations used for two and three dimensions when computing segment mass or center of gravity. In chapter 7, segment mass is calculated using regression equations developed by Vaughan, Davis, and O'Connor (1992). These equations require the anthropometric measures listed in table 3.5 and illustrated in figure 3.8. Essentially, the segment masses of the three segments of the lower extremity were computed from these equations:

$$m_{foot} = 0.0083\,m_{total}$$
$$+ 254.5\left(l_{foot}h_{malleolus}w_{malleolus}\right) \tag{3.27}$$
$$- 0.065$$

$$m_{leg} = 0.0226\,m_{total} \atop + 31.33\left(l_{leg}c_{leg}^2\right) + 0.016 \tag{3.28}$$

$$m_{thigh} = 0.1032\,m_{total} + 12.76\left(l_{thigh}c_{midthigh}^2\right) \atop - 1.023 \tag{3.29}$$

where m_{total} is the total body mass; the ls represent the segment lengths; $h_{malleolus}$ and $w_{malleolus}$ are the malleolus

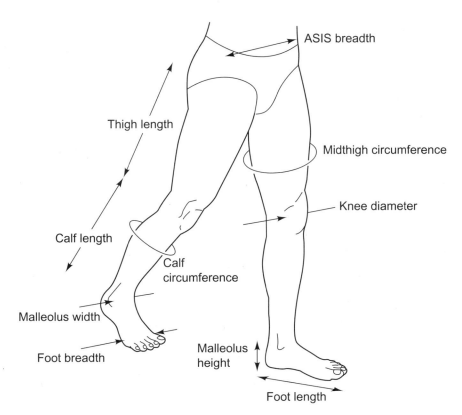

○ **Figure 3.8** Anthropometric measurements needed for the body segment parameter equations of Vaughan, Davis, and O'Connor (1992).

Table 3.6 Anthropometric Data Required for Body Segment Parameter Equations of Vaughan, Davis, and O'Connor

Number	Anthropometric measure	Units
1	Total body mass	kg
2	Anterior-superior iliac spine	m
3	Thigh length	m
4	Midthigh circumference	m
5	Leg (calf) length	m
6	Leg circumference	m
7	Knee diameter	m
8	Foot length (heel-toe)	m
9	Malleolus height	m
10	Malleolus width	m
11	Foot breadth	m

Reprinted, by permission, from C.L. Vaughan, B.L. Davis, and J.C. O'Connor, 1992, *Dynamics of Human Gait* (Champaign, IL: Human Kinetics).

height and width, respectively, c_{leg} is the leg (calf) circumference; and $c_{midthigh}$ is the midthigh circumference. Note that all masses are in kilograms and all other measurements are in meters.

To compute segment center of gravity, an additional equation, identical to those presented previously for planar mechanics, is applied. That is, the center of gravity in the Z direction (z_{cg}) is

$$z_{cg} = z_{proximal} + R_{proximal} (z_{distal} - z_{proximal}) \qquad (3.30)$$

where $z_{proximal}$ and z_{distal} are the Z coordinates of the proximal and distal ends, respectively, and $R_{proximal}$ is the distance from the proximal end of the segment to the center as a proportion of the segment's length.

Vaughan, Davis, and O'Connor (1992) used the proportions (*R*-values) reported by Chandler et al. (1975). These are listed in table 3.4. That is, $R_{proximal}$(foot) = 0.4485, $R_{proximal}$(leg) = 0.3705, and $R_{proximal}$(thigh) = 0.3719. Note that these researchers provided body segment parameter information only for the lower extremity.

SEGMENT MOMENT OF INERTIA

Computing segmental moments of inertia in 3-D requires the same basic approach that is used for planar (2-D) mechanics, but instead of using a single scalar quantity for moment of inertia *(I)*, a moment of inertia tensor *([I])* is used. This tensor is a 3 × 3 matrix:

$$\begin{vmatrix} I_x & P_{xy} & P_{xz} \\ P_{yx} & I_y & P_{yz} \\ P_{zx} & P_{zy} & I_z \end{vmatrix} \qquad (3.31)$$

where the diagonal elements $(I_x, I_y,$ and $I_z)$ are called the principal mass moments of inertia and the off-diagonal elements are called the *mass products of inertia* (the *Ps*).

In general, all nine elements of this tensor must be measured, computed, or estimated to define, for example, the resultant moment of force $(\vec{M}_R)$. That is, $\vec{M}_R = [\bar{I}]\vec{\alpha}$, where $\vec{\alpha}$ is the angular acceleration vector of the rigid body or segment and $[\bar{I}]$ is the centroidal (at the center of gravity) moment of inertia tensor. This situation is simplified by using an axis system that places one axis along the longitudinal axis of the segment reducing the inertia tensor to a diagonal matrix in which all the products of inertia are zero. That is, the inertia tensor is reduced to

$$\begin{vmatrix} I_x & 0 & 0 \\ 0 & I_y & 0 \\ 0 & 0 & I_z \end{vmatrix} \qquad (3.32)$$

In this way, only the elements along the principal diagonal need to be computed. Details about 3-D mechanics are presented in chapter 7. This chapter, however, is concerned only with computing the three principal moments of inertia.

As mentioned previously, Hanavan in 1964 (see also Miller and Morrison 1975) laid down the most commonly used method for modeling human body segments in three dimensions. He used the empirically derived data provided by Dempster (1955) and then measured various anthropometric measures. These were then used as parameters to mathematically derive the principal mass moments of inertia based on equations derived from integral calculus. Figure 3.9 illustrates Hanavan's mathematical model of the body. Table 3.7 shows equations for computing the principal mass moments of inertia of various geometric, uniformly dense solids.

Later, other researchers developed additional methods of estimating the principal moments of inertia as well as mass proportions and segmental centers of gravity. As stated earlier, Zatsiorsky and Seluyanov (1983, 1985) used gamma mass scanning

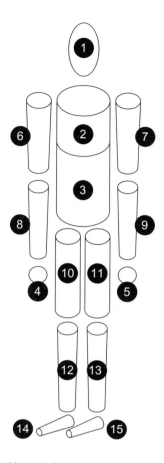

° Figure 3.9 Hanavan's geometrical model of the body.

to acquire tables of body segment parameters for young males and females. These authors divided the body into 15 segments, 3 for the trunk and 3 for each extremity. They also used bony landmarks, as Clauser, McConville, and Young (1969) and Chandler and colleagues (1975) had, to define segment endpoints, in contrast to Dempster's (1955) use of joint centers of rotation. Paolo de Leva (1996) provided methods that convert Zatsiorsky and Seluyanov's data to the form used by Dempster and Hanavan (i.e., segment endpoints defined by their joint centers of rotation).

In chapter 7, equations for computing the 3-D kinetics of the lower extremities are presented. These equations (table 3.5) use the regression equations developed by Vaughan, Davis, and O'Connor (1992) to compute the body segment parameters. They require 11 anthropometric measures (20 if both lower extremities are analyzed) to derive the body segment parameters of the three segments of the lower extremity (the foot, leg, and thigh). These parameters are shown graphically in figure 3.8 and listed in table 3.6.

Table 3.7   Principal Mass Moments of Inertia of Solid Geometrical Shapes

	I_x	I_y	I_z	
Slender rod	0	$1/12ml^2$	$1/12ml^2$	
m = mass; l = length of rod				
Rectangular plate	$1/12m(b^2 + c^2)$	$1/12mc^2$	$1/12mb^2$	
m = mass; b = height of plate; c = width of plate				
Thin disk	$1/2mr^2$	$1/4mr^2$	$1/4mr^2$	
m = mass; r = radius of disk				
Rectangular prism	$1/12m(b^2 + c^2)$	$1/12m(a^2 + c^2)$	$1/12m(a^2 + b^2)$	
m = mass; a = depth (x); b = height (y); c = width (z)				
Circular prism	$1/2mr^2$	$1/12m(3r^2 + l^2)$	$1/12m(3r^2 + l^2)$	
m = mass; l = length of cylinder; r = radius				
Elliptical cylinder	$1/12m(3c^2 + l^2)$	$1/12m(3b^2 + l^2)$	$1/4m(b^2 + c^2)$	
m = mass; l = length of cylinder (x); b = height/2 (y); c = width/2 (z)				

(continued)

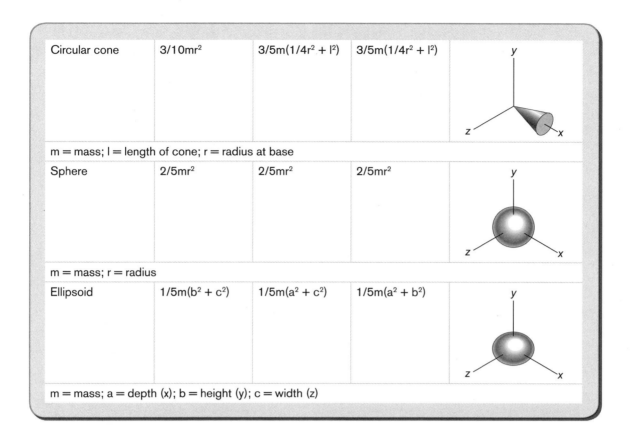

| Circular cone | $3/10mr^2$ | $3/5m(1/4r^2 + l^2)$ | $3/5m(1/4r^2 + l^2)$ | |

m = mass; l = length of cone; r = radius at base

| Sphere | $2/5mr^2$ | $2/5mr^2$ | $2/5mr^2$ | |

m = mass; r = radius

| Ellipsoid | $1/5m(b^2 + c^2)$ | $1/5m(a^2 + c^2)$ | $1/5m(a^2 + b^2)$ | |

m = mass; a = depth (x); b = height (y); c = width (z)

SUMMARY

Two methods for computationally obtaining body segment parameters were outlined. The first method, based primarily on the techniques and data derived from Dempster (1955) or Clauser, McConville, and Young (1969) and Chandler et al. (1975), were outlined for 2-D analyses. This method, based on proportions and measured anthropometrics, can estimate the body segment parameters for linked-segment models of the body. The researcher decides on the number of segments according to the complexity of the motion being investigated. Proportions can also be obtained from various databases, depending on the population (male/female, young/old, and so on) being investigated.

The second method, based primarily on techniques defined by Vaughan, Davis, and O'Connor (1992), was outlined for 3-D analysis of the lower extremities. This method requires the measurement of specific anthropometric dimensions so that 3-D body segment parameters can be estimated. Again, the equations presented can be extended to other segments by resorting to the research literature.

Note that although accurate body segment parameters are desirable, errors in these parameters may have little effect on kinetic measurements, especially when the body is in contact with the environment. In such cases, large errors in the body segment parameters have little influence on the computation of, for example, joint moments of force or moment powers. This is because the relative magnitudes of the inertial forces ($-ma$) and especially the moments of force ($-I\alpha$) are small compared with the moments caused by the GRF.

SUGGESTED READINGS

Jensen, R.K. 1993. Human morphology: Its role in the mechanics of motion. *Journal of Biomechanics* 26(Supp. no. 1): 81-94.

Krogman, W.M., and F.E. Johnston. 1963. *Human Mechanics: Four Monographs Abridged*. AMRL Technical Document Report 63-123. Wright-Patterson Air Force Base, OH.

Miller, D.I., and R.C. Nelson. 1973. *Biomechanics of Sport*. Philadelphia: Lea and Febiger.

Nigg, B.M., & W. Herzog. 1999. *Biomechanics of the Musculoskeletal System*. 2nd ed. Toronto: John Wiley & Sons.

Plagenhoef, S. 1971. *Patterns of Human Motion: A Cinematographic Analysis*. Englewood Cliffs: Prentice Hall.

Winter, D.A. 1990. *Biomechanics and Motor Control of Human Movement*. 2nd ed. Toronto: John Wiley & Sons.

Zatsiorsky, V.M. 2002. *Kinetics of Human Motion*. Champaign, IL: Human Kinetics.

CHAPTER 4

Forces and Their Measurement

Graham E. Caldwell, D. Gordon E. Robertson,
and Saunders N. Whittlesey

The field of mechanics is partitioned into the study of motion (kinematics) and the study of the causes of motion (kinetics). Chapters 1 and 2 covered important aspects of the kinematics of human movement, and we now turn our attention to the underlying kinetics. In this chapter, we

- introduce the concepts of force, which causes linear motion, and torque, also called moment of force, or simply moment, which causes angular motion;

- discuss the effect of applied forces and moments of force through the consideration of laws and equations set forth by Newton and Euler;

- explain how to create and use free-body diagrams;

- identify various forces encountered in biomechanical investigations;

- define the mechanical concepts of impulse and momentum, which dictate the effect of changing levels of force and moment applied over a duration; and

- describe how to measure force and moment for human biomechanics research.

FORCE

The term *force* is common in everyday language. Curiously, in its use in physics, force can only be defined by the effect that it has on one or more objects. More specifically, force represents the action of one body on another. In our study of human biomechanics, we are interested in two specific effects related to force. The first is the effect that a force has on a particle or a perfectly rigid body, while the second is the effect that a force has on a deformable body or material. The effects of forces on particles or rigid bodies can be assessed using Newton's laws and are critical to understanding the causes of the kinematic data described in earlier chapters. A rigid body is one in which the constituent particles have fixed positions relative to each other. The effects of forces on a deformable body or material are important when considering internal forces within biological tissues and in understanding how it is possible to measure external forces applied to or by humans.

Force is a vector quantity defined by its magnitude, direction, and point of application. For the linear motion of particles or rigid bodies, only the force magnitude and direction are important. However, the point of application is critical if angular motion is also under consideration. Furthermore, although the concept of a force vector leads one to consider a "point" of application, in reality most external forces are generated by direct contact spread over a finite area rather than at a single point. This introduces the concept of pressure, that is, force distributed over an area of contact. Within the realm of human biomechanics, kinetic analyses usually concern either external forces and pressures applied through direct contact with the ground or an object (e.g., tool, machine, ball, bicycle, or keyboard) or internal forces within muscles, ligaments, bones, and joints. Although kinetic parameters sometimes can be measured directly, often they must be calculated or estimated based on measurement of the observed kinematics.

NEWTON'S LAWS

In 1687, Isaac Newton published a book on "rational mechanics" titled *Philosophiae Naturalis Principia Mathematica,* or *Mathematical Principles of Natural Philosophy,* with two revised editions appearing before his death in 1727. Although our knowledge of the physics of motion has evolved substantially since its publication, the importance of the *Principia* can be appreciated by the fact that its basic tenets are still taught in high schools and universities throughout the world more than 300 years after its first appearance.

Although Einstein's 1905 theory of relativity demonstrated that Newton's mechanics were incomplete, Newton's laws are still valid for all but the movement of subatomic particles and situations in which velocity approaches the speed of light; they remain the principal tools of biomechanics and engineering. The laws that Newton set forth in the *Principia* define the relationship between force and the linear motion of a particle or rigid body (referred to henceforth as a *body* or *object)* to which it is applied. Three such relationships are described here. Readers should note that these laws may be described in slightly different forms in other texts as a result of differing translations from the original Latin, changes in common English over the past 300 years, and varying levels of background preparation in the target student audience.

Newton's first law, the law of inertia, describes how a body moves in the absence of external forces, stating that "a body will remain in its current state of motion unless acted upon by an external force." The "state of motion" is described by the body's momentum *(p),* defined as the product of its mass and linear velocity *(p = mv).* Simply put, if no force is applied to a body, its momentum will remain constant. For many situations in biomechanics, mass will be constant, so the absence of force means that the body will remain

at constant linear velocity. Juxtaposed with this situation describing the absence of force is the law of acceleration (Newton's second law), which describes how a rigid body moves when an external force is applied. It states, "the force will cause the body to accelerate in direct proportion to the magnitude of the force and in the same direction as the force." This proportionality can be stated as an equality with the introduction of the body's mass, resulting in the famous Newtonian equation $F = ma$. The International System unit for force is the newton (N), with 1 N being the force needed to cause a mass of 1 kg to accelerate by 1 m/s^2. Simply put, an applied force causes a rigid body to accelerate by changing its velocity, direction of motion, or both.

Newton's third law describes how two masses interact with each other. The law of reaction states that "when one body applies a force to another body, the second body applies an equal and opposite reaction force on the first body." A common example in biomechanics involves a human in contact with the surface of the Earth, as when running (figure 4.1). When the runner's foot strikes the ground, the runner applies a force to the Earth. This force can be represented as a vector having a certain magnitude and direction. At the same time, the Earth applies to the runner a reaction force of equal magnitude but opposite direction. Other examples include a person making contact with a ball, bat, tool, or other handheld implement. Because of the opposition in

○ **Figure 4.1**   The physical interaction of two bodies results in the application of *action* and *reaction* forces. In this example, the runner is pushing against the ground. The figure on the left depicts the force acting on the Earth resulting from the runner's muscular efforts. On the right, the equal but opposite reaction force that acts on the runner is shown, giving rise to the term *ground reaction force.*

direction between the so-called action and reaction forces, be careful in these situations to correctly recognize which force acts on which body when assessing the effect caused by the force.

In many situations, more than one force acts on a body at a given point in time. This situation is easily handled within Newton's laws through the concept of a resultant force vector. Because each force is a vector quantity, a set of forces acting on a body can be combined through vector summation into a single resultant force vector. Newton's laws can then be considered using this single resultant force. A useful tool when dealing with kinetics problems is the free-body diagram (FBD), which is a simple sketch of the body that includes all of the forces acting on it. Drawing an FBD reminds researchers of the existence of each force and helps them to visualize the direction of reaction forces that are acting on the body in question.

FREE-BODY DIAGRAMS

Diagrams usually are helpful in visualizing mechanics problems. The FBD is a formal means of presenting the forces, moments of force, and geometry of mechanical systems. This diagram is all-important; the first step in the solution of a mechanical problem is to draw the FBD. The second step is to use the FBD to derive the equations of motion of the object. Finally, known numerical values are substituted and the equations are solved for the unknown terms. There are several formalities to FBD construction:

- Draw the object of interest in minimalist form (either an outline or even just a single line), free of the environment and other bodies.

- Write out the coordinates of the object to completely specify its position.

- Indicate the object's center of mass with a marker; it is from here that the accelerations are drawn.

- Draw and label all external reaction forces and moments of force. Base the directions for these forces and moments by how the object experiences them. For example, direct vertical ground reaction forces upward and frictional forces opposite to the direction of motion of the contacting surfaces.

- Draw all unknown forces and moments with positive coordinate system directions. Unknown forces must be applied wherever the body is in contact with the environment or other bodies (or body segments).

- It is also desirable to draw and label global coordinate system (GCS) axes off to the side of the diagram indicating which are the positive directions.

Figure 4.2a shows a runner crossing a force plate. In this case, because of the relative complexity of the body shape, we chose to draw a stick figure of the runner. At the mass center we drew the force of gravity (mg) and the force of the wind at the center of the frontal area. The ground reaction forces are included at the subject's stance foot. Note that these are the reaction forces acting on the runner, rather than the forces the runner is applying to the Earth. Figure 4.2b is an FBD of a bicycle crank showing the known pedal forces. The crank is represented as a single line and its mass center is indicated with a dot. At the distal end of the crank are the horizontal and vertical forces of the pedal; we measured these forces and labeled their magnitudes. Because the pedal has an axle with smooth bearings, we assumed that there was no moment of force exerted about the pedal axle. At the proximal end of the crank, we drew the reaction forces on the crank's axle (R_{Ax}, R_{Ay}). These are unknown, so we gave them names and drew them in the positive GCS directions. We also drew a resultant moment (M_A) about the crank axle, which we gave a positive counterclockwise direction because its magnitude is unknown. This moment is nonzero because of the resistance of the chain driving the bicycle.

EXAMPLE 4.1

1. Draw the FBD for a rowing oar.
2. Draw the FBD for a running human, including wind resistance. What is the problem with drawing the wind resistance?

 ➤ *See answer 4.1 on page 271.*

TYPES OF FORCES

At present, physicists contend that there are four fundamental forces in nature: the "strong" and "weak" nuclear forces, electromagnetic force, and gravitational force. Of these forces, only gravitational and electromagnetic forces concern the biomechanist. All forces experienced by the body are some combination of these two.

Another of Newton's substantial contributions to our understanding of motion was his description of how masses interact even when they are not in contact. His universal law of gravitation states that "two bodies attract each other with a force that is

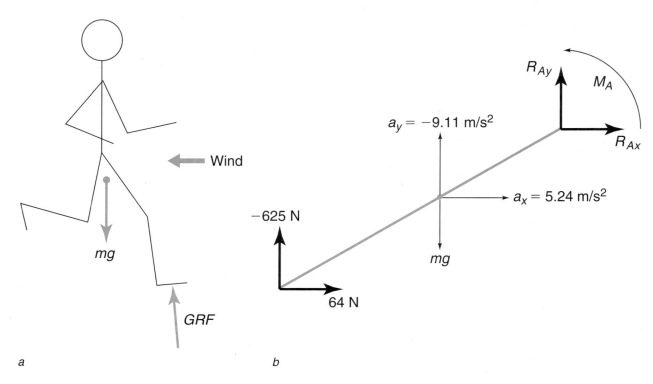

° **Figure 4.2**   Free-body diagrams of a runner *(a)* and a bicycle crank *(b)*. It is assumed that there is no moment of force at the pedal but there is a moment at the crank axle (M_A) due to the chain ring.

proportional to the product of their masses and inversely proportional to the square of the distance between them." With the introduction of the universal constant of gravitation *(G)*, we can quantify the magnitude of the gravitational force: $F_g = G(m_1 \times m_2)/r^2$, where m_1 and m_2 are the masses of the two bodies and *r* is the distance between their mass centers. Whereas Newton considered this concept of gravity in his quest to understand planetary motion, in the realm of biomechanics its most important application is for the effect that the Earth's gravity has on bodies near its surface. In this case, m_1 is the mass of the Earth, m_2 is the mass of an object on the surface of the Earth, and *r* is the specific distance, *R*, from the body's center to the center of the Earth. If we consider the gravitational force acting on mass m_2, the effect of this force is described by the law of acceleration $(F = m_2 a)$. By substitution into the gravitational equation, we see that $m_2 a = G(m_1 \times m_2)/R^2$. The body's mass, m_2, appears on both sides of the equation and therefore drops out, leaving an expression for the acceleration of the body caused by the Earth's gravitational force, $a = Gm_1/R^2$. Because all of the terms on the right-hand side are constants, the acceleration, *a*, is also a constant. This is the acceleration resulting from Earth's gravity, com-

monly called *g* and approximately equal to 9.81 m/s². Furthermore, the gravitational force is given the specific name **weight,** with the equation $W = mg$ being a specific form of the more general $F = ma$. (Note that body weight is a force and therefore a vector. In contrast, body mass is a scalar quantity.) Therefore, on an FBD for an object on or near the Earth's surface, the weight force should be drawn acting in the direction toward the Earth's center, usually designated as vertically down.

If a person stands quietly while waiting for a bus, his weight force vector is acting to accelerate him downward at 9.81 m/s². However, this acceleration does not occur, as a person's vertical velocity remains constant at zero. The reason for this lack of acceleration is Newton's third law, which governs the forces associated with the contact between the person's feet and the pavement. The person is applying a force to the Earth equal to his weight, while the Earth is applying an equal and opposite reaction force on the person. On an FBD of the person (figure 4.3), we would draw a force vector pointing upward (away from the Earth's center) called the **ground reaction force** (GRF or F_g). Therefore, there are two forces acting on the person in the vertical direction: the weight force downward (negative) and the GRF upward (positive). In this quiet-stance situation, these two forces have the same magnitude. Vector summation results in a resultant vertical force of zero $(W + F_g = 0)$, which is the reason the person's vertical velocity remains constant.

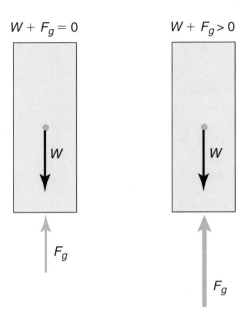

○ **Figure 4.3** An FBD of a person standing on the Earth must show two forces. The first is the weight force vector *(W)* acting downward (negative) because of the gravitational attraction of the Earth. The second force is the GRF *(F_g)* acting upward (positive) because of the physical contact between the person and the Earth. For a person in a quiet stance (left panel), these two opposing forces are approximately equal in magnitude, and therefore the person's center of mass undergoes no acceleration $(\Sigma F = 0)$. If the person uses muscular effort to push down on the ground (right panel), the resulting increase in F_g will cause an upward acceleration because F_g will now be larger than W (i.e., $\Sigma F > 0$).

Although equal to the person's weight in this example, in general the magnitude of the GRF can vary, and therefore can be greater than or less than the magnitude of the person's weight, which is a constant $(W = mg)$. If a person standing on the ground activates his leg extensor muscles (i.e., pushes down on the ground), the GRF will rise above the magnitude of his weight (figure 4.3). The resultant force vector will therefore be nonzero and directed upward $(F_g > W,$ therefore $W + F_g > 0)$. The law of acceleration dictates that this positive resultant force causes acceleration in the upward direction, the magnitude of which depends on the person's mass $(F = ma,$ rearranged as $a = F/m)$. If the person had been standing

quietly before activating his muscles, the initial vertical velocity would have been zero. The positive resultant force and acceleration result in a change in the velocity from zero to an upwardly directed positive velocity. In contrast, if the person had just jumped off a step, his initial velocity at ground contact would have been downward (negative). The positive resultant force and acceleration act to reduce this negative velocity (slow the downward motion). If large enough and applied long enough, the positive resultant force reduces the velocity to zero, stopping the downward movement, or even reversing the movement from downward to upward.

Although this example refers to the vertical direction, the GRF is a 3-D vector that can have components in the horizontal plane that are often referenced to body position, with anterior/posterior (A/P) and medial/lateral (M/L) components (figure 4.4). The 3-D direction of the GRF vector depends on how the person applies the force to the ground and dictates the relative size of the vertical, A/P, and M/L components. For example, a soccer player trying to initiate forward motion pushes downward and backward on the ground. The GRF therefore is directed upward and forward, resulting in positive vertical and anterior GRF components. A similar push-off to initiate a forward and lateral movement results in a lateral GRF component, as well. The ability to generate these horizontal components depends on the nature of the foot/ground

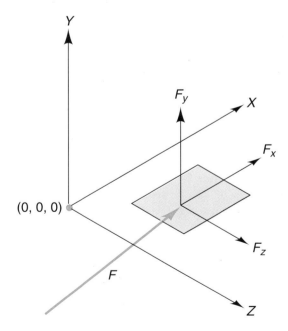

○ **Figure 4.4** The GRF vector F can be resolved into its three components, F_x, F_y, and F_z.

contact. Recall that in the vertical direction, the GRF was the result of the resistance to the gravitational attraction provided by the Earth's solid surface. In the horizontal directions, the foot/ground interface must provide a similar resistance; the person must be able to push on the Earth to generate a GRF. A force known as friction can provide this resistance. If a person stands on a truly frictionless surface, it is impossible to generate A/P or M/L ground reaction force components.

Friction is a specific force that acts whenever two contacting surfaces slide over each other. Frictional forces are directed parallel to the surfaces and always oppose the relative motion of the two surfaces. In some cases, the frictional force is large enough to prevent movement; this is known as static friction. In other cases, applied forces are great enough to cause movement, and then kinetic friction acts to resist the motion. Imagine that a block sitting on a level surface is subjected to a slowly increasing applied horizontal force, $F_{applied}$ (figure 4.5). When $F_{applied}$ is small, the frictional force $(F_{friction})$ is able to resist movement, so the block remains stationary $(F_{applied} = -F_{friction}; F_{applied} + F_{friction} = 0)$. As $F_{applied}$ increases, $F_{friction}$ continually grows to match it in magnitude to keep the block stationary. However, at some point, $F_{friction}$ will reach its maximal static value, and further increases in $F_{applied}$ will not be met by equal increases in $F_{friction}$. It is at this point that movement of the block commences. The maximal value of $F_{friction}$ in static conditions is known as the limiting static friction force, and it is calculated using the equation $F_{maximum} = \mu_{static} N$, where μ_{static} is the coefficient of static friction and N is the normal force acting across the two surfaces (figure 4.5). As the magnitude of $F_{applied}$ exceeds that of $F_{maximum}$, the block is set in motion, with $F_{applied}$ tending to accelerate it, whereas the kinetic frictional force $F_{kinetic}$ opposes the motion (i.e., tends to slow it down). The magnitude of $F_{kinetic}$ is somewhat less than that of $F_{maximum}$ and is approximately constant despite the magnitude of $F_{applied}$ or the velocity attained by the block. The kinetic friction force can be calculated with the formula $F_{kinetic} = \mu_{kinetic} N$, where $\mu_{kinetic}$ is the coefficient of kinetic friction and N is the normal force. The normal force is the force perpendicular to the surfaces that keeps the surfaces in contact. Note that μ_{static} and $\mu_{kinetic}$ both depend on the nature of the two surfaces involved and that μ_{static} is always slightly greater than $\mu_{kinetic}$.

Readers will encounter other classes of forces throughout the biomechanics literature. Internal forces usually refer to those generated or borne by tissues within the body, such as muscles, ligaments, tendons, cartilage, or bone. In contrast, external

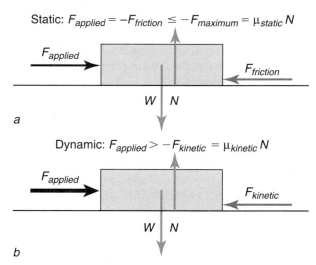

Static: $F_{applied} = -F_{friction} \leq -F_{maximum} = \mu_{static} N$

a

Dynamic: $F_{applied} > -F_{kinetic} = \mu_{kinetic} N$

b

○ **Figure 4.5**   When a force, $F_{applied}$, is applied in a direction that would tend to cause an object to slide across a surface, a frictional force vector, $F_{friction}$, opposes the applied force. A static situation is shown in the top panel, in which $F_{applied}$ (and $F_{friction}$) are less than the limiting static friction force $F_{maximum} = \mu_{static} N$, where μ_{static} is the coefficient of static friction and N is the normal force acting across the two surfaces. In this static case, $F_{applied} = -F_{friction}$. If $F_{applied}$ becomes greater than $F_{maximum}$, the object will slide resisted by the dynamic frictional force $F_{kinetic} = \mu_{kinetic} N$, where $\mu_{kinetic}$ is the coefficient of dynamic friction (bottom panel).

forces are those imposed on the body by contact with other objects, such as the reaction forces described earlier. *Inertia forces* are those associated with accelerating bodies, and they arise from a slightly different expression of Newton's law of acceleration $(F = ma)$. If the right-hand side is subtracted from both sides of the equation, the expression becomes $(F - ma = 0$; this formulation is known as d'Alembert's principle). The term $-ma$ is called the inertial force; it is dimensionally equivalent to other forces that constitute the resultant force F on the left-hand side of the equation. This so-called pseudo-force is felt when an elevator rapidly slows as it approaches its destination floor. The body's inertia wishes to continue upward, and the inertial force results in a decrease in the reaction force between the feet and the floor. Another example is the *g*-force experienced during rapid acceleration in a car or plane; in these cases, the body wants to remain stationary while the vehicle moves rapidly forward. In these situations, a person feels like she is being pushed into the seat but the real force is the seat pushing her forward.

Another type of pseudo-force arises when considering objects rotating about an axis and is associated with the ever-changing linear direction of particles within the rotating rigid body. In this circumstance, an outwardly directed centrifugal force represents the inertial tendency for the particles to continue moving away from the axis of rotation, while the inwardly directed radial or centripetal force acts to prevent such an occurrence. A third pseudo-force, the Coriolis force, occurs when considering the rotations of reference systems within a given system. Readers are directed to physics or engineering texts (e.g., Beer and Johnston 1977) for a more complete description and the computation of these forces.

MOMENT OF FORCE, OR TORQUE

Earlier we noted that the point of application of a force vector is important only if angular motion of a rigid body is under consideration. By definition, angular motion takes place around an axis of rotation (see chapters 1 and 3). If a force vector $\vec{F}$ is applied to a rigid body so that its line of action passes directly through the axis of rotation, no angular motion is induced (figure 4.6). However, if force vector $\vec{F}$ is moved to a parallel location so that its line of action falls some distance from the axis, the force tends to cause rotation. Forces that do not pass through the axis of rotation are known as eccentric (off-center) forces. If a displacement vector, $\vec{r}$, is defined from the axis of rotation to the point of force application, the vector cross product of the force and displacement vector is known as the moment of force $(\vec{M})$ so that $\vec{M} = \vec{r} \times \vec{F}$. The perpendicular distance from the axis of rotation to the force line of action is known as the moment arm (d) of the force. Figure 4.6 illustrates that $d = r \sin \theta$, where θ is the angle formed by the lines of action of the displacement vector $\vec{r}$ and force vector $\vec{F}$. Using the perpendicular distance, d, the magnitude of the moment of force can be computed as $M = Fd$. The unit for a moment of force (or simply "moment") is the newton meter (N·m). Clearly, the point of application of a force vector dictates the magnitude of the moment, because altering the application point changes the force moment arm.

A single eccentric force produces both linear and rotational effects. The force itself causes the object to accelerate according to Newton's law of acceleration regardless of whether the force is directed through the axis of rotation or not. A purely rotational motion, with no linear acceleration, can be produced by two forces acting as a force couple. A force couple consists of two noncollinear but parallel forces of equal magnitude acting in opposite directions. For example, in figure 4.6, the equal parallel

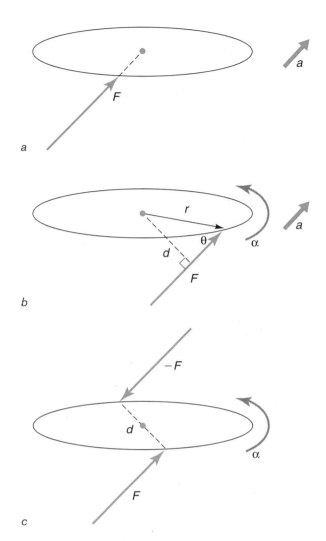

Figure 4.6 In (a), the force F acts through the body's center of mass, which also is its axis of rotation. Therefore, the force causes only translation. If F is applied so that it does not act through the axis of rotation, a moment of force, M, is created ($M = Fd$) and the body undergoes both linear and angular acceleration (b). (c) illustrates a force couple consisting of parallel forces F and $-F$, separated by the moment arm d. The force couple causes angular acceleration only.

forces F and $-F$ are separated by a perpendicular distance, d, and thereby form a force couple that applies a moment equal to Fd. Because F and $-F$ are in opposite directions, they sum to zero, and thus the resultant force applied to the object is zero. This results in an absence of linear acceleration.

Another term commonly used instead of moment of force is torque. Some physics and engineering texts distinguish between the two terms, associating torque with either force couples or "twisting" movements where it is difficult to identify a single force vector and point of application. Within the biomechanics community, the terms are used interchangeably by most, and we use both terms throughout this text.

The exact rotational kinematic effect of an applied torque is dictated by the angular version of Newton's laws of motion, the first two of which deal with angular momentum in the absence or presence of torque. In the absence of an applied torque, a rotating body continues to rotate with constant angular momentum, analogous to the linear law of inertia. Angular momentum, L, is defined[1] as the product of the body's mass moment of inertia, I, and its angular velocity, ω, so that $L = I\omega$. The units for angular momentum are therefore kg·m^2/s. When a moment of force is applied to a rigid body, the angular momentum changes so that $M = dL/dt$, where dL/dt is the time derivative of angular momentum. This is known as Euler's equation, after the famous 16th century Swiss mathematician Leonhard Euler. It is the angular equivalent of $F = ma$, which Newton originally expressed as $F = dp/dt$, where dp/dt is the time derivative of the linear momentum (p) of a particle or rigid body. In the linear case, a change in velocity (acceleration) is the only possibility because the mass of a rigid body is a constant.[2] In the angular case, if the mass moment of inertia, I, is a constant, Euler's equation becomes $M = I\alpha$, where α is the rotating body's angular acceleration. However, when measuring the human body, changes in configuration are possible, and thus the moment of inertia can change. Therefore, the more general form of Euler's equation dictating changes in angular momentum (rather than acceleration) is useful, even though the body's mass does not change. Note that Euler's full equations for 3-D motion are more complex and will not be dealt with here. Consult an engineering mechanics text such as Beer and Johnston (1977) for a complete description of the 3-D case.

LINEAR IMPULSE AND MOMENTUM

Newton's second law, $F = ma$, can be applied instantaneously or when an average force is considered. When a researcher wants to know the influence of a force that varies over its duration of application, the impulse-momentum relationship becomes useful. This relationship is directly derivable from Newton's second law; as noted previously, it was originally written as a relationship between force and momentum. At the time, the term *momentum* was not used;

Newton referred instead to the quantity of motion. A mathematical derivation of the impulse-momentum relationship for forces begins with Newton's law of acceleration,

$$F = ma = m\frac{dv}{dt} \qquad (4.1)$$

Next, rearrange the equation by multiplying both sides by dt.

$$Fdt = mdv \qquad (4.2)$$

Finally, integrating both sides of the equation yields the impulse-momentum relationship.

$$\int Fdt = mv_{final} - mv_{initial} \qquad (4.3)$$

where the left-hand side is the linear impulse of the resultant force, F, and the right-hand side represents the change in linear momentum of mass, m. The terms mv_{final} and $mv_{initial}$ are the final and initial linear momenta of the body, respectively. The units of linear impulse are newton seconds (N·s), which are dimensionally equivalent to the units for linear momentum (kg·m/s).

Thus, the linear impulse of a force is defined as the integral of the force over its period of application, and this impulse changes the body's momentum. Graphically, linear impulse is the area under a force history. Figure 4.7 illustrates that (a) increasing the amplitude of the force, (b) increasing the duration of the force, (c) increasing both amplitude and dura-

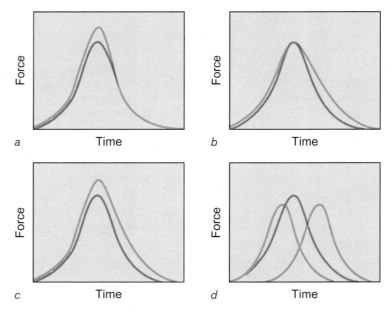

○ Figure 4.7   Increasing impulses by (a) increasing amplitude of the force, (b) increasing the duration of the force, (c) increasing both the amplitude and duration, and (d) increasing the frequency or number of impulses. Note that the gray curve is identical in all frames.

tion, and *(d)* increasing the number of impulses (i.e., the frequency of impulses) increase the impulse on a body.

MEASURED FORCES, LINEAR IMPULSE, AND MOMENTUM

The impulse-momentum relationship can be used to evaluate the effectiveness of a force in altering the momentum or velocity of a body. For example, a person performing a start in sprinting (Lemaire and Robertson 1990b) or swimming (Robertson and Stewart 1997) tries to apply horizontal reaction forces to initiate horizontal motion, and sensors capable of recording these forces can quantify the effectiveness of the start. Figure 4.8 shows the horizontal impulses of a start from instrumented track starting blocks (Lemaire and Robertson 1990b), a force platform mounted on swimmer's starting blocks (Robertson and Stewart 1997), and a force platform imbedded in an ice surface (Roy 1978). Integrating the areas shown in figure 4.8 and dividing by the mass of the athlete permit determination of the change in the athlete's horizontal velocity. Note that it is assumed

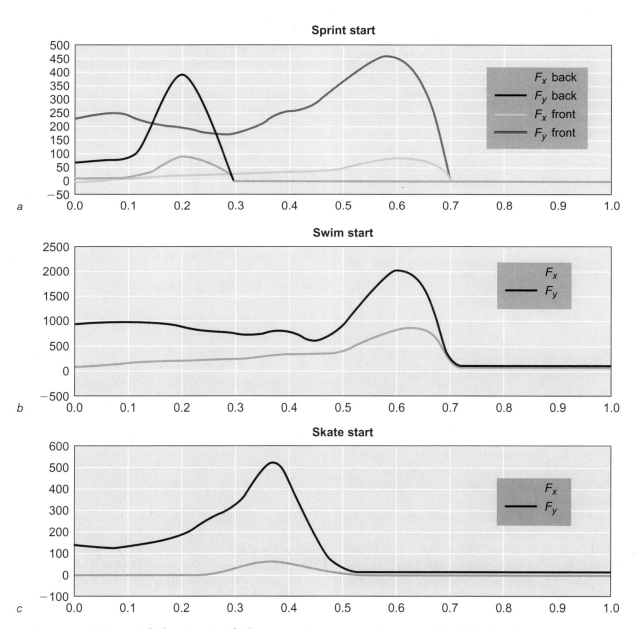

Figure 4.8   Horizontal (F_x) and vertical (F_y) impulses of a start from the front and back blocks of instrumented track starting blocks *(a)*, from a force platform mounted on a swimmer's starting platform *(b)*, and a force platform imbedded in an ice surface *(c)*. Abscissa is time in seconds and ordinate is force in newtons.

with these types of skills that the initial velocity is zero, which is required for these skills; otherwise, the start is considered *false,* for which the athlete is penalized. This requirement is not applicable to relay starts, where the athlete is allowed to have a *running start.* In such a situation, the researcher has to measure the athlete's velocity before the start of force application. Thus, for a given impulse, the velocity of the person after the impulse is

$$v_{final_x} = \frac{\int_{t_{initial}}^{t_{final}} F_x dt}{m} + v_{initial_x} \qquad (4.4)$$

where m is the person's mass, $v_{initial_x}$ is the initial velocity (zero if the motion starts from rest), and the numerator is the impulse of the horizontal force—that is, the area under the horizontal reaction force history—from time $t_{initial}$ to t_{final}.

A similar application concerns research on vertical jumping or landing from a jump, in which vertical force platform signals are integrated over time to obtain the changes in the jumper's vertical momentum. In the case of standing jumps, the athlete's initial velocity is zero; therefore the takeoff velocity of the jumper can be computed directly from the force platform signals (i.e., the change in velocity is equivalent to the takeoff velocity). There is a slight difference in the equation for the vertical impulse because gravity must be excluded to obtain the vertical velocity. The equation for the vertical velocity is therefore

$$v_{final_y} = \frac{\int_{t_{initial}}^{t_{final}} (F_y - W) dt}{m} + v_{initial_y} \qquad (4.5)$$

where W is the person's weight in newtons, $v_{initial_y}$ is the initial vertical velocity (zero if the motion starts from rest), and F_y is the vertical GRF. Of course, this equation applies only if all forces act against the force platform. For example, if one of the person's feet is off the platform, the vertical velocity will be underestimated. Similarly, it is assumed that no other body part acts against the ground or the environment. If so, additional force-measuring instruments must be used.

Other applications for the impulse-momentum relationship include activities such as rowing, canoeing, golf, batting, and cycling, in which the forces applied by the hands or feet can be measured by force-sensing elements to quantify their impulses. Again, the effectiveness of the force can be directly quantified by integrating the applied force over time. The simplest method for computing these integrals is to use Riemann integration. With this method, the force signals collected from an analog-to-digital

(A/D) converter of a computer are added and then the sum is multiplied by the sampling interval (Δt). If the sampling rate of the force signal is 100 Hz, the sampling interval is 0.01 s. The equation for a Riemann integral is

$$impulse = \Delta t \sum_{i=1}^{n} F_i \qquad (4.6)$$

where F_i is the sampled forces and n is the number of force samples. Another, more accurate integral uses trapezoidal integration:

$$impulse = \Delta t \left(\frac{F_1 + F_n}{2} + \sum_{i=2}^{n-1} F_i \right) \qquad (4.7)$$

Essentially, this is half the sum of the first and last forces plus the sum of the remaining forces times the sampling duration. Other, more sophisticated integrals are also possible, such as Simpson's rule integration, but with a sufficiently high sampling rate there is little difference in the resulting integrals. Readers are directed to college calculus or numerical-analysis textbooks for more information on these integration techniques. Also important in this process is the selection of an appropriate sampling rate and smoothing function used during data collection and reduction (see chapter 11 for data-smoothing functions). The sampling rate should be selected so that it is neither too low, in which the true peaks and valleys in the force history are clipped, nor too high, which increases errors due to the integration process. A suitable sampling rate for many jumping and starting situations is 100 Hz. After calculating the impulse, compute the change in velocity by dividing the impulse by the body's mass.

A slightly different approach can be used to observe the instantaneous changes in velocity resulting from force application. We begin by converting the GRFs into acceleration histories by dividing by the person's mass at each instant in time. The vertical GRF must be reduced by subtracting the person's weight *(W):*

$$a_x = F_x / m$$
$$a_y = (F_y - W) / m \qquad (4.8)$$

Note, that W must be very accurate, otherwise there is going to be an ever-increasing error during the integration process. The best solution is to record a brief period immediately before the impulse starts from which the person's weight can be determined. It happens that a person's weight as registered by a force platform varies slightly depending on where the feet are placed.

These acceleration patterns have the same shape as the force profiles because they have merely been scaled by the constant mass. The acceleration histories are then integrated to obtain the velocity histories by iteratively adding successive changes in velocity. The velocity history is computed by repeatedly applying this integration equation:

$$v_i = v_{i-1} + a_i(\Delta t) \qquad (4.9)$$

where v_i is the velocity at time i, v_{i-1} is the previous time interval's velocity, a_i is the acceleration, and Δt is the sampling time interval. The first initial velocity, called a constant of integration in calculus, must be known. If the activity starts statically, the first initial velocity is zero; if not, then the researcher must compute or measure the initial velocity with another system, such as videography. In a similar manner, the second integral of force theoretically can yield the displacement of the body. The integration process is repeated using the computed velocity signal to obtain the displacement history, which introduces another constant representing the initial position of the person. For simplicity, we can set the initial position to zero and then determine displacement (s_i) that occurs after starting the integration. That is, $s_i = s_{i-1} + v_i(\Delta t)$, where v_i is the velocity computed from the previous iteration of the equation.

Note that this integration process can become unstable if instrumentation problems arise. If the force signal drifts (i.e., low frequency changes in the baseline) as the subject stands motionless on the force plate, the computed displacement signal rapidly becomes unrealistic. This drift is a characteristic of piezoelectric force plates. It is advisable, therefore, to conduct this type of measurement by minimizing the integration time and calculating body weight from the force recording at the instant immediately before the integration starts, when the person is standing motionless. Small errors in weight determination occur when a person stands on a force platform in response to exact foot placement and other environmental factors. These slight inaccuracies in the subject's weight cause large errors in the displacement record because of the double integration of the force signal (Hatze 1998). This is the inverse of the situation that occurs when deriving acceleration from displacement, in which small, high-frequency displacement errors create large acceleration errors (see chapters 1 and 11).

SEGMENTAL AND TOTAL BODY LINEAR MOMENTUM

When analyzing human motion, it is not always possible to directly measure external forces that act on the body. Instead, momentum can be computed indirectly from the kinematics of body markers and computed segment centers of gravity (chapter 3). Once the segment centers are known, it is a relatively simple matter of multiplying the velocity vectors by the segment masses (see chapter 3). That is,

$$\vec{p} = m\vec{v} \quad \text{or}$$
$$p_x = mv_x, \ p_y = mv_y, \ p_z = mv_z \qquad (4.10)$$

Total body momentum is also easy to compute with the necessary input data, because momenta can be summed vectorially. The total body linear momentum is therefore the sum of its segmental momenta. That is,

$$\vec{p}_{total} = \sum_{s=1}^{S} m_s \vec{v}_s \qquad (4.11)$$

where m_s is the segment masses, $\vec{v}_s$ is the velocity vectors of the segment centers, and S is the total number of segments. The scalar versions are:

$$p_{total\,x} = \sum_{s=1}^{S} m_s v_{sx}$$
$$p_{total\,y} = \sum_{s=1}^{S} m_s v_{sy} \qquad (4.12)$$
$$p_{total\,z} = \sum_{s=1}^{S} m_s v_{sz}$$

This measurement is not often used in biomechanics because it requires recording the kinematics of all the body's segments, which are typically difficult to obtain, especially in three dimensions. However, this technique has been used for the study of airborne dynamics, such as long jumping (Ramey 1973a, 1973b), high jumping (Dapena 1978), diving (Miller 1970, 1973; Miller and Sprigings 2001), and trampolining (Yeadon 1990a, 1990b). In these situations, conservation of linear momentum occurs in the horizontal directions and momentum decreases predictably in the vertical direction because of gravity.

ANGULAR IMPULSE AND MOMENTUM

Angular impulse and angular momentum are the rotational equivalents of linear impulse and linear momentum. They are derived from Euler's equation $(M = I\alpha)$ in a fashion similar to the linear $F = ma$. Recall that Euler expressed his equation as $M = dL/dt$, where dL/dt is the time derivative of angular momentum $(L = I\omega)$. Rearranging Euler's equation results in $M\Delta t = \Delta I\omega$. Given a resultant moment of force, M_R, acting on a body, integrating over time yields the angular impulse applied to the body. That is,

$$\text{angular impulse} = \int_{t_{initial}}^{t_{final}} M_R \, dt \qquad (4.13)$$

where $t_{initial}$ and t_{final} define the duration of the impulse in seconds. Given that the angular momentum of a system of rigid segments is L, the angular impulse-momentum relationship can be written

$$L_{final} = L_{initial} + \text{angular impulse} \qquad (4.14)$$

That is, the angular momentum of the system after an angular impulse is equal to the angular momentum of the system before the impulse plus the angular impulse. Note that the system's moment of inertia may have a different value before and after the duration of the impulse, depending upon segmental configuration. Unlike the constant mass assumption of linear motion, the rotational moment of inertia can be quite different from one instant to another. For example, a diver or gymnast in the layout position can have a 10-fold greater moment of inertia than when in the tuck position. Thus, the effects of an angular impulse vary with changes in the body's moment of inertia. This factor is taken into consideration by segmenting the body into a linked system of rigid bodies and computing each segment's contribution to the total body's angular momentum. Each segment contributes two terms to the angular momentum of the whole body: one term sometimes called the local angular momentum and another called the moment of momentum or remote angular momentum. The first term describes rotation of the segment about its own center of gravity, whereas the second, the moment of momentum, corresponds to the angular momentum created by the segment's center of gravity rotating about the total body's center of gravity. These terms are defined next.

SEGMENTAL ANGULAR MOMENTUM

Whereas the linear momentum of a segment is the product of its mass and linear velocity, the angular momentum of a segment (L_s) rotating about its center of gravity is the product of its moment of inertia and its angular velocity:

$$L_s = I_s \omega_s \qquad (4.15)$$

where I_s is the moment of inertia (in kg·m²) of the segment about its center of gravity and ω_s is the angular velocity of the segment (in rad/s). Of course, segments rarely rotate solely about their own center of gravity. To determine the angular momentum of a segment about another axis (e.g., the total body center of gravity or the proximal end of the segment), the moment of momentum (L_{mofm})

is needed. This term, based on the parallel axis theorem, is defined as

$$L_{mofm} = \left[\vec{r}_s \times m_s \vec{v}_s \right]_z = m_s \left(r_x v_y - r_y v_x \right) \qquad (4.16)$$

where (r_x, r_y) is the position vector that goes from the axis of rotation to the segment's center of gravity, m_s is the segment's mass, and (v_x, v_y) is the linear velocity vector of the segment. Note that the symbol $\times$ means that the two vectors are multiplied as a cross or vector product and the symbols $[\vec{r}_s \times m_s \vec{v}_s]_z$ mean that only the scalar component about the Z-axis is to be considered.

TOTAL BODY ANGULAR MOMENTUM

To obtain the angular momentum of a whole body, several different approaches may be taken. If the body is made up of a series of interconnected segments (as the human body is), then the total body angular momentum is the sum of all the segment angular momenta plus their associated moments of momentum. For example, to calculate the total body angular momentum (L_{total}) about the total body center of gravity for planar analyses, this equation applies:

$$L_{total} = \sum_{s=1}^{S} I_s \omega_s + \sum_{s=1}^{S} \left[\vec{r}_s \times m_s \vec{v}_s \right]_z \qquad (4.17)$$

where $\vec{r}_s$ represents the position vector connecting the total body center of gravity to the center of gravity of the segment, that is, $(x_s - x_{total}, y_s - y_{total})$; $\vec{v}_s$ is the velocity of the segment's center; and S represents the number of body segments.

In general, moment of momentum terms are larger than segment angular momentum terms because a segment's moment of inertia is usually less than 1 (in kg·m²) whereas the mass of a segment is greater than 1 (in kg). Furthermore, the position vectors for the least massive segments can be quite large, and consequently their cross products with velocity are relatively large compared to the segment's rotational velocity.

ANGULAR IMPULSE

An alternative way of determining total body angular momentum makes use of the external forces and moments of force acting on the body and the angular impulses they produce. Figure 4.9 shows four examples of external forces that produce angular impulses and consequently affect the angular momentum of the bodies. Notice that in all cases the lines of action of the external forces do not pass through the bodies' centers of gravity. To determine how much angular impulse is produced, one must measure the forces

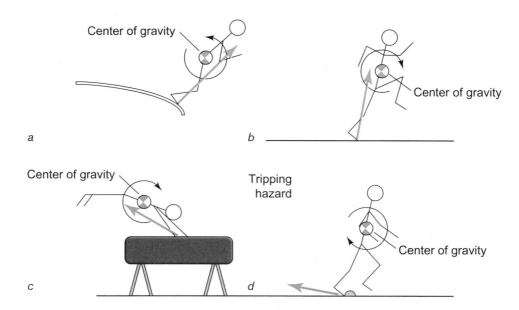

a   b

c   Tripping hazard   d

Figure 4.9 Examples of forces that cause angular impulses and angular momenta. Curved lines show the direction of the applied angular impulse. Example *a* is a diver, *b* is a long jumper at takeoff, *c* is a gymnast, and *d* is someone tripping while walking.

over time and, simultaneously, record the body's center of gravity trajectory. In addition, all other external forces must be quantified to determine their magnitude, direction, and points of application on the body. The only external force that does not need to be measured is gravity, because it is a **central force** that passes through the body's center of gravity and therefore causes no rotational momentum.

Recall that angular impulse is the time integral of the resultant moment of force or eccentric force acting on the body (figure 4.10), whereas angular momentum is the quantity of rotational motion of the body. In mathematical form,

$$\text{angular impulse} = \int_{t_i}^{t_f} M_R \, dt \qquad (4.18)$$

or, if the moment of force, M_R, is constant, then

$$\text{angular impulse} = M_R \Delta t \qquad (4.19)$$

where Δt is the duration of the impulse.

Alternatively, if there is a single external force acting on the body (figure 4.11), the angular impulse can be quantified:

$$\text{angular impulse} = \int \left(\vec{r} \times \vec{F} \right) dt \qquad (4.20)$$

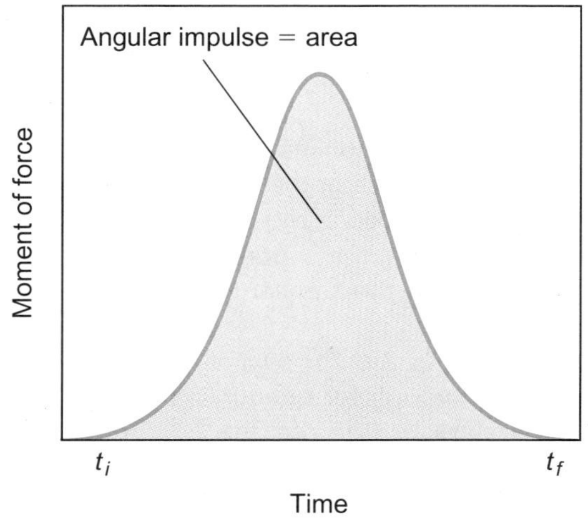

Moment of force

Angular impulse = area

t_i t_f

Time

Figure 4.10 Angular impulse is defined as the area under the moment of force versus time curve.

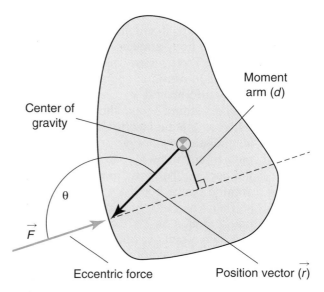

Center of gravity

Moment arm (d)

θ

$\vec{F}$

Eccentric force   Position vector $(\vec{r})$

Figure 4.11 Eccentric forces are capable of producing angular impulses.

where $\vec{F}$ is the applied force and $\vec{r}$ is the position vector from the total body center of gravity to the force's point of application.

Note that $\vec{r} \times \vec{F}$ is the cross product of the force and position vectors. In two dimensions, the magnitude of this product is $r_x F_y - r_y F_x$ (see appendix D for vector products in three dimensions). If the applied force is a central force such as gravity, it causes no angular impulse and therefore causes no change in the angular momentum of the body. This important principle, called the *law of conservation of angular momentum,* applies when the body is airborne and the only external force is gravity (i.e., ignoring air resistance).

Other situations also follow this law if any rotational friction can be neglected, such as standing on a frictionless turntable or spinning on an icy surface. Thus, the person is assumed to spin with constant angular momentum, but not necessarily constant angular velocity. The person's spin rate (angular velocity) may be increased or decreased, respectively, by decreasing or increasing the total body's moment of inertia through movements of body segments. The law of conservation of angular momentum says that the total body angular momentum (L_{total}) stays constant about any axis whenever the applied moments are zero and the resultant force is a central force:

$$L_{total} = \text{constant} \qquad (4.21)$$

The majority of researchers who have attempted to quantify angular momentum have been concerned with the study of athletic airborne motions, for example, diving (Miller 1970, 1973; Murtaugh and Miller 2001), figure skating (Albert and Miller 1996), long jumping (Lemaire and Robertson 1990a; Ramey 1973a, 1973b, 1974), gymnastics (Gervais and Tally 1993; Kwon 1996), and trampolining (Yeadon 1990a, 1990b).

MEASUREMENT OF FORCE

There are many tools available to the biomechanist for the measurement of force and moment of force. Although all of these tools can be called transducers, we separate them into force platforms, pressure distribution sensors, internally applied force sensors, and isokinetic devices.

From the Scientific Literature

Yu, B., and J.G. Hay. 1995. Angular momentum and performance in the triple jump: A cross-sectional analysis. *Journal of Applied Biomechanics* **11:81-102.**

This project quantified the segmental and total body angular momenta for four airborne phases of the triple jump—the last stride, the hop, the step, and the jump. Elite-level triple jumpers (n = 13) were videotaped during a competition and only their longest effort was analyzed. Three-dimensional coordinates of the segments and segmental and total body angular momenta were computed for the airborne portions of the four phases. The average values of the angular momentum for each phase and each axis of rotation were then normalized by mass and height squared and correlated to actual jump performance.

The authors found that there was a statistically significant nonlinear correlation between the mediolateral angular momentum of the support phase of the step and the jump distance achieved $(r = 0.86)$. They concluded that to achieve the required angular momentum at the step, that momentum must be obtained during the support phase of the hop. Furthermore, the step phase should minimally change the angular momentum created by the hop.

This was the first paper to report the angular momenta in 3-D for such a complicated motion as the triple jump. Not only was it a breakthrough for quantifying such a dynamic motion, but it also was done on four different skills (the run, the hop, the step, and the jump) performed by elite athletes in a highly competitive situation (U.S. Olympic trials).

FORCE TRANSDUCERS

Much of our discussion has dealt with the effects caused by forces applied to a particle or rigid body. In fact, modeling any object or body segment as "rigid" in the presence of an applied force is only an approximation of reality. Because all objects deform to a certain extent, our definition of a rigid body (all particles having fixed positions relative to each other) is not strictly correct. In many cases, the deformation and errors associated with nonrigidity are small. Furthermore, this deformation can be useful to biomechanists because it allows them to measure the applied forces using force transducers.

Sensing elements of different types can be adhered to or built into deformable materials. When a force is applied, the sensing elements register the amount the material deforms. Typically, the sensing elements have electrical properties and constitute part of an electronic circuit. For example, *resistive* or *piezoresistive* elements can serve as resistors within a circuit such as a Wheatstone bridge. Deformation causes structural and geometric changes in resistors that alter their electrical resistance (e.g., a thin piece of metal becomes thinner when stretched, altering its ability to transmit electric charge). This change in resistance alters the voltage drop in an appropriate electronic circuit. Piezoresistive elements are based on semiconductor materials such as silicon and are more sensitive than ordinary resistive materials. Because force, deformation, resistance, and voltage are directly related, knowledge of these relationships allows one to calculate the force applied by measuring the voltage change in such a circuit. A different type of sensing element is made from piezoelectric crystal, a naturally occurring mineral that produces electric charge in response to deformation from an applied force. The piezoelectric crystals must be connected to a *charge amplifier*, and the associated electronic circuit is much different from that in the piezoresistive case. However, the concept is the same: An applied force causes deformation that leads to a measurable electrical response, and the magnitude of the electrical response is directly related to the magnitude of the force. Readers are directed to appendix C for more information on electronic circuits.

The quality of a force transducer depends on the relationships between the applied force, the deformation, the electrical characteristics of the sensing element, and the measured electrical output. If we ignore the underlying electrical details, the relationship that is of interest is the one between the applied force input and the registered voltage output (figure 4.12). The responses shown in these graphs are derived from static measurements, for which a constant force is applied to a transducer, and the resulting (and, we hope, constant) voltage is registered. Each data point represents a given force level and the voltage response. By changing the magnitude of the constant applied force, voltage responses across a range of forces are recorded. If the resulting relationship is linear (figure 4.12a), the slope of the relationship ($\Delta F / \Delta V$) represents a *calibration coefficient,* called the sensitivity, which can be used to convert from measured voltage into force in newtons (e.g., 6.5 volts [V] $\times$ 63 N/V = 409.5 N). If the relationship is nonlinear (figure 4.12b), the data can be fitted to a *calibration equation* (e.g., second- or third-order polynomial) that can be used for the conversion to force.

Another consideration is the *range* of forces a transducer will measure before its response changes markedly or it is damaged. Transducers are rated for a particular force range, over which their response is linear; if higher forces are applied, the voltage output may saturate at a given level. A related issue is the *sensitivity* of the transducer. A force transducer should be matched to the range of forces one wishes to measure; it should be sensitive enough to detect small changes in applied force, yet still have enough range. Another point of concern is hysteresis (figure 4.12c), in which a different force-to-voltage relation is found when the force is incrementally increased compared to when it is incrementally decreased. This is undesirable because, in theory, different calibration coefficients or equations should be used in loading and unloading situations.

Static characteristics are important and easy to assess, but in most biomechanics applications, the applied forces continually change in magnitude. Therefore, the *dynamic response* capabilities of the transducer are equally important. The *frequency response* characteristics of the transducer should be matched to those of the applied force. The physical characteristics of the transducer's construction permit it to respond to a limited range of input force frequencies. If the input force changes too rapidly, the transducer may be unable to respond quickly enough to faithfully register the true time history of the force. This is analogous to a low-pass filter, which attenuates or eliminates the higher-frequency components of the input force signal (see chapter 11, on signal processing). On the other hand, the transducer's construction may cause the unwanted amplification of some frequencies within the input force. Any physical structure or system will respond to a forced vibration in a characteristic way based on its internal mass, elasticity, and damping. The mass and elasticity dictate the *natural frequency* of a structure,

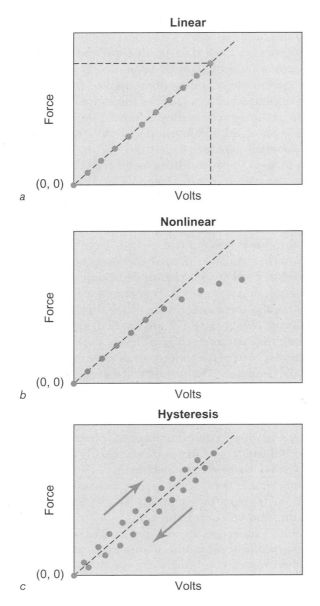

o Figure 4.12 Three possible relations between an input force and the output response voltage for a force transducer. *(a)* shows a *linear* response, in which the slope of the force-voltage relation can be used as a *calibration coefficient*. The relation is *nonlinear* in *(b)*, whereas *(c)* illustrates the concept of *hysteresis*.

and the structure resonates when an external vibration at or above the natural frequency is imposed. This concept is useful in the construction of tuning forks, which oscillate at their natural frequency when struck with an applied force. In biomechanics, this response is undesirable, because an applied force that contains significant energy at the natural frequency of a force transducer causes the transducer to vibrate on its own, therefore exaggerating the force response at that frequency.

Transducers can be purchased based on their inherent response properties, such as range, linearity, sensitivity, frequency response, and natural frequency. However, transducers can have very different characteristics when they are mounted or installed as one component in a typical measurement system. For example, a transducer with a high natural frequency (e.g., 2000 Hz) vibrates excessively at lower frequencies if it is attached to a wood frame with springs. One area of particular concern involves the construction of measurement devices with multidirectional transducers that measure force in three orthogonal directions. The transducers may have excellent isolation of forces applied in the three directions, but the measurement device in which they are installed may have considerable elasticity that transmits forces across these orthogonal axes, leading to *crosstalk*. For example, the application of a force purely in the F_x direction may result in the transducer responding in the F_y and F_z directions as well. Such a response can be dealt with by using a *calibration matrix* that relates the response of the transducer in all three directions to given input forces.

FORCE PLATFORMS

The most commonly used type of force transducer in biomechanics is the **force platform** (or **force plate**), which is an instrumented plate installed flush with the ground for the registration of GRF. Early models used springs (Elftman 1934) and inked rubber pyramids to display pressure patterns. Currently, two types of sensors are used in commercially available force platforms: strain gauges and piezoelectric crystals. The strain-gauge models are less expensive and have good static capabilities, but do not have the range and sensitivity of the piezoelectric models. The piezoelectric platforms have high-frequency response, but must have special electronics to enable measurement of static force.

Early force platforms were designed with a single instrumented central column or pedestal, whereas recent designs usually have four instrumented pedestals located near the corners of the plate (figure 4.14). Researchers have abandoned the single-pedestal design because forces tend to become more inaccurate the farther an applied force moves away from the pedestal. With the four-pedestal models, forces are accurate whenever the contact force is applied within the area bounded by the pedestals. The pedestals are instrumented to measure forces and bending moments that result from all forces and moments applied to the platform's top plate. Most commercial platforms are instrumented to measure in three dimensions—vertical (Z), along the length of the plate (Y), and across the width of the plate (X). Force platforms operate on the principle that no matter how many objects apply force to different locations on the top plate, there is one resultant force vector—the GRF—that is numerically and physically equivalent to all the applied forces.

Any single 3-D force vector applied to a force plate can be described by nine quantities. The three orthogonal components of the force vector are designated as F_x, F_y, and F_z, whereas three spatial coordinates, x, y, and z, designate the force vector location with respect to the *plate reference system* (PRS) origin. Because the force vector usually results from a distribution of forces to an area of contact on the surface of the plate, its location is often called the *center of pressure* (COP). The final three quantities are orthogonal moments M_x, M_y, and M_z, taken with respect to the PRS origin. The exact location of the PRS origin depends on the specific location of the force-sensing pedestals, but in general the origin is located in the middle of the plate slightly beneath the level of the top surface (figure 4.13). Thus, the COP coordinate z is a constant equal to the depth of the PRS origin beneath the top of the plate.

Although these nine parameters represent the force applied to the plate, we generally are interested in only six quantities with respect to the reaction force vector applied by the force plate to the human performer. This change in emphasis comes from our interest in the motion of the human, not the force plate itself. These six quantities are the three GRF components $(R_x, R_y,$ and $R_z)$, the COP location of the reaction force vector in a GCS (x,y), and a moment known as the **free moment**, M_z'. Note that the GCS has its Z-axis in the vertical direction, the Y-axis is horizontal in the direction of progression, and the horizontal X-axis is aligned roughly in the medial/lateral direction. Only the x and y COP coordinates need to be computed, because the vertical coordinate is on the force platform's top plate, usually designated as $z = 0$ within the GCS. The free moment M_z' represents the reaction to a twisting moment applied by the subject about a vertical axis located at the COP coordinates. The free moments about the X- and Y-axes are assumed to be zero, because these can occur only if there is a direct connection between the shoe and plate (as with glue). Note that these six quantities are related to, but not equal to, the nine force platform parameters described in the previous paragraph, for three reasons:

- the PRS and GCS axes' directions may not coincide,
- our interest is in the reaction forces $(R_x, R_y,$ and $R_z)$ rather than the applied forces $(F_x, F_y,$ and $F_z)$, and
- we need to know the COP location in the GCS rather than in the PRS.

In the next section, we investigate how these six quantities are computed from the nine force platform measures.

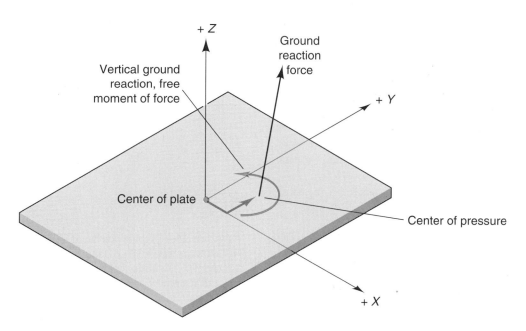

○ Figure 4.13 Force platform with its reaction to an applied force and vertical moment of force.

Processing Force Platform Signals

Each type of force platform has its own unique set of equations from which the six measures of the ground reaction are computed. Note that each force platform manufacturer may use a unique PRS, and that these reference systems do not necessarily coincide with the GCS established for the kinematic data-capture system (see chapters 1 and 2). For consistency, we will use the PRS system established in the previous section (figure 4.13). Example equations are given for two commercial brands, the strain-gauge AMTI plate (AMTI, Watertown, MA) and the piezoelectric-based Kistler platform (Kistler AG, Winterthur, Switzerland). Other manufacturers may have slightly different configurations, but we should still be able to compute the six measures. The equations for an AMTI force platform are derived from output signals labeled F_x, F_y, F_z, M_x, M_y, and M_z (figure 4.14). The six AMTI equations are:

$$F_x' = F_x f_x \quad F_y' = F_y f_y \quad F_z' = F_z f_z$$
$$x = -(M_y g_y + F_x'z) / F_z'$$
$$y = (M_x g_x - F_y'z) / F_z' \tag{4.22}$$
$$M_z' = M_z g_z + F_x'y - F_y'x$$

where $(F_x', F_y',$ and $F_z')$ are the components of the GRF, $(x, y, 0)$ are the coordinates of the COP, and M_z' is the free moment of force, $f_x, f_y,$ and f_z are scale factors that convert the forces from voltages to newtons, and g_x, g_y, and g_z are scale factors that convert the bending moments from voltages to newton meters. The six output signals from an AMTI plate are fed to an amplifier unit with selectable gain levels that determine the exact values of the f_i and g_i scale factors. Each AMTI plate has a unique, factory-calibrated value for z, the distance from the top of the plate to the PRS origin. Note that the applied moments $(M_x, M_y,$ and $M_z)$ about the PRS origin are used to only calculate the COP location.

Kistler force platforms have 12 piezoelectric force sensors, packaged as three orthogonal cylinders in each of the four pedestals (figure 4.14). Horizontal, but not vertical, sensors are summed in pairs so that the following eight signals are output: F_{x12}, F_{x34}, F_{y14}, F_{y23}, F_{z1}, F_{z2}, F_{z3}, and F_{z4}. The equations for computing the six quantities of the GRF are:

$$F_x' = (F_{x12} + F_{x34}) f_{xy}$$
$$F_y' = (F_{y14} + F_{y23}) f_{xy}$$
$$F_z' = (F_{z1} + F_{z2} + F_{z3} + F_{z4}) f_z$$
$$x = -[a(-F_{z1} + F_{z2} + F_{z3} - F_{z4})f_z - F_x'z] \tag{4.23}$$
$$y = [b(F_{z1} + F_{z2} - F_{z3} - F_{z4})f_z + F_y'z]F_z'$$
$$M_z' = b(-F_{x12} - F_{x23})f_{xy} + a(F_{y14} - F_{y23})f_{xy} + xF_y' + yF_x'$$

where (F_x', F_y', F_z') are the components of the GRF, $(x, y, 0)$ are the coordinates of the COP, M_z' is the free moment of force, f_{xy} and f_z are scale factors that convert the forces from voltages to newtons, and a and b are the distances between the sensors and the plate center in the X and Y directions, respectively. The eight output signals are fed to a charge amplifier unit that includes selectable gain levels for each channel that determines the exact values of the f_{xy}

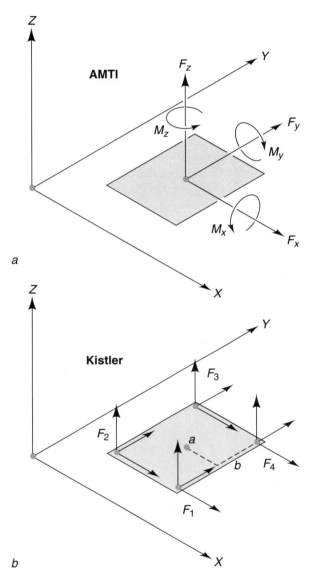

○ **Figure 4.14** Outputs from an AMTI force plate (top panel) include three forces, F_x, F_y, and F_z, and three moments, M_x, M_y, and M_z. The bottom panel shows the four piezoelectric sensors in a Kistler force plate, spaced equidistant from the plate center (lengths a and b). The three force components from each of the four sensors are used either alone (F_z) or in summed pairs (F_x, F_y) in the calibration equations (see text).

and f_z scale factors. As with the AMTI plates, each Kistler platform has its own factory-calibrated specific value for z, the distance from the top of the plate to the PRS origin.

The equations for both the AMTI and Kistler force plates result in the computation of six quantities, but they are not yet in the form needed for analysis of their effects on the human subject. The three force components $(F_x', F_y',$ and $F_z')$ are in reference to the plate and must be multiplied by –1 to provide the reaction forces (R_x, R_y, R_z) applied to the subject. The same reversal in sign applies to the free moment M_z'. The next potential adjustment occurs because the three axes of the PRS and the GCS may be labeled differently or misaligned. The GCS origin and axes' directions are set when calibrating the imaging system used in kinematic data capture. For example, in locomotor studies, the subject usually progresses in such a way that her foot will land on the force plate along its greatest length, Y, in our PRS. In the GCS defined above, this may also correspond to the Y direction. However, the Y directions of the GCS and PRS may be opposite in polarity (figure 4.15). In addition, for 2-D studies, the direction of progression may be designated as the X-axis (see chapter 1). If the force plate data are to be combined with the kinematic data in an inverse dynamics analysis (chapter 5), the researcher must ensure that force data from individual PRS directions are correctly transformed into the GCS directions.

The same PRS/GCS alignment issue also applies to the COP coordinate data, which is a matter of spatial synchronization. Recall that the calculated x and y coordinates are in the PRS, but we want to express them in relation to the subject's kinematics in the GCS. Thus, we need to know the location of the PRS origin with respect to the GCS origin. An easy solution is to locate the GCS origin at the very center of the plate so that the origin x and y coordinates coincide (the vertical origins will be separated by the distance z). Another solution is to place a reflective marker at a known coordinate location in the PRS (e.g., on one corner of the plate). Knowledge of that marker's location in both the GCS and PRS will allow the coordinates x and y to be transformed into the GCS by a simple linear transformation.

A more complex problem occurs when the PRS and GCS axes are misaligned so that the two horizontal PRS axes are rotated by an angle, ϕ, with respect to the two GCS axes (figure 4.15). If the angle ϕ is known, a rotational transformation as described in chapter 2 can be performed to mathematically align

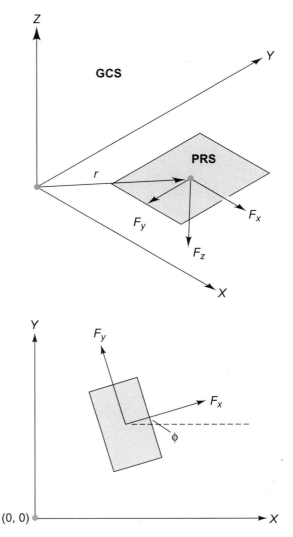

° **Figure 4.15** Spatial alignment of the force plate PRS and GCS is critical for inverse dynamics analyses in which kinematic and force data are combined. The top panel illustrates an example of differences in the origin location and axes' orientation of the PRS and GCS. The position vector r describes the location of the PRS origin within the GCS. The bottom panel shows an overhead view of angular misalignment between the PRS and GCS in the horizontal plane. If the angle ϕ is known, this misalignment can be corrected using rotational transformation techniques.

the two reference systems. This situation can usually be avoided by carefully selecting the GCS axes' directions in relation to the force platform at the time of calibration.

Another issue relevant to the combination of force and kinematic data is the temporal synchronization of the two data sets. If the kinematic data-capture system is separate from the analog-to-digital converter used to collect the force plate data, the same event must be recorded on both systems to ensure proper time synchronization of the two systems. Synchronization units that use the vertical channel from the force plate to trigger an LED in the view of the imaging system are commonly used. When a small threshold vertical force is applied (such as when the subject first touches the plate), the LED is turned on; this identifies the instant of first foot contact in both data-collection systems. If the motion studied begins with the subject on the plate (e.g., jumping), the point of takeoff can be established as the time the LED turns off. Some commercial products (e.g., Vicon, Qualisys, or Simi) ensure correct time synchronization with analog-to-digital conversion modules operated within the kinematic capture process. These systems capture the force and kinematic data simultaneously and allow the kinematic data to be sampled at a rate lower than for the analog data. One caveat is that the analog sampling rate must be an even multiple of the kinematic data-capture rate (e.g., 1000 Hz force to 200 Hz kinematics, or multiples of 60 Hz for video capture systems).

Center of Pressure on the Foot

Once the COP has been located within the GCS through the calculation of its (x,y) coordinate pair, it is important to examine how this location relates to the position of the subject. During the stance phase of walking or running, the portion of either foot in contact with the ground continually changes, and the COP constantly moves along the surface of the plate. When only one foot is on the plate, the COP lies within the part of the foot or shoe that contacts the platform, and it moves beneath the foot in a characteristic manner typically from the heel to the toes (Cavanagh 1978). When more than one foot is in contact with the force platform (a situation that normally should be avoided), the COP lies within the areas formed by either foot or the area between the two feet. If a kinematic analysis is done simultaneously with the force measurement, both COP- and foot-location data will be in the GCS, and their relative positions can be computed. Note that foot position is usually designated based on the locations of several markers on the foot (see chapters 1 and 2).

Several difficulties arise when comparing the COP and foot locations. The location of the COP is highly variable, especially during periods when the foot is initially contacting or leaving the force plate (i.e., the landing and liftoff phases). This variability comes from the equations used to calculate the COP coordinates (x,y), which are unstable when the vertical forces are small, near the beginning and end of the stance phase. Also, slippage may occur during loading and unloading, causing further errors in the COP computation. Often, unrealistic COPs must be discarded, such as any positions that are outside the area of the foot that is in contact with the platform. However, the foot-marker data do not define the outline of this contact area, so it is difficult to implement such a checking algorithm. Frequently, the researcher must do a visual inspection of the individual marker coordinate data (e.g., the heel marker, ankle marker, fifth metatarsal head marker) to determine which portion of the foot is on the ground at a particular time and check that the COP coordinates are realistic. Finally, averaging the COPs is difficult because the subject will land at a different place on the force platform each time. However, Cavanagh (1978) provided a means of averaging COP paths in relation to the outside edges of the shoe during gait (figure 4.16).

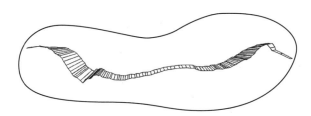

○ **Figure 4.16** Center of pressure path relative to a foot print. Line segments represent directions of force vectors in the transverse plane. Line closest to medial side of foot is path of center of pressure.

Combining Force Platform Data

Occasionally, the GRFs from two or more force platforms need to be combined to yield a single force. Gerber and Stuessi (1987) provided the equations necessary to perform this operation.

Figure 4.17 shows the forces from a single footfall in which the rear foot landed on one force platform and the forefoot on another, a setup used in a two-segment foot model to improve accuracy in modeling the foot dynamics during gait (Cronin and Robertson 2000). Also shown is the combined single GRF. This process enables the computation of the moments of force across the metatarsal-phalangeal joint, which is usually modeled as a rigid body.

Presentation of Force Data

The six quantities of the GRF can be displayed in many different ways. As with kinematic data presentation (chapter 1), a common method is to present the GRF, COP coordinates, and free moment as

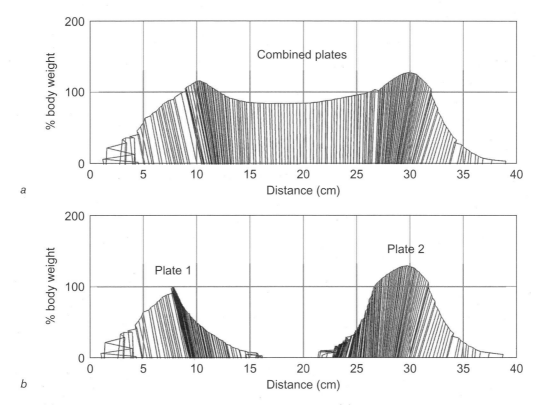

° **Figure 4.17**   (a) Combined ground reaction force vectors from walking. (b) Same ground reaction forces from force platforms under the rear foot and forefoot. Line segments indicate direction and relative magnitudes of GRFs. The lower ends of the segments correspond to the location of the center of pressure in the A/P direction.

time-series or history plots (figure 4.18). To facilitate the comparison of the COP and the foot location, it is sometimes helpful to plot their location data together (figure 4.16). Another useful graph is a trajectory plot of the COP *(x,y)* coordinates; again, inclusion of the foot-marker data can be helpful. Sample trajectory plots are shown in figure 4.19 for a person walking (tracing *a*) and standing (tracing *b),* and are frequently used for posture and balance research.

A *force signature,* vector, or *butterfly* graph combines the three force components and COP location into one 2-D image that can be rotated to view in three dimensions (Cavanagh 1978). Figure 4.20 shows force signatures for walking and running. A typical walking signature has two humps and often a spike immediately after initial foot contact. This spike has been attributed to the passive characteristics of the subject's footwear and the anatomical characteristics of the subject's foot, because its characteristics exhibit too rapid a change to have been caused by the relatively slow-acting muscles. The double hump disappears during running and the peak vertical force increases to approximately three times the body weight. During walking, the peak vertical

forces fluctuate around the body-weight line by approximately 30% of body weight.

The force signature is an especially useful tool if an inverse dynamics study (chapters 5 and 7) is conducted. Inverse dynamics determines the internal forces and moments that act across the joints in response to such external forces as GRFs. By displaying the motion pattern and the force signature on the same display, the researcher can ensure that the two data-collection systems (motion and force) and their reference axes are synchronized temporally and spatially. Figure 4.21 shows four different sagittal views of the force signatures for walking from left to right. Only tracing *a* shows the correct shape of the force signatures. In the other three, the directions of the horizontal GRF, the COP, or both are reversed. By walking across a force platform and examining the resulting force signature, the researcher can check that the axes of the force platform match the axes of the motion-collection system. Displaying stick figures of the subject's motion along with the synchronized force signature can also be used as a validation check. The force signature's vectors should appear to be synchronized in time and space beneath the subject's foot. Figure 4.22 illustrates

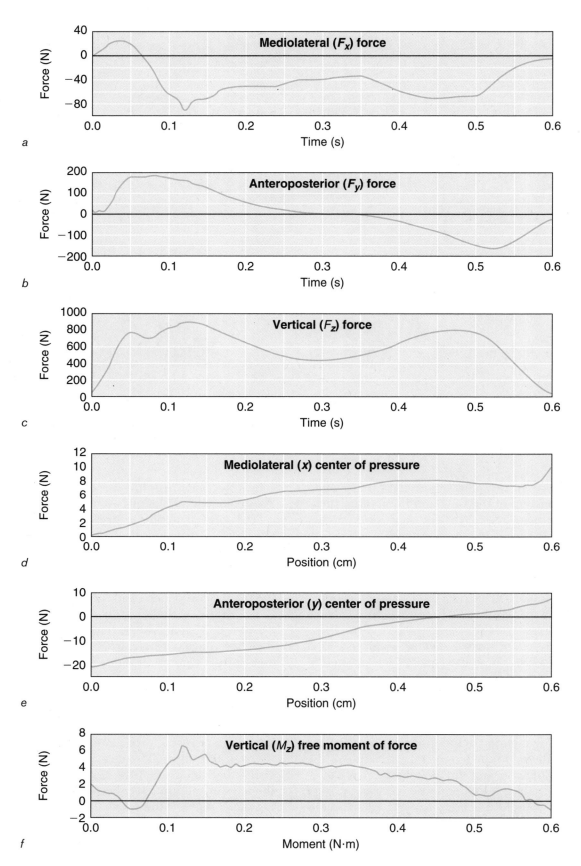

● Figure 4.18   (a, b, c) Histories (time series) of a typical ground reaction force, (d, e) its center of pressure location, and (f) its vertical (free) moment of force during walking.

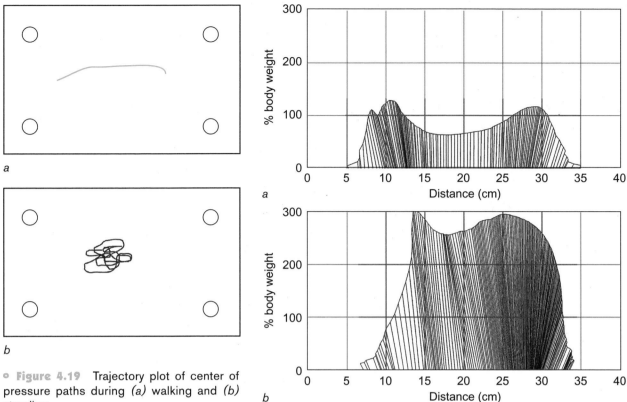

○ **Figure 4.19** Trajectory plot of center of pressure paths during *(a)* walking and *(b)* standing.

○ **Figure 4.20** Force signature of *(a)* walking and *(b)* running. Notice the initial spike at the beginning of each signature and notice that the walking trial has two major peaks (excluding the initial spike).

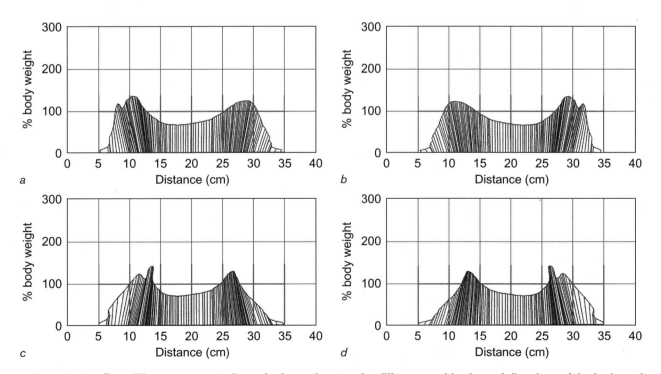

○ **Figure 4.21** Four different representations of a force signature for different combinations of directions of the horizontal force and center of pressure. The correct pattern is *(a)*; both the horizontal forces and centers of pressure are reversed in *(b)*; only the horizontal forces are reversed in *(c)*; and in *(d)* only the horizontal centers of pressure are reversed.

correct and incorrect spatial synchronization of the forces during walking.

Another way of combining GRFs is to average a series of signals from the same subject or across several subjects. Averaging the forces is not difficult, although each force time series is of a different dura-tion. By standardizing the time base to percentages of the footfall duration, often called the *stance phase*, the various trials can be averaged (see chapter 1 for details on ensemble averaging). Of course, the resulting averages are no longer time-based, they are instead percentages of the stance phase.

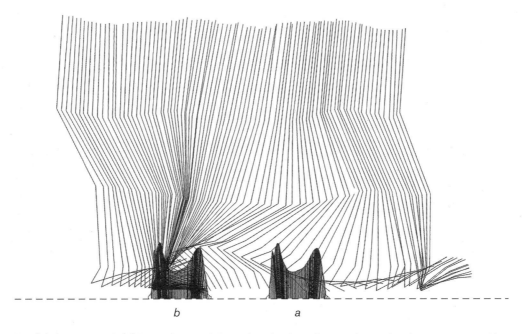

b　　　　　a

◦ Figure 4.22 *(a)* Correct and *(b)* incorrect spatial synchronization of ground reaction force vectors with a stick figure representation of the lower-extremity and trunk motion patterns.

From the Literature

Caldwell, G.E., L. Li, S.D. McCole, and J.M. Hagberg. 1998. Pedal and crank kinetics in uphill cycling. *Journal of Applied Biomechanics* **14:245-59.**

Research on the biomechanics and neural control of cycling has benefited greatly from the advent of instrumented pedals that measure the force vector applied during vari-ous cycling conditions. This study used a pedal based on piezoelectric crystal technology (Broker and Gregor 1990) to examine pedal and crank kinetics during both seated and standing uphill cycling. These authors measured forces acting both normal and tangential to the surface of the pedal as elite cyclists climbed a simulated 8% grade in a laboratory. A unique aspect of measuring pedal forces is that the transducer orientation changes as the subject manipulates his foot position, unlike a force platform rigidly embedded in a laboratory floor. By measuring the angle of the pedal with kinematic markers, the normal and tangential forces were transformed into a GCS (i.e., vertical and horizontal force components). The forces were further transformed into components that acted perpendicularly and parallel to the crank arm as it rotated through a complete 360° revolution. The perpendicular force components were considered effective because they

(continued)

contributed to crank torque $(T = Fd$, where d represents the length from the pedal to the crank axle) to propel the bicycle. The magnitude and pattern of crank torque were altered substantially when the subjects changed from a seated to a standing posture (figure 4.23). This was attributed to more effective use of gravitational force in parts of the crank cycle associated with upward and forward movement of the subject's center of mass as he rose from the saddle.

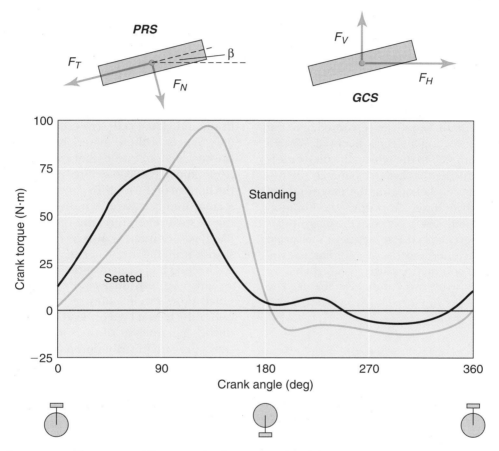

○ **Figure 4.23**   The top panel illustrates that instrumented pedals measure forces relative to a pedal reference system. In a sagittal view, the forces are measured tangential (F_T) and normal (F_N) to the pedal surface. By accounting for the pedal angle, β, these forces can be transformed into the GCS for registration of vertical (F_V) and horizontal (F_H) force components. The pedal forces can be used to calculate the torque exerted about the crank, which is the propelling force for the bicycle. The crank torque pattern over one complete crank arm revolution (360°) is shown in the bottom panel for a subject climbing a hill while seated and while standing.

PRESSURE DISTRIBUTION SENSORS

We have discussed force as a vector quantity characterized by magnitude, direction, and point of application. This viewpoint is in fact a mathematical construct useful for the description and application of Newton's laws of motion. Usually, a force is distributed over an area of contact rather than concentrated at one specific point of application. For example, a GRF vector has a COP point location, but in reality an ever-changing area of the shoe's sole is in contact with the ground at any given time during a stance. The force vector is in fact distributed over this contact area, and its distribution can be analyzed using the concept of pressure, defined as the force per unit area and expressed in newtons per square meter (N/m^2); also called pascals (Pa, although usually kilopascals [kPa] are more common). This can be conceptualized as an array of force vectors distributed evenly across the area of contact, each one applied to a unit surface area (e.g., 1 mm^2). Some forces within this array are larger than others; the overall pattern of these force vectors constitutes the force distribution across the contact area. The summation of these distributed forces equals the magnitude of the overall force vector measured with a force plate.

Measurement of pressure distribution over an area is possible with the use of *capacitance* or *conductance* sensors. These sensors consist of multilayered materials that can constitute part of an electronic circuit, in much the same manner as was described earlier for force transducers. Capacitance sensors consist of two electrically conductive sheets separated by a thin layer of a nonconducting dielectric material. When a normal force is applied to the sensor, the dielectric is compressed, reducing the distance between the two conducting sheets and changing the magnitude of a measured electric charge. Appropriate calibration establishes the relationship between the magnitudes of the applied force and the subsequent measured current. Conductance sensors are similar in construction, but a conductive rather than dielectric material separates the two conductive sheets. The

separating material has electrical properties different from those of the two sheets and offers some resistance to the flow of current between the two sheets. Application of a normal force causes compression of the middle layer, changing its electrical resistance. This alteration in resistance can be used to invoke a measurable change in voltage that will be related to the applied force magnitude. Appendix C gives more details on electronic circuits. Note that both types of sensors are responsive to normal forces, and therefore horizontal shear forces cannot be detected.

Capacitance and conductance materials are manufactured so that independent cells of equal area are formed, and the circuit is designed to measure the pressure within each cell. The resolution of the pressure distribution is dictated by the size of the individual cells, typically on the order of 0.5 cm^2. Sheets of these materials can be formed into pressure mats or insoles that are placed inside shoes. Approximately 400 individual pressure cells may line the bottom of the foot. To extract the data from such a large number of cells, individual wires would be impractical. Therefore, the insoles are constructed like flexible circuit boards by using thin conductive strips within the insole to carry the signals to a small connecting box worn near the subject's ankle. Other systems use individual *piezoceramic* sensors that are glued onto specific sites of interest (e.g., directly under the lateral heel or head of the first metatarsal during gait analysis of the foot). Figure 4.24 illustrates pressure-distribution patterns measured with pressure insoles that were taken from two specific points within the stance phase of a runner.

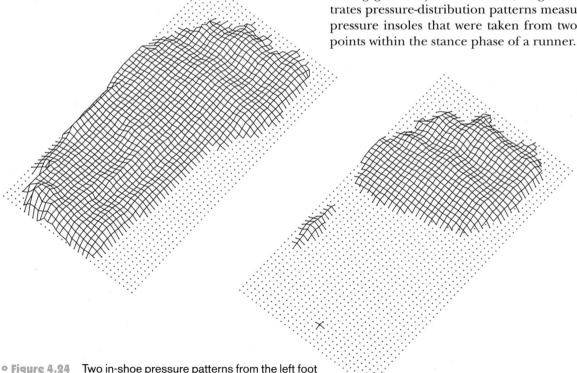

○ Figure 4.24   Two in-shoe pressure patterns from the left foot at different times in the stance phase of running.

INTERNAL FORCE MEASUREMENT

In many cases, it is of interest to know the internal forces acting on individual ligaments, tendons, and joints. Unfortunately, measurement of these internal forces requires highly invasive techniques, although they are commonly used in animal studies (e.g., Herzog and Leonard 1991) and have been used in a limited number of human studies (e.g., Gregor et al. 1991). Although it is doubtful that these invasive protocols will ever be widely adopted for human research, a few options are mentioned here. A less-invasive technique is to use musculoskeletal modeling to estimate these forces during human movements (see chapter 9).

The most common technique for measuring force in tendons (or ligaments) is the buckle transducer, which consists of a small rectangular frame that mounts on the tendon and a transverse steel beam that rests against the tendon (figure 4.25). Strain gauges change their electrical resistance when the tendon transmits force and deforms the transverse beam. For animal preparations, the transducer can be surgically inserted and the wound allowed to heal, with the thin electrical wires used to carry the signals through the skin to associated external circuitry still exposed. A similar approach has been used in the few human studies, although little healing time was permitted; the transducer insertion, experimental data collection, and transducer removal were performed all in 1 day (Gregor et al. 1991). Other transducer designs include foil strain gauges or liquid metal gauge transducers, which are small tubes filled with a conductive liquid metal that can be inserted within a ligament. Axial strain in the ligament causes the tube to elongate and become thinner, thereby altering the electrical resistance afforded by the liquid metal (Lamontagne et al. 1985). Another device is the Hall-effect transducer, which consists of a small, magnetized rod and a tube containing a Hall generator. As a force is imposed on a tendon, the rod moves with respect to the tube, causing the Hall generator to produce a voltage. Finally, the use of fiber-optic transducers has enabled measurement of forces in large, superficial tendons (Komi et al. 1996).

ISOVELOCITY DYNAMOMETERS

One of the most commonly used strength-testing devices is the *isokinetic dynamometer,* which tests the muscular strength capabilities of patients and athletes in different stages of recovery, training, and rehabilitation. These machines were designed to measure the torque a subject produces at a single joint under such controlled kinematic conditions as isometric (a fixed joint angular position) and isovelocity (a predetermined constant angular velocity). In general, the torque produced during these muscular efforts changes continuously, although the term *isokinetic* means *same force* or *same torque.* The term *isovelocity (same velocity)* is therefore preferred (Bobbert and van Ingen Schenau 1990; Chapman 1985). The dynamic isovelocity conditions are further categorized into concentric, when the muscles crossing the joint shorten, and eccentric, when the muscles lengthen.

Although there are several commercial dynamometers, some research laboratories have chosen to build their own to give them better control over the machine's specifications (e.g., its maximal rotational velocity). In either case, the machines permit a subject to apply torque to the dynamometer using a single, isolated joint. The subject is usually seated with the more distal segment of the joint attached to a solid metal bar that rotates about a fixed axis of rotation (figure 4.26). The subject either grasps

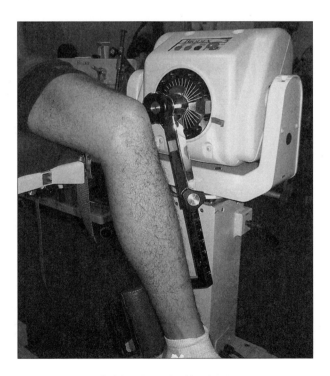

° **Figure 4.26** Subject in an isokinetic dynamometer.

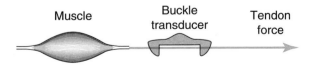

° **Figure 4.25** Buckle-style, muscle force transducer. Note that the muscle and transducer are not drawn to scale; muscle is much larger.

the bar by its handle or applies force to a padded plate attached to it. A series of straps firmly affixes the distal segment to the padded plate and the proximal segment and trunk to the seat. This prevents extraneous movement and ensures that only the muscles surrounding the tested joint generate the force applied to the bar. It is important to position the subject with the joint and machine arm axes of rotation aligned, because misalignment can cause errors in the measurement of the joint torque (Herzog 1988). In some cases, the torque is measured with a torque transducer element built into the physical axis of rotation of the bar. In other designs, a force sensor is placed between the bar and the padded plate, and torque is calculated from the recorded force and the perpendicular distance from the plate to the axis of rotation. A potentiometer built into the axis of rotation measures the angle, and in some machines a tachometer independently measures angular velocity.

Within the research community, dynamometers are used most often to study the *torque–angle (T–θ)* and *torque–angular velocity (T–ω)* relationships for individual joints. The isometric *T–θ* relation is generated from a series of muscular efforts performed with the dynamometer bar fixed (i.e., at velocity equal to zero) in various angular positions. Beginning from rest in each position, the subject is encouraged to produce maximal joint torque against the padded plate of the machine bar so that the applied torque will increase from zero to a more or less steady plateau level (figure 4.27). By averaging a section of the torque plateau, a single point in the isometric *T–θ* relation is generated (i.e., the maximal isometric torque value at that specific angle). Repeating this procedure at different angular positions generates a series of points that define the joint's maximal isometric capabilities across the range of motion (figure 4.28). The reasons that the maximal isometric torque varies with joint angle include the muscular force-

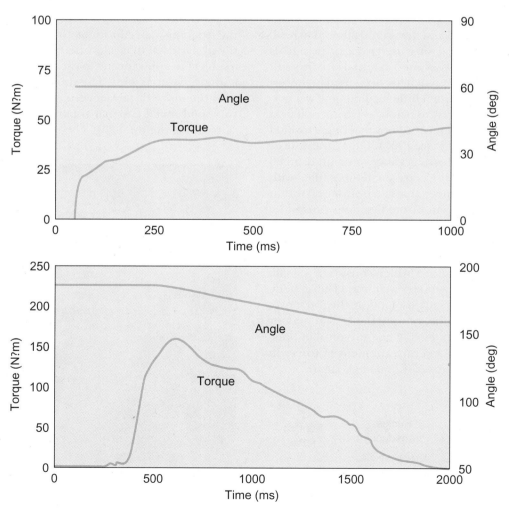

◦ Figure 4.27   The top panel illustrates a typical *isometric* maximal effort contraction at a fixed angular position ($\theta = 60°$). The bottom panel shows a typical *dynamic* maximal effort contraction at a fixed angular velocity ($\omega = 60°/s$).

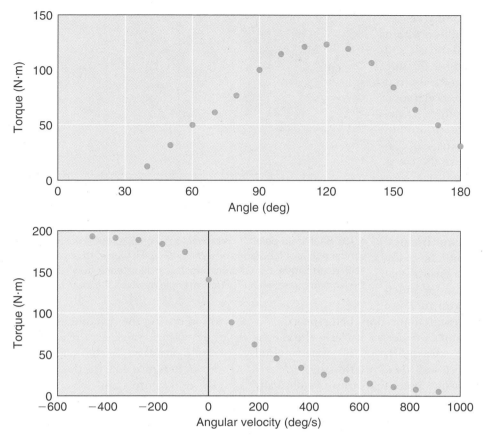

○ **Figure 4.28** The top panel illustrates a typical $T–\theta$ relation for the extensor muscles of the knee joint. Each data point (T) was taken from a maximal effort isometric contraction at a fixed angular position (θ). The bottom panel shows a typical dynamic $T–\omega$ relation for the knee extensors. Each data point (T) represents the peak torque from a maximal effort isovelocity contraction at a fixed angular velocity (ω).

length relationship (see chapter 9), the change in muscle moment arm with angle, and the ability to completely activate muscles in all joint positions. Also, passive structures such as ligaments and bony contact points may produce torque at the extremes of joint motion, adding to the active contributions of the muscles themselves.

A similar protocol is used to determine the dynamic torque–angular velocity relationship at a joint. In this case, the subject performs a series of maximal muscular efforts while the dynamometer bar is constrained to move at or below a predetermined angular velocity (e.g., 30°/s). Beginning at a set angle from rest, the subject exerts maximal effort against the bar. The applied torque causes the bar to rotate, and if the subject exerts enough torque the bar attains the predetermined velocity. The torque profile here is quite different from in the isometric case (figure 4.27), because it rises to a peak value and then falls. Recording the peak torque and the angular velocity gives one point for the $T–\omega$ relation, and the complete joint $T–\omega$ capability can be

ascertained by performing repeated trials across a wide range of angular velocities (figure 4.28). As with the isometric $T–\theta$ relation, several reasons can account for the variation in torque potential with angular velocity, the most important of which is the force-velocity characteristics of the muscles crossing the tested joint (see chapter 9).

A number of issues must be considered when collecting data with an isovelocity dynamometer (see Herzog 1988). If the movement takes place within a sagittal plane, the distal segment and dynamometer arm will change their orientations with respect to gravity as the movement unfolds. The body segment and dynamometer arm both have a finite weight and will affect the measured joint torque (Winter, Wells, and Orr 1981). Most commercial dynamometers include a *gravity correction* option for handling this situation. Another consideration is the speed at which the dynamometer bar is moving. Although moving the segment and bar at a constant angular velocity is the goal, the movement by necessity begins and ends with a velocity of zero. This means that

there are periods of acceleration and deceleration within
each experimental trial, and
only a finite length of time
and movement at the desired
constant velocity. Acceleration
periods can produce unde-
sirable inertial loads and an
overshoot phenomenon. Furthermore, because the bar is
constrained to move at *or below*
a set velocity, in some trials
the subject may be unable
to achieve the desired target
velocity. The experimental
data records should be examined carefully to ensure that
the conditions under which
the torque is attained are
correct for the research questions being posed.

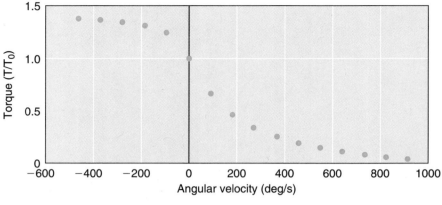

○ **Figure 4.29**   Standardized dynamic $T–\omega$ relation for the knee extensors (from figure 4.28). Each data point (T) represents the peak torque from a maximal effort isovelocity contraction at a fixed angular velocity $(-\omega)$. However, to account for the occurrence of each peak torque at a different angular position, each "raw" torque has been divided by the isometric maximum (T_o) for the angle at which the peak occurred.

A second group of concerns involves the meaning of the data extracted from the experimental
records. The $T–\theta$ and $T–\omega$ relationships clearly
illustrate that torque potential varies isometrically
as a function of angle and dynamically as a function
of velocity. Although these relationships are found
using two different test protocols (i.e., one static
and one dynamic), it has been noted that underlying muscular properties are "in effect" during all
experimental trials. The torque–angle relationship
should be taken into account when assessing the
dynamic torque–velocity relation to ensure that
variation in peak torque at different velocities is
not partly the result of peaks occurring at different angular positions. Indeed, in plantarflexion,
peak torque occurs at progressively greater angles
as concentric velocity increases (Bobbert and van
Ingen Schenau 1990). One method used to prevent
this problem is to measure the torque at one specific
angle, regardless of whether a peak occurred in the
torque time record (e.g., Froese and Houston 1985).
Although this seems sensible, it is flawed because
of the passive series elasticity that resides in muscle
(Hill 1938). This series elastic component changes
length if the torque either rises or falls when the
angle-specific measurement is taken, meaning that
the force-generating structures within the muscles
are not at the same velocity as registered by the
dynamometer (Caldwell, Adams, and Whetstone
1993). A better approach is to establish the isometric $T–\theta$ relation and then express each peak torque

found during the dynamic trials as a percentage of
the isometric value for the angle at which the peak
torque occurred. Such a scaled $T–\omega$ relationship is
shown in figure 4.29 for the "raw" $T–\omega$ relationship
shown in figure 4.28.

SUMMARY

In this chapter, we introduced the conceptual basics
for understanding Newton's laws of motion and how
these laws can be used to study human motion. Of
importance are the concepts of force applied to
particles and forces and moments of force applied
to rigid bodies, and the resulting linear and angular
effects. Free-body diagrams were introduced as a way
of visualizing systems in which multiple forces act. The
measurement of forces was discussed, including the
types of force transducers most commonly used in biomechanics. Readers should now be able to understand
how forces are measured in biomechanics research
and be able to undertake their own force and torque
measurements if the proper equipment is available.

SUGGESTED READINGS

Beer, F.P., and E.R. Johnston, Jr. 1977. *Vector Mechanics
for Engineers: Statics and Dynamics.* 3rd ed. Toronto:
McGraw-Hill.

Hamill, J., and K.M. Knutzen. 1995. *Biomechanical Basis of
Human Movement.* Baltimore: Williams & Wilkins.

Nigg, B.M., and W. Herzog. 1994. *Biomechanics of the Musculo-skeletal System.* Toronto: John Wiley & Sons.

---

[1]L is the accepted SI abbreviation for angular momentum. Many textbooks use H.

[2]Note that scientists in the aerospace industry must use Newton's original formula because much of a rocket's mass is the
fuel used for propulsion; therefore, the mass of the rocket continually changes.

CHAPTER 5

Two-Dimensional Inverse Dynamics

Saunders N. Whittlesey
and D. Gordon E. Robertson

Inverse dynamics is the specialized branch of mechanics that bridges the areas of kinematics and kinetics. It is the process by which forces and moments of force are indirectly determined from the kinematics and inertial properties of moving bodies. In principle, inverse dynamics also applies to stationary bodies, but usually it is applied to bodies in motion. It derives from Newton's second law, where the resultant force is partitioned into known and unknown forces. The unknown forces are combined to form a single *net force* that can then be solved. A similar process is done for the moments of force so that a single *net moment of force* is computed. This chapter

- defines the process of inverse dynamics for planar motion analysis,

- presents the standard method for numerically computing the internal kinetics of planar human movements,

- describes the concept of general plane motion,

- outlines the method of sections for individually analyzing components of a system or segments of a human body,

- outlines how inverse dynamics aids research of joint mechanics, and

- examines applications of inverse dynamics in biomechanics research.

Inverse dynamics of human movement dates to the seminal work of Wilhelm Braune and Otto Fischer between 1895 and 1904. This work was later revisited by Herbert Elftman for his research on walking (1939a, 1939b) and running (1940). Little follow-up research was conducted until Bresler and Frankel (1950) conducted further studies of gait in three dimensions (3-D) and Bresler and Berry (1951) expanded the approach to include the powers produced by the ankle, knee, and hip moments during normal, level walking. Because Bresler and Frankel's 3-D approach measured the moments of force against a Newtonian or absolute frame of reference, it was not possible to determine the contributions made by the flexors or extensors of a joint vs. the abductors and adductors.

Again, few inverse dynamics studies of human motion were conducted until the 1970s, when the advent of commercial force platforms to measure the ground reaction forces (GRFs) during gait and inexpensive computers to provide the necessary processing power spurred new research. Another important development has been the recent propagation of automated and semiautomated motion-analysis systems based on video or infrared camera technologies, which greatly decrease the time required to process the motion data.

Inverse dynamics studies have since been carried out on such diverse movements as lifting (McGill and Norman 1985), skating (Koning, de Groot, and van

Ingen Schenau 1991), jogging (Winter 1983a), race walking (White and Winter 1985), sprinting (Lemaire and Robertson 1989), jumping (Stefanyshyn and Nigg 1998), rowing (Robertson and Fortin 1994; Smith 1996), and kicking (Robertson and Mosher 1985), to name a few. Yet inverse dynamics has not been applied to many fundamental movements, such as swimming and skiing because of the unknown external forces of water and snow, or batting, puck shooting, and golfing because of the indeterminacy caused by the two arms and the implement (the bat, stick, or club) forming a closed kinematic chain. Future research may be able to overcome these difficulties.

PLANAR MOTION ANALYSIS

One of the primary goals of biomechanics research is to quantify the patterns of force produced by the muscles, ligaments, and bones. Unfortunately, recording these forces directly (a process called dynamometry) requires invasive and potentially hazardous instruments that inevitably disturb the observed motion. Some technologies that measure internal forces include a surgical staple for forces in bones (Rolf et al. 1997) and mercury strain gauges (Brown et al. 1986; Lamontagne et al. 1985) or buckle force transducers (Komi 1990) for forces in muscle tendons or ligaments. While these devices enable the direct measurement of internal forces, they have been used only to measure forces in single tissues and are not suitable for analyzing the complex interaction of muscle contractions across several joints simultaneously. Figure 5.1 (Seireg and Arvikar 1975) shows the complexity of forces a biomechanist must consider when trying to analyze the mechanics of the lower extremity. In figure 5.2, the lines of action of only the major muscles of the lower extremity have been graphically presented by Pierrynowski (1982). It is easy to imagine the difficulty of and the risks associated with attempting to attach a gauge to each of these tendons.

Inverse dynamics, although incapable of quantifying the forces in specific anatomical structures, is able to measure the *net* effect of all of the internal forces

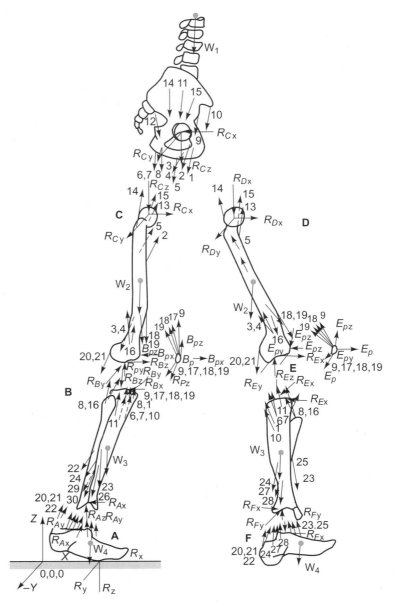

○ Figure 5.1 Free-body diagrams of the segments of the lower extremity during walking.

Reprinted, by permission, from A. Seireg and R.J. Arvikar, 1975, "The prediction of muscular load sharing and joint forces in the lower extremities during walking," *Journal of Biomechanics* 18: 89-102.

and moments of force acting across several joints. In this way, a researcher can infer what total forces and moments are necessary to create the motion and quantify both the internal and external work done at each joint. The steps set out next clarify the process for reducing complex anatomical structures to a solvable series of equations that indirectly quantify the kinetics of human or animal movements.

Figure 5.3 shows the space and free-body diagrams of one lower extremity during the push-off phase of running. Three equations of motion can

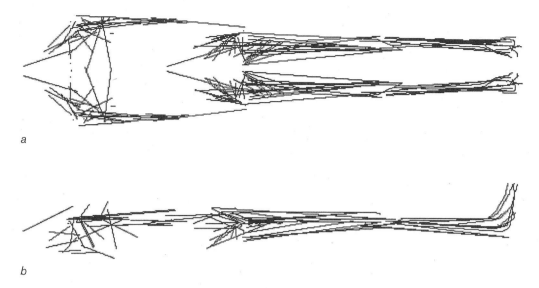

a

b

° **Figure 5.2** Lines of action of the muscle forces in the lower extremity and trunk: *(a)* front view, *(b)* side view.

Adapted from data, by permission, from M.R. Pierrynowski, 1982, "A physiological model for the solution of individual muscle forces during normal human walking," (PhD diss., Simon Fraser University).

be written for each segment in a two-dimensional (2-D) analysis, so for the foot, three unknowns can be solved (figure 5.4). Unfortunately, because there are many more than three unknowns, the situation is called *indeterminate*. Indeterminacy occurs when there are more unknowns than there are independent equations. To reduce the number of unknowns, each force can be resolved to its equivalent force and moment of force at the segment's endpoint. The process starts at a terminal segment, such as the foot or hand, where the forces at one end of the segment are known or zero. They are zero when the segment is not in contact with the environment or another object. For example, the foot during the swing phase of gait experiences no forces at its distal end; when it contacts the ground, however, the GRF must be measured by, for example, a force platform.

A detailed free-body diagram (FBD) of the foot in contact with the ground is illustrated in figure 5.4. Notice the many types of forces crossing the ankle joint, including muscle and ligament forces and bone-on-bone forces; many others have been left out (e.g., forces from skin, bursa, and joint capsule). Furthermore, the foot is assumed to be a "rigid body," although some researchers have modeled it as having two segments (Cronin and Robertson 2000; Stefanyshyn and Nigg 1998). A rigid body is an object that has no moving parts and cannot be deformed. This state implies that its inertial properties are fixed values (i.e., that its mass, center of gravity, and mass distribution are constant).

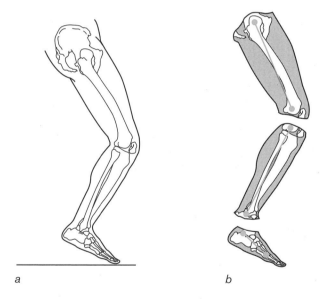

a   b

° **Figure 5.3** *(a)* Space and *(b)* free-body diagrams of the foot during the push-off of running.

Figure 5.5 shows how to replace a single muscle force with an equivalent force and moment of force about a common axis. In this example, the muscle force exerted by the tibialis anterior muscle on the foot segment is replaced by an equivalent force and moment of force at the ankle center of rotation. Assuming that the foot is a "rigid body," a force $(\vec{F}^*)$ equal in magnitude and direction to the muscle force $(\vec{F})$ is placed at the ankle. Because this would

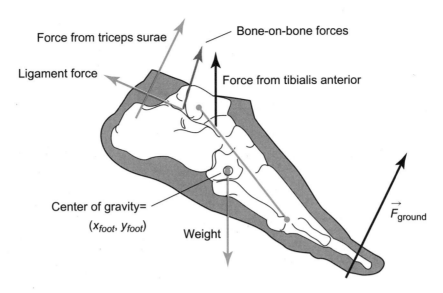

FBD of the foot showing anatomical forces.

unbalance the free body, a second force $(-\vec{F}^{*})$ is added to maintain equilibrium (figure 5.5b). Next, the force couple ($\vec{F}^{*}$ and $-\vec{F}^{*}$) is replaced by the moment of force ($M_{F}\vec{k}$). The resulting force and moment of force in figure 5.5c have the same mechanical effects as the single muscle force in figure 5.5a, assuming that the foot is a rigid body.

The first step to simplifying the complex situation shown in figure 5.4 is to replace every force that acts across the ankle with its equivalent force and moment of force about a common axis. Figure 5.6 shows this situation. Note that forces with lines of action that pass through the ankle joint center produce no moment of force around the joint. Thus, the major structures that contribute to the net moments

of force are the muscle forces. The ligament and bone-on-bone forces contribute mainly to the net force experienced by the ankle and only affect the ankle moment of force when the ankle is at the ends of its range of motion.

Muscles attach in such a way that their turning effects about a joint are enhanced, and most have third-class leverage to promote speed of movement. Thus, muscles rarely attach so that they cross directly over a joint axis of rotation because that would eliminate their ability to create a moment about the joint. Ligaments, on the other hand, often cross joint axes, because their primary role is to hold joints together rather than to create rotations of the segments that they connect. They do, however, have a role in pro-

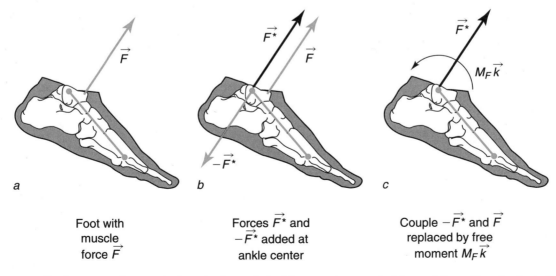

| Foot with muscle force $\vec{F}$ | Forces $\vec{F}^{*}$ and $-\vec{F}^{*}$ added at ankle center | Couple $-\vec{F}^{*}$ and $\vec{F}$ replaced by free moment $M_{F}\vec{k}$ |

○ **Figure 5.5**   Replacement of a muscle force by its equivalent force and moment of force at the ankle axis of rotation.

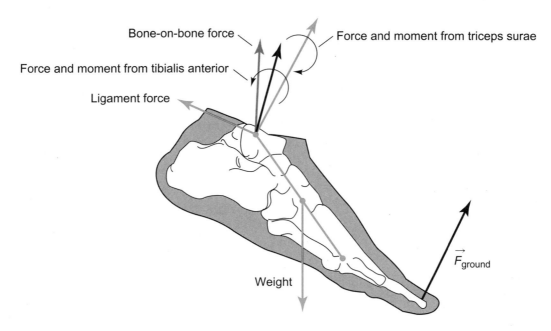

Figure 5.6   FBD of the foot showing the muscle forces replaced by their equivalent force and moment about the ankle.

ducing moments of force when the joint nears or reaches its range-of-motion limits. For example, at the knee, the collateral ligaments prevent varus and valgus rotations and the cruciate ligaments restrict hyperextension. Often, the ligaments and bony prominences produce force couples that prevent excessive rotation, such as when the olecranon process and the ligaments of the elbow prevent hyperextension of the elbow.

To complete the inverse dynamics process for the foot, every anatomical force, including ligament and bone-on-bone (actually, cartilaginous) forces, must be transferred to the common axis at the ankle. Note

that only forces that act across the ankle are included in this process. Internal forces that originate and terminate within the foot are excluded, as are external forces in contact with the sole of the foot. Figure 5.7 represents the situation after all of the ankle forces have been resolved. In this figure, the ankle forces and moments of force are summed to produce a single force and moment of force, called the *net force* and *net moment of force*, respectively. They are also sometimes called the *joint force* and *joint moment of force*, but this is confusing because there are many different joint forces included in this sum, such as those caused by the joint capsule, the ligaments, and

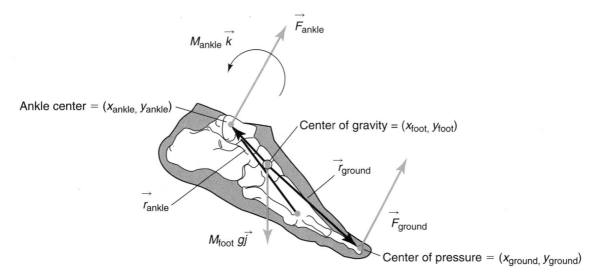

Figure 5.7   Reduced FBD showing net force and moment of force.

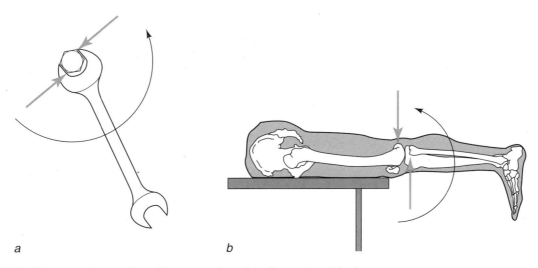

a   b

○ **Figure 5.8**   Force couples produced by a wrench and the ligaments of the knee.

the articular surfaces (cartilage). Another confusing term is *resultant joint force* and *resultant moment of force,* because these terms may be confused with the resultant force and moment of force of the foot segment itself. Recall that the resultant force and moment of force of a rigid body are the sums of all forces and moments acting on the body. These sums are *not* the same as the net force and moment of force just defined. The resultant force and moment of force concern Newton's first and second laws.

The term *moment of force* is often called *torque* in the scientific literature. In engineering, torque is usually considered a moment of force that causes rotation about the longitudinal axis of an object. For example, a torque wrench measures the axial moment of force when tightening nuts or bolts, and a torque motor generates spin about an engine's spin axis. In the biomechanics literature, however, as stated in chapter 4, torque and moment of force are used interchangeably.

Another term related to moment of force is the *force couple.* A force couple occurs when two parallel, noncollinear forces of equal magnitude but opposite direction act on a body. The effect of a force couple is special because the forces, being equal but opposite in direction, effect no translation on the body when they act. They do, however, attempt to produce a pure rotation, or torque, of the body. For example, a wrench (figure 5.8) causes two parallel forces when applied to the head of a nut. The nut translates because of the threads of the screw, but turns around the bolt because of the rotational forces (i.e., moment of force, force couple, or torque) created by the wrench.

Another interesting characteristic of a force couple, or *couple* for short, is that, when the couple

is applied to a rigid body, the effect of the couple is independent of its point of application. This makes it a *free moment,* which means that the body experiencing the couple will react in the same way wherever the couple is applied as long as the lines of axis of the force are parallel. For example, a piece of wood that is being drilled will react the same no matter where the drill contacts the wood as long as the drill bit enters the wood from the same parallel direction. Of course, how the wood actually reacts will depend on friction, clamping, and other forces, but the drill will inflict the same rotational motion on the wood no matter where it enters.

The work done by the net moments of force quantifies the mechanical work done by only the various tissues that act across and contribute a turning effect at a particular joint. All other forces, including gravity, are excluded from contributing to the net force and moment of force. More details about how the work of the moment of force is calculated are delineated in chapter 6.

Net forces and moments are not real entities; they are mathematical concepts and therefore can never be measured directly. They do, however, represent the summed or net effect of all the structures that produce forces or moments of force across a joint. Some researchers (e.g., Miller and Nelson 1973) have called the source of the net moment of force a "single equivalent muscle." They contend that each joint has two single, equivalent muscles that produce the net moments of force about each joint—for example, one for flexion and the other for extension—depending on the joint's anatomy. Others have called the net moments of force "muscle moments," but this nomenclature should be avoided because, even though muscles are the main contributors to the net moment, other structures also contribute, especially at the ends of the range of motion. An illustration of this situation is when the knee reaches maximal flexion during the swing phase of sprinting. Lemaire and Robertson (1989) and others showed that although a very large moment of force occurs, the likely cause is not an eccentric contraction of the extensors; instead it is the result of the calf and thigh bumping together. On the other hand, the same cannot be said to occur for the negative work done by the knee extensors during the swing phase of walking, because the joint does not fully flex and therefore muscles must be recruited to limit knee flexion (Winter and Robertson 1979).

NUMERICAL FORMULATION

This section presents the standard method in biomechanics for numerically computing the internal kinetics of planar human movements. In this process, we use body kinematics and anthropometric parameters to calculate the *net* forces and moments at the joints. This process employs three important principles: Newton's second law ($\sum \vec{F} = m\vec{a}$), the principle of *superposition,* and an engineering technique known as the *method of sections.* The principle of superposi-

tion holds that in a system with multiple factors (i.e., forces and moments), given certain conditions, we can either sum the effects of multiple factors or treat them independently. In the method of sections, the basic idea is to imagine cutting a mechanical system into components and determining the interactions between them. For example, we usually section the human lower extremity into a thigh, leg, and foot. Then, via Newton's second law, we can determine the forces acting at the joints by using measured values for the GRFs and the acceleration and mass of each segment. This process, called the *linked-segment* or *iterative Newton-Euler method,* is diagrammed in figure 5.9. The majority of this chapter explains how this method works. We will begin with kinetic analysis of single objects in 2-D, then demonstrate how to analyze the kinetics of a joint via the method of sections, and, finally, explain the general procedure diagrammed in figure 5.9 for the entire lower extremity.

Note the diagram conventions used in this chapter: Linear parameters are drawn with straight arrows and angular parameters, with curved arrows. Known kinematic data (linear and angular accelerations) are drawn with blue arrows. Known forces and moments are drawn with black arrows. Unknown forces and moments are drawn with gray arrows. These conventions will assist you in visualizing the solution process.

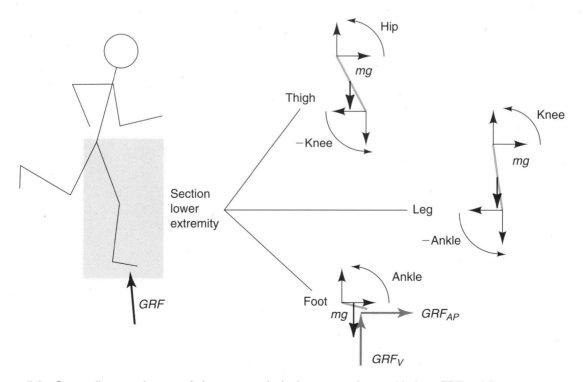

Figure 5.9 Space diagram of a runner's lower extremity in the stance phase, with three FBDs of the segments.

GENERAL PLANE MOTION

General plane motion is an engineering term for 2-D movement. In this case, an object has three degrees of freedom (DOF): two linear positions and an angular position. Often, we draw these as translations along the x- and y-axes and a rotation about the z-axis. As discussed in chapter 2, many lower-extremity movements can be analyzed using this simplified representation, including walking, running, cycling, rowing, jumping, kicking, and lifting. However, despite the simplification to 2-D analysis, the resulting mechanics can still be complicated. For example, a football punter has three lower-extremity segments that swing forward much like a whip, kick the ball, and elevate. Even the ball has a somewhat complicated movement, translating both horizontally and vertically and rotating. To determine the kinetics of such situations, the fundamental principle is that the three DOF are treated independently. That is, we exploit the fact that an object accelerates in the vertical direction only when acted on by a vertical force and accelerates in the horizontal direction only when acted on by a horizontal force. Similarly, the body does not rotate unless a moment (torque) is applied to it. The principle of superposition states that when one or more of these actions occur, we can analyze them separately. We therefore separate all forces and moments into three coordinates and solve them separately.

To illustrate, let us consider the football example. In figure 5.10d, a football being kicked is subjected to the force of the punter's foot. The ball moves horizontally and vertically and also rotates. Our goal is to determine the force with which the ball was kicked. We cannot measure the force directly with an instrumented ball or shoe. However, we can film the ball's movement and measure its mass and moment of inertia. Given these data, we are left with an apparently confusing situation to analyze: The single force of the foot has caused all three coordinates to change. However, the situation becomes simpler when we employ superposition. The horizontal and vertical accelerations of the ball's mass center must be proportional to their respective forces, and the angular acceleration must be proportional to the applied moment. Consider the examples in figure 5.10. Figure 5.10a and b are rather obvious, but are presented for the sake of demonstration. In figure 5.10a, a horizontal force is applied to the ball through its center

of mass, and the ball will accelerate horizontally; it will not accelerate vertically because there is no vertical force. Similarly, in figure 5.10b, the ball will accelerate only vertically because there is no horizontal force. In figure 5.10c, the force acts at a 45° angle, and therefore the ball accelerates at a 45° angle. This is just a superposition of the situations in figure 5.10a and b. We do not deal with the force at this angle; rather, we measure the accelerations in the horizontal and vertical directions, and therefore we can determine the forces in the horizontal and vertical directions. In Figure 5.10d, the applied force is not directed through the center of mass. In this case, the force is the same as it is in figure 5.10c, so the ball's center of mass has the same acceleration. However, there is also an angular acceleration proportional to the product of the force, *F,* and the distance, *d,* between its line of action and the center of mass. The acceleration, *a,* in this case is the same as in figure 5.10c. However, the ball will also rotate.

To reiterate, a force causes a body's center of mass to accelerate in the same direction as that

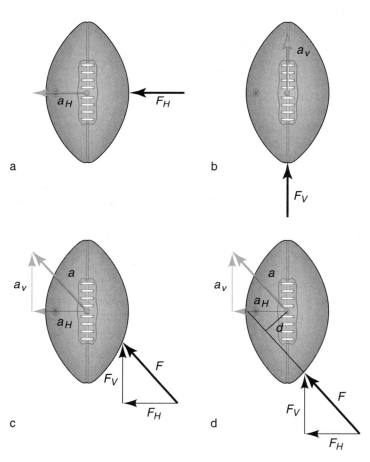

○ **Figure 5.10** Four FBDs of a football experiencing four different external forces.

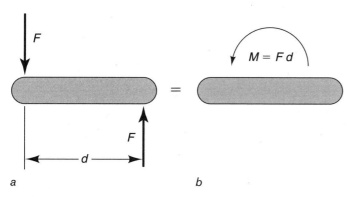

force. A force does not cause a body to rotate; only the moment of a force causes a body to rotate. These principles derive from Newton's first and second laws. If a ball is kicked, a resulting force is imparted on the ball. This force *is* the reaction of the ball being accelerated. Whether or not the ball was kicked through its center of mass, causing it to rotate, is irrelevant. Rotation is affected in proportion to the distance by which the applied force and mass center are out of line.

Let us explore moments further. Referring to figure 5.11, a moment can be defined as the effect of a force-couple system, that is, two forces of equal magnitude and opposite direction that are not collinear (figure 5.11a). In this system, the sum of the forces is zero. However, because the two forces are noncollinear, they cause the body to rotate. This is drawn diagrammatically as a curved arrow (figure 5.11b).

Returning to the football in figure 5.10, there is a force couple in all four diagrams, the applied force, *F*, and the reaction, *ma*. In figure 5.10a through c, this force couple is collinear, so there is no moment. However, in figure 5.10d, the force and reaction are noncollinear. There is a perpendicular distance between the line of action of the force and the reaction of the mass center, and the result is a moment that rotates the ball.

Let us formalize the present discussion. Given a 2-D FBD, the process is to employ Newton's second law in the horizontal, vertical, and rotational directions:

$$\Sigma F_x = ma_x \quad \Sigma F_y = ma_y \quad \Sigma M = I\alpha \quad (5.1)$$

The mass, *m*, and moment of inertia, *I*, for the object in question are determined beforehand. The linear and angular accelerations are determined from camera data. The sum of the forces/moments on the left side of each equation (i.e., each Σ term) can combine many forces/moments, but it should contain only one unknown—a net force or moment—to solve for. Because of this, the two forces usually must be solved for before solving for the unknown moment.

When putting together the sums of forces and moments, it is all-important to adhere to the sign conventions established by the FBD. Inverse dynamics problems often require careful bookkeeping of positive and negative signs. As the examples in this chapter show, the FBD takes care of this as long as we honor the sign conventions that have been drawn. For example, in many cases we solve for a force or moment even though we are uncertain of its direction. This is not a problem with FBDs: We merely draw the force or moment with an assumed direction. If, in fact, the force points in the opposite direction, our calculations simply return a negative numerical value.

A sum of moments must be calculated about a single point on the object in question. There is no right or wrong point about which to calculate; however, some points are simpler than others. If we sum moments about a point where one or more forces act, then the moments of those forces will be zero because their moment arms are zero. Therefore, sometimes in human movement, it is convenient to calculate about a joint center. However, in most cases we calculate moments about the mass center; there, we can neglect the reaction force *(ma)* and gravity terms because their moment arms are zero.

This text uses the convention that a counterclockwise moment is positive, also called the *right-hand rule*. Coordinate system axes on FBDs establish their positive force directions. When solving a problem from an FBD, the proper technique is to first write the equations in algebraic form, honoring the sign conventions, and then to substitute known numerical values with their signs once all of the algebraic manipulations have been carried out. These procedures are only learned by example, so we now present a few of them.

EXAMPLE 5.1

5.1a—Suppose that in our football example, the ball was kicked and its mass center moved horizontally. The horizontal acceleration was −64 m/s², the angular acceleration was −28 rad/s², the mass of the ball was 0.25 kg, and its moment of inertia was 0.05 kg·m². What was the kicking force? How far from the center of the ball was the force's line of action?

➤ *See answer 5.1a on page 271.*

5.1b—Now solve the same problem assuming that there was a force plate under the football and that the tee the football rested on resisted the kicking force. The force of the tee was 4 N in the horizontal direction; its center of pressure (COP) was 15 cm below the mass center of the ball.

➤ *See answer 5.1b on page 272.*

EXAMPLE 5.2

A commuter is standing inside a subway car as it accelerates away from the station at 3 m/s². She maintains a perfectly rigid, upright posture. Her body mass of 60 kg is centered 1.2 m off the floor, and the moment of inertia about her ankles is 130 kg·m². What floor forces and ankle moment must the commuter exert to maintain her stance?

➤ *See answer 5.2 on page 272.*

EXAMPLE 5.3

A tennis racket is swung in the horizontal plane. The racket's mass center is accelerating at 32 m/s² and its angular acceleration is 10 rad/s². Its mass is 0.5 kg and its moment of inertia is 0.1 kg·m². From the base of the racket, the locations of the hand and racket mass center are 7 and 35 cm, respectively. Ignoring the force of gravity, what are the net force and moment exerted on the racket? Given that the hand is about 6 cm wide, interpret the meaning of the net moment (i.e., what is the actual force couple at the hand?).

➤ *See answer 5.3 on page 273.*

In these examples, we provided distances between various points on the FBDs. These distances are not the data we measure with camera systems and force plates. Rather, these instruments measure the locations of points, specifically the locations of joint centers and the GRF (the COP). We need these points to calculate the moment arms of the various forces in the FBD. This is simply a matter of subtracting corresponding positions in the global coordinate system (GCS) directions. However, a commonly made error is that the moment arm of a force in the x direction is a distance in the perpendicular y direction, and vice versa. This is a very important point that we again illustrate with examples.

EXAMPLE 5.4

Given the FBD of the arbitrary object following, calculate the reactions R_x, R_y, M_z at the unknown end. Its mass is 8.0 kg and its moment of inertia is 0.2 kg·m² about the mass center.

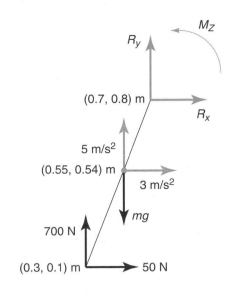

➤ *See answer 5.4 on page 273.*

EXAMPLE 5.5

Given the following data for a bicycle crank, draw the FBD and calculate the forces and moments at the crank axle: pedal force x = −200 N; pedal force y = −800 N; pedal axle at (0.625, 0.310) m; crank axle at (0.473, 0.398) m; crank mass center at (0.542, 0.358) m. The crank mass is 0.1 kg and its moment of inertia is 0.003 kg·m². Its accelerations are −0.4 m/s² in the x direction and −0.7 m/s² in the y direction; its angular acceleration is 10 rad/s².

➤ *See answer 5.5 on page 274.*

Having discussed general plane motion for a single object, we now turn to the solution technique for multiobject systems like arms and legs. The procedure for each body segment is exactly the same as for single-body systems. The only difference is that the method of sections is applied to manage the interactive forces between segments. In fact, this technique was applied in the previous example of the bicycle crank: We had to "section" the crank, that is, imagine cutting it off at the axle, to determine the forces and moment at the axle. Let us explore this technique in detail.

METHOD OF SECTIONS

Engineering analysis of a mechanical device usually focuses on a limited number of key points of the structure. For example, on a railroad bridge truss, we normally study the points where its various pieces are riveted together. The same is true when we analyze the kinetics of human movement. We generally do not concern ourselves with a complete map of the forces and moments within the body. Rather, we study specific points of the body—most commonly, the joints. We therefore "section" the body at the joints and calculate the reactions between adjacent segments that keep them from flying apart. These forces and moments at a sectioned joint are unknown. Therefore, when constructing our FBD, we must draw a reaction for each DOF—that is, a horizontal force, a vertical force, and a moment. In some calculations, it is possible for one or two of these reactions to be zero, but the method of sections requires that each be drawn and solved for because they are unknowns.

The method of sections is straightforward:

- Imagine cutting the body at the joint of interest.
- Draw FBDs of the sectioned pieces.
- At the sectioned point of one piece, draw the unknown horizontal and vertical reactions and the net moment, honoring the *positive* directions of the GCS.
- At the sectioned point of the other piece, draw the unknown forces and moment in the *negative* directions of the GCS. This is Newton's third law.
- Solve the three equations of motion for one of the sections.

SINGLE SEGMENT ANALYSIS

In many cases, we may be interested in only one of the sectioned pieces. Let us start with such an example and then follow with a complete example.

EXAMPLE 5.6

Consider an arm being held horizontally. What are the shoulder reaction forces and joint moment? Assume that the arm is stationary and rigid. The weights of the upper arm, forearm, and hand are 4, 3, and 1 kg, respectively, and their mass centers are, respectively, 10, 30, and 42 cm from the shoulder.

➤ See answer 5.6 on page 275.

EXAMPLE 5.7

Suppose the hand is holding a 2 kg weight. What are the shoulder reaction forces and joint moment now?

➤ See answer 5.7 on page 275.

EXAMPLE 5.8

The elbow is 22 cm from the shoulder. What are the reactions at the elbow on the forearm? On the upper arm? Solve again for the reactions at the shoulder using the FBD of the upper arm.

➤ See answer 5.8 on page 276.

MULTISEGMENTED ANALYSIS

Complete analysis of a human limb follows the procedure used in the previous examples. We simply have to formalize our solution process. This process has a specific order: We start at the most distal segment and continue proximally. The reason for this is that we have only three equations to apply to each segment, which means that we can have only three unknowns for each segment—one horizontal force, one vertical force, and one moment. However, we can see (return to figure 5.9, if necessary) that if we were to section and analyze either the thigh or the leg, we would have six unknowns—two forces and one moment at each joint. The solution is to start with a segment that has only one joint (i.e., the most distal), and from there to proceed to the adjacent segment. For this we use Newton's third law, applying the negative of the reactions of the segment that we just solved, as shown for the lower extremity in figure 5.12. Of the lower-extremity segments, only the foot has the requisite three unknowns, so we solve this segment first. Note how the actions on the ankle of the foot have corresponding equal and opposite reactions on the ankle of the leg. Then we can calculate the unknown reactions at the knee. These are drawn in reverse in the FBD for the thigh, and then we solve for the hip reactions.

One final, but important, detail about this process is that the sign of each numerical value does not change from one segment to the next. From Newton's third law, every action has an equal but opposite reaction. At the joints, therefore, the forces on the distal end of one segment must be equal but opposite to those on the proximal end of the adjacent segment. However, we never change the signs of numerical values. The FBDs take care of this. Note that in figure 5.12, the knee forces and moments are drawn in opposing directions. Following the procedures shown previously, first construct the equations for a segment from the FBD without considering the numeric values. Once the equations are constructed, the numeric values with their signs are substituted into these equations. Note how this process is carried out in the examples that follow.

We provide example calculations for each of the two distinct phases of human locomotion, swing and stance. When solving the kinetics of a swing limb, the process is almost exactly the same as for stance. The only difference is that the GRFs are zero, and thus we can neglect their terms in the equations of motion. The following procedure is virtually identical to the calculations that would be carried out in computer code for one frame of data.

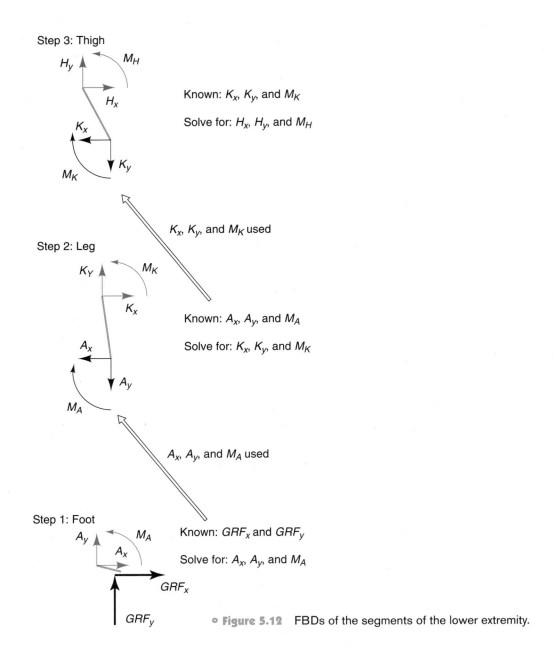

Figure 5.12   FBDs of the segments of the lower extremity.

EXAMPLE 5.9

Determine the joint reaction forces and moments at the ankle, knee, and hip given the following data. These occurred during the swing phase of walking, so the GRFs are zero.

	Mass (kg)	I **(kg·m²)**	a_x **(m/s²)**	a_y **(m/s²)**	α **(rad/s²)**	**C. M. (m)**
Foot	1.2	0.011	−4.39	6.77	5.12	(0.373, 0.117)
Leg	2.4	0.064	−4.01	2.75	−3.08	(0.437, 0.320)
Thigh	6.0	0.130	6.58	−1.21	8.62	(0.573, 0.616)

The ankle is at (0.303, 0.189) m, the knee is at (0.539, 0.420) m, and the hip is at (0.600, 0.765) m.

➤ *See answer 5.9 on page 276.*

EXAMPLE 5.10

Determine the joint reaction forces and moments at the ankle, knee, and hip given the following data. These occurred during the stance phase of walking, so the ground reaction forces are nonzero. The solution process is almost identical.

	Mass (kg)	I (kg·m^2)	a_x (m/s^2)	a_y (m/s^2)	α (rad/s^2)	C. M. (m)
Foot	1.2	0.011	−5.33	−1.71	−20.2	(0.734, 0.089)
Leg	2.4	0.064	−1.82	−0.56	−22.4	(0.583, 0.242)
Thigh	6.0	0.130	1.01	0.37	8.6	(0.473, 0.566)

The ankle is at (0.637, 0.063) m, the knee at (0.541, 0.379) m, and the hip at (0.421, 0.708) m. The horizontal GRF is −110 N, the vertical GRF is 720 N, and its center of pressure is (0.677, 0.0) m.

➤ *See answer 5.10 on page 278.*

HUMAN JOINT KINETICS

What exactly are these forces and moments that we have just calculated? The answer, which was given earlier in this chapter, deserves revisiting: The net forces and moments represent the sum of the actions of all the joint structures. Commonly made errors are thinking of a joint reaction force as the force on the articular surfaces of the bone and a joint moment as the effect of a particular muscle. These interpretations are incorrect because joint reaction forces and moments are more abstract than that; they are sums, net effects. We calculated in the two preceding examples *relative* measures to compare the efforts of the three lower-extremity joints in producing the movement and forces that were measured. We do not even have estimates of the activities of the quadriceps, triceps surae, or any other muscle. We do not have estimates of the forces on the articular surfaces of the joints or any other anatomical structure. Let us explain this further.

As depicted in figure 5.2, many muscle-tendon complexes, ligaments, and other joint structures bridge each joint. Each of these structures exerts a particular force depending on the specific movement. In figure 5.2, we neglected all friction between the articular surfaces as well as between all other adjacent structures. It cannot be overemphasized that joint reaction forces and moments should be discussed without reference to specific anatomical structures. There are several reasons for this. Equal joint reactions may be carried by entirely different structures. Consider, for example, the elbow joint of a gymnast: When hanging from the rings, the elbow is subjected to a tensile force that must be borne by the various tendons, ligaments, and other structures crossing the elbow. In contrast, when the gymnast performs a handstand, many of these same structures can be lax, because much of the load is shifted to the articular cartilage. From the sectional analysis presented in this chapter, we would calculate equal but opposite joint reaction forces in each case; both would equal one-half of body weight minus the weight of the forearms. However, the distribution of the force among the joint structures is completely different in these two cases.

Even if we are not analyzing an agile gymnast, the fact remains that we do not know from net joint forces and moments how loading is shared between various structures. This situation, in which there are more forces than equations, is said to be *statically indeterminate*. In plain language, it is a case in which we know the total load on a system, but we are not able to determine the distribution of the load without considering the specific properties of the load-bearing structures. This is analogous to a group of people moving a heavy object such as a piano: We know that they are carrying the total weight of the piano, but without a force plate under each individual, we cannot know how much weight each person bears.

Consider another specific example, a human standing quietly with straight lower extremities. We could calculate a joint reaction force at the knee. Assuming that the limbs share body weight evenly, the knee reaction force would equal one-half of the body weight minus the weight of the leg and foot. If we asked the subject to clench his muscles as much as possible, the joint reaction forces would not change. However, the tensions in the tendons would increase, as would the compressive forces on the bones. The changes are equal but opposite, and thus do not express themselves as an external, joint reaction force change.

Before discussing patterns of joint moments during human movements, we detail the limitations of these measures. As stated earlier, they are somewhat abstract. However, the purpose of the previous discussion was simply to delineate the limits of joint moments. This being done, these data can be discussed appropriately.

LIMITATIONS

Aside from the intrinsic limitations of 2-D kinematics discussed in chapter 2, there are several other important limitations in our foregoing analysis of 2-D kinetics.

Effects of friction and joint structures are not considered. The tensions of various ligaments become high near the limits of joints, and thus moments can occur when muscles are inactive. Also, while the frictional forces in joints are very small in young subjects, this is often not the case for individuals with diseased joints. Interested readers should refer to Mansour and Audu (1986), Audu and Davy (1988), or McFaull and Lamontagne (1993, 1998). Segments are assumed to be rigid. When a segment is not rigid, it attenuates the forces that are applied to it. This is the basis for car suspension systems: the forces felt by the occupant are less than those felt by the tires on the road. Human body segments with at least one full-length bone such as the thigh and leg are reasonably rigid and transmit forces well. However, the foot and trunk are flexible, and it is well documented that the present joint moment calculations are slightly inaccurate for these structures (see, for example, Robertson and Winter 1980). In the foot, for example, various ligaments stretch to attenuate the GRFs. It is for this reason that individuals walking barefoot on a hard floor tend to walk on their toes: the calcaneus-talus bone structures are much more rigid than the forefoot.

The present model is sensitive to its input data. Errors in GRFs, COPs, marker locations, segment inertial properties, joint center estimates, and segment accelerations all affect the joint moment data. Some of these problems are less significant than others. For example, GRFs during locomotion tend to dominate stance-phase kinetics; measuring them accurately prevents the majority of accuracy problems. However, during the swing phase of locomotion, data-treatment techniques and anthropometric estimates are critical. Interested readers can refer to Pezzack, Norman, and Winter (1977), Wood (1982), or Whittlesey and Hamill (1996) for more information. Exact comparisons of moments calculated in

different studies are not appropriate; we recommend allowing at least a 10% margin of error.

Individual muscle activity cannot be determined from the present model. We do not know the tension in any given muscle simply because muscle actions are represented by moments. Moreover, we do not even know the moment of a single muscle because multiple muscles, ligaments, and other structures cross each joint. Muscle forces are estimated using the musculoskeletal techniques discussed in chapter 9. An important aspect of this is that *cocontraction* of muscles occurs in essentially all human movements. Thus, for example, if knee extensor moments decrease under a certain condition, we do not know whether the decrease occurs because of decreased quadriceps activity or increased hamstring activity. As another example, a subject asked to stand straight and clench her lower-extremity muscles would have joint moments close to zero even though her muscles were fully activated; their actions would work against each other.

Two-joint muscles are not well represented by the present model. Although the moments of two-joint muscles are included in joint moment calculations, the segment calculations effectively assume that muscles only cross one joint. Again, this is a problem that is addressed by musculoskeletal models (see chapter 9).

People with amputations require different interpretations than other individuals. For example, an ankle moment can be measured for prostheses even when the leg and foot are a single, semirigid piece. Prosthetic knees have stops to prevent hyperextension and other components, such as springs and frictional elements, to control their movement. These cause knee moments that require completely different interpretations from those seen in intact subjects. Similar considerations also apply to subjects with braces such as ankle-foot orthoses.

It was a conscious choice on the authors' part to present the interpretation of joint moments after this discussion of their limitations. Limitations do not invalidate model data, but they do limit the extent of interpretation. Note in the following discussion that no reference is made to specific muscles or muscle groups and that the magnitudes of different peaks are not referenced to the nearest 0.1 of a newton meter.

RELATIVE MOTION METHOD VS. ABSOLUTE MOTION METHOD

The equations presented above are a way of computing the net forces and moments. Plagenhoef (1968, 1971) called this approach the *absolute motion method* of inverse dynamics because the segmental kinematic data are computed based on an absolute or fixed frame of reference. An alternative approach outlined by Plagenhoef is the *relative motion method*. This method quantifies the motion of the first segment in a kinematic chain from an absolute frame of reference, but all other segments are referenced to moving axes that rotate with the segment. Thus, each segment's axis, except that of the first, moves relative to the preceding segment. This method has the advantage of showing how one joint's moment of force contributes to the moments of force of the other joints in the kinematic chain. The drawback is that the level of complexity of the analysis increases with every link added to the kinematic chain. Furthermore, the method requires the inclusion of Coriolis forces, which are forces that appear when an object rotates within a rotating frame of reference. These fictitious forces—sometimes called pseudo-forces—only exist because of their rotating frames of reference. From an inertial (i.e., fixed or absolute) frame of reference, they do not exist.

From the Scientific Literature

Winter, D.A. 1980. Overall principle of lower limb support during stance phase of gait. *Journal of Biomechanics* **13:923-7.**

This paper presents a special way of combining the moments of force of the lower extremity during the stance phase of gait. During stance, the three moments of force—the ankle, knee, and hip—combine to support the body and prevent collapse. The author found that by adding the three moments in such a way that the extensor moments had a positive value, the resulting "support moment" followed a particular shape. The support moment $(M_{support})$ was defined mathematically as

$$M_{support} = -M_{hip} + M_{knee} - M_{ankle} \qquad (5.2)$$

Notice that the negative signs for the hip and ankle moments change their directions so that an extensor moment from these joints makes a positive contribution to the support moment. A flexor moment at any joint reduces the amplitude of the support moment. Figure 5.13 shows the average support moment of normal subjects and the support moment and hip, knee, and ankle moments of a 73-year-old male with a hip replacement.

This useful tool allows a clinical researcher to monitor a patient's progress during gait rehabilitation. As the patient becomes stronger or coordinates the three joints more effectively, the support moment gets larger. People with one or even two limb joints that cannot adequately contribute to the support moment will still be supported by that limb if the remaining joints' moments are large enough to produce a positive support moment.

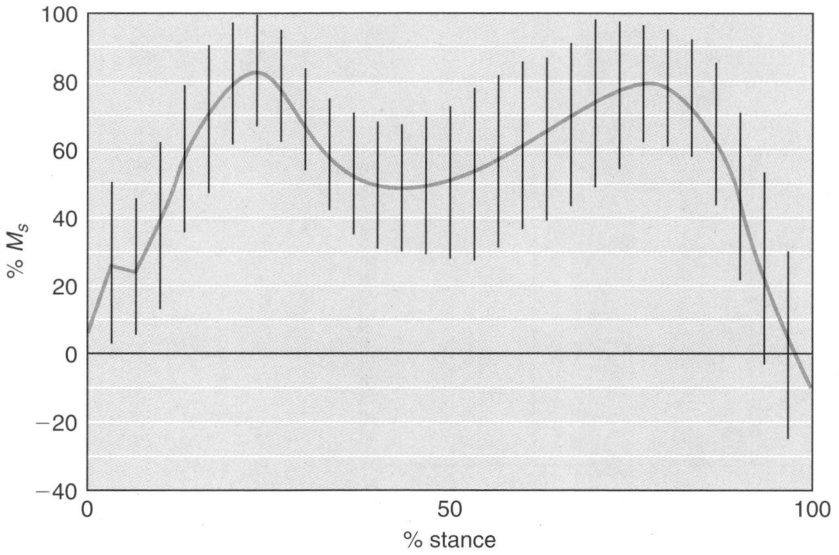

○ Figure 5.13a   Averaged support moment of subjects with intact joints.

Reprinted from Winter, D.A. Overall principle of lower limb support during stance phase of gait. *Journal of Biomechanics* 13:923-7, 1980, with permission from Elsevier.

(continued)

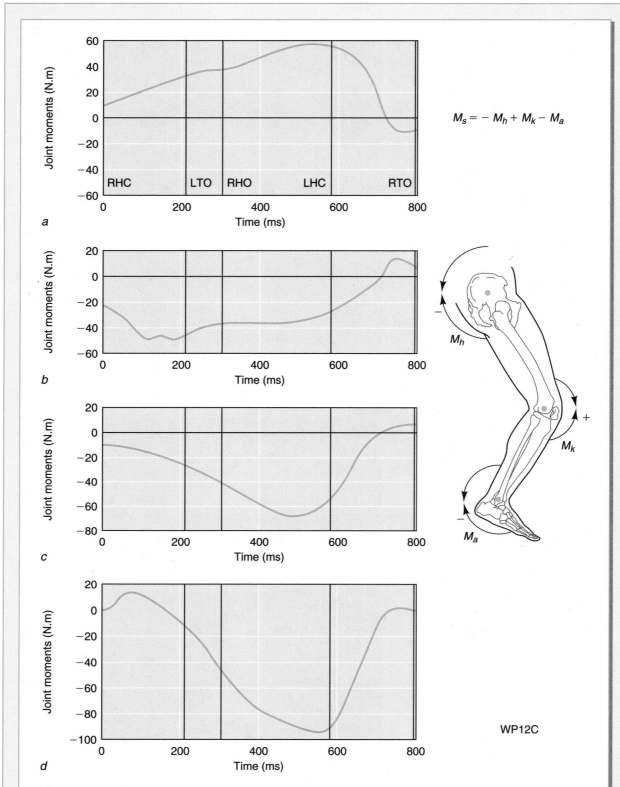

$M_s = -M_h + M_k - M_a$

a

b

c

d

WP12C

○ Figure 5.13b The support moment and hip, knee, and ankle moments of a 73-year-old male with a hip replacement during the stance phase of walking.

Reprinted from Winter, D.A. Overall principle of lower limb support during stance phase of gait. *Journal of Biomechanics* 13: 923-7, 1980, with permission from Elsevier.

Rarely have researchers tried to compare the two methods. However, Pezzack (1976) did compare both methods using the same coordinate data and found that the relative motion method was less accurate, especially as the kinematic chain got longer (had more segments). For short kinematic chains (two or three segments), both methods yielded similar results. Most researchers have adopted the absolute motion method, because most data-collection systems measure segmental kinematics with respect to axes affixed to the ground or laboratory floor.

APPLICATIONS

There are many uses for the results of an inverse dynamics analysis. One application sums the extensor moments of a lower extremity during the stance phase of walking and jogging to find characteristic patterns and predict if people with artificial joints or pathological conditions have a sufficient support moment to prevent collapse (Winter 1980, 1983a; see previous From the Scientific Literature). Others have used the net forces and moments in musculoskeletal models to compute the compressive loading of the base of the spine for research on lifting and low back pain (e.g., McGill and Norman 1985). An extension of this approach is to calculate the compression and shear force at a joint. To do this, the researcher must know the insertion point of the active muscle acting across a joint and assume that there are no other active muscles.

Computing the force in a muscle requires several assumptions to prevent indeterminacy (i.e., too few equations and too many unknowns). For example, if a single muscle can be assumed to act across a joint and no other structure contributes to the net moment of force, then, if the muscle's insertion point and line of action are known (from radiographs or estimation), the muscle force (F_{muscle}) is defined:

$$F_{muscle} = M/(r \sin \theta) \tag{5.3}$$

where M is the net moment of force at the joint, r is the distance from the joint center to the insertion point of the muscle, and θ is the angle of the muscle's line of action and the position vector between the joint center and the muscle's insertion point. Of course, such a situation rarely occurs because most joints have multiple synergistic muscles with different insertions and lines of action as well as antagonistic muscles that often act in cocontraction. By monitoring the electrical activity (with an electromyograph) of both the agonists and antagonists, one can reduce these problems, but even an inactive muscle can create forces, especially if it is stretched beyond its resting length. The contributions of other tissues to the net moment of force can also be minimized as long as the motion being analyzed does not include the ends of the range of motion, when these structures become significant.

Once the researcher has estimated the force in the muscle, the muscle stress can be computed by determining the cross-sectional area. The cross-sectional areas of particular muscles can be derived from published reports or measured directly from MRI scans or radiographs. Axial stress (σ) is defined as axial force divided by cross-sectional area. For pennate types of muscles in which the force is assumed to act along the line of the muscle, the stress is defined as $\sigma = F_{muscle}/A$, where F_{muscle} is the muscle force in newtons and A is the cross-sectional area in square meters. The units of stress are called pascals (Pa), but because of the large magnitudes, kilopascals (kPa) are usually used. Of course, true stress on the muscle cannot be quantified because of the difficulty of directly measuring the actual muscle force.

The following sections present and discuss the patterns of planar lower-extremity joint moments during walking and running. The convention for

From the Scientific Literature

McGill, S., and R.W. Norman. 1985. Dynamically and statically determined low back moments during lifting. *Journal of Biomechanics* **18:877-85.**

This study presents a method for computing the compressive load at the L4/L5 joint from data collected on the motion of the upper body during a manual lifting task. Three different methods for computing the net moments of force at L4/L5 were compared. A conventional dynamic analysis was performed using planar inverse dynamics to compute the net forces and moments at the shoulder and neck, from which the net forces

(continued)

and moments were calculated for the lumbar end of the trunk (L4/L5). Second, a static analysis was done by zeroing the accelerations of the segments. A third, quasidynamic approach assumed a static model but utilized dynamical information about the load. Once the lumbar net forces and moments were calculated, it was assumed that a "single equivalent muscle" was responsible for producing the moments of force and that the effective moment arm of this muscle was 5 cm. The magnitude of the compressive force $(F_{compress})$ on the L4/L5 joint was then computed by measuring the angle of the lumbar spine (actually, the trunk), θ. The equation used was

$$F_{compress} = \frac{M}{r} + F_x \cos\theta + F_y \sin\theta \qquad (5.4)$$

where M is the net moment of force at the L4/L5 joint, r is the moment arm of the single equivalent extensor muscle across L4/L5 (5 cm), (F_x, F_y) is the net force at the L4/L5 joint and θ is the angle of the trunk (the line between L4/L5 and C7/T1).

The researchers showed that there were "statistically significant and appreciable difference(s)" among the results of the three methods for the pattern and peak values of the net moment of force at L4/L5. In general, the dynamic method yielded larger peak moments than the static approach but smaller values than the quasidynamic method. Comparisons of a single subject's lumbar moment histories and the averaged histories of all subjects showed that each approach produced very different patterns of activity. They also showed that while the static approach produced compressive loads that were less than the 1981 National Institute for Occupational Health and Safety (NIOSH) lifting standard, the more accurate dynamic model produced loads greater than the "maximum permissible load." The quasidynamic approach produced even higher compressive loads. Clearly, then, one should apply the most accurate method to obtain realistic conclusions.

presenting these figures is that extensor moments are presented as positive and flexor moments as negative. This is in agreement with the engineering standard that mechanical actions that *lengthen* a system are positive (positive strain) and actions that *shorten* the system are negative (negative strain). In figure 5.9, hip flexor moments and dorsiflexor moments were calculated as positive. Therefore, these two moments are usually presented as the negative of what is calculated.

WALKING

Joint moments during walking have many typical features. Sample data are presented in figure 5.14 for the ankle, knee, and hip joint moments. These data are presented as percentages of the gait cycle; the vertical line at 60% of the cycle represents toe-off and the 0 and 100% points of the cycle reflect heel contact. On the vertical axes, the joint moments are presented in N·m. Sometimes these values are scaled to the percent of body mass or percent of body mass times leg length to assist in comparing

different subjects. We keep these data in N·m for continuity with the preceding examples. In general, joint moments do scale up and down with body size. They also change magnitude with the speed of movement.

Referring to the ankle moment in figure 5.14 shows that there was a dorsiflexor moment after heel contact that peaked at about 15 N·m. This moment prevents the foot from rotating too quickly from heel contact to foot-flat, a condition known clinically as foot-slap. Although 15 N·m is a relatively small moment on the scale of this graph, this peak is nonetheless a very common feature of normal walking. Thereafter, we see a large plantar flexor moment that peaked at about 160 N·m near 40% of the stride duration. This reflects the effort necessary to effect push-off. As this peak diminishes, the limb becomes unloaded. As toe-off occurs, we again see a small dorsiflexor moment of about 10 N·m. This action, although small, is important because it lifts the toes out of the way of the ground. Individuals with dorsiflexor dysfunction have a problem with their toes catching the ground at this part of the

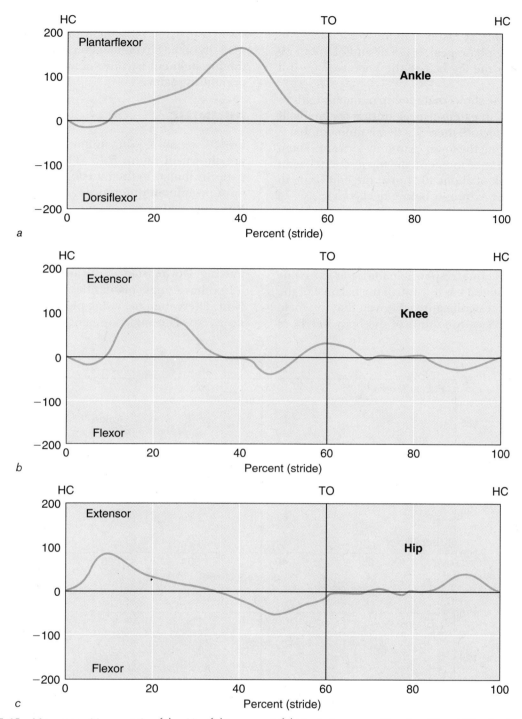

° Figure 5.14   Moments of force at the *(a)* ankle, *(b)* knee, and *(c)* hip during normal level walking. HC = heel contact; TO = toe-off.

gait cycle. For the rest of the swing phase, the ankle moment is near zero.

In figure 5.14b, there are four distinct peaks of the knee moment. The largest peak during stance is extensor, typically peaking around 100 N·m. During this peak, the limb is loaded and thus the extensor moment acts to prevent collapse of the limb.

Note that this peak occurs slightly earlier than the peak of the ankle moment. Often we see a smaller knee flexor peak before toe-off as the leg is pulled through the remainder of the stance. During swing, the first peak is extensor; it limits the flexion of the knee that occurs because the lower extremity is being swung forward from the hip. Without this peak, the knee would reach a highly flexed position, especially at faster walking velocities. The second swing-phase peak is flexor and of about 30 N·m; it slows the leg before the knee reaches full extension.

Figure 5.14c shows that the hip moment tends to have an 80 N·m extensor peak during the first half of stance. At toe-off, there is a flexor moment that is needed to swing the lower extremity forward. Then, similar to the knee moment, the hip moment has an extensor peak of about 40 N·m at the end swing to slow the lower extremity before heel contact.

The hip moment is in fact the most variable of the three lower-extremity joint moments. The foot moment tends to be the least variable because the segment is constrained by the ground. The hip, in contrast, is not only responsible for lower-extremity control, but it also has to control the balance of the torso. This is a significant task, given that the torso comprises at least two-thirds of the body weight of an average individual. Winter and Sienko (1988) showed that the movement of the torso reflects a majority of the hip moment variability. In this regard, the hip moment becomes somewhat hard to interpret during stance.

RUNNING

Lower-extremity joint moments during running are shown in figure 5.15. These moments have patterns similar to their analogues in walking. The most prominent differences are their magnitudes. Running is a more forceful activity than walking; thus, just as GRFs are larger during running, so, too, are the joint moments. The stance phase is a smaller percentage of the running cycle than during walking. During stance, the ankle moment (figure 5.15a) has a single plantar flexor peak of about 200 N·m. Thereafter, the swing-phase ankle moment is near zero. The ankle moment at heel contact does

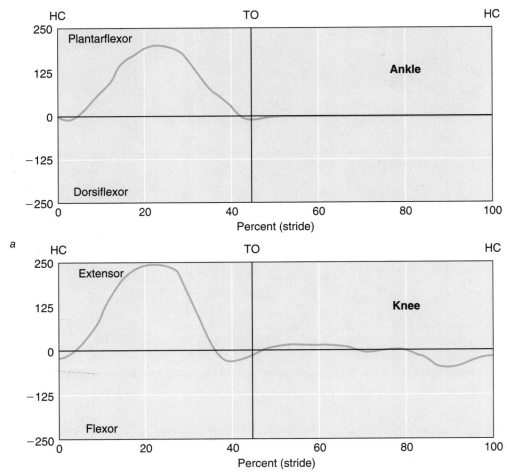

○ Figure 5.15 Moments of force at the *(a)* ankle and *(b)* knee during normal level running. HC = heel contact; TO = toe-off.

(continued)

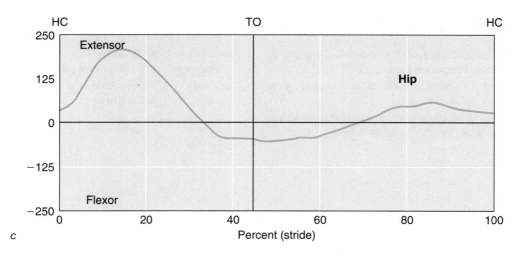

HC   TO   HC

○ **Figure 5.15**   (continued) Moments of force at the (c) hip during normal level running. HC = heel contact; TO = toe-off.

vary slightly depending on running style. Runners with a heel-toe footfall pattern exhibit a small dorsiflexor peak at heel contact, much like in walking. Individuals who run foot-flat or on their toes do not have this peak because there is no need to control foot-slap.

The stance-phase knee moment (Figure 5.15b), like the ankle moment, consists primarily of a single extensor peak of about 250 N·m. Around toe-off, there is usually a flexor moment of about 30 N·m. This action flexes the knee rapidly before the lower extremity is swung forward. During the swing, there is an extensor phase in which the leg is swung forward. Finally, there is a flexor phase before heel contact to slow the leg.

The hip moment (figure 5.15c) is extensor for much of stance, peaking at over 200 N·m. We then see a shift to net flexor activity around toe-off to swing the lower extremity forward. Then there is an extensor action to slow the thigh before heel contact.

SUMMARY

This chapter focused heavily on proper technique for inverse dynamics problems. This focus is necessary simply because the technique clearly has many steps and potential pitfalls (Hatze 2002). Students are encouraged to practice such problems until they can solve them without referring to this book. Students should be able to draw an FBD for any segment, construct the three equations of motion, and solve them.

Students are also encouraged to be mindful of the limitations of joint moments. Joint moments are only a summary representation of human effort, one step more advanced than the information obtained from a force plate. Joint moments are not an end-all statement of human kinetics; rather, they are convenient standards for evaluating the *relative* efforts of different joints and movements. Researchers interested in specific muscle actions must employ either electromyography (chapter 8), musculoskeletal models (chapter 9), or both. It is noteworthy that the great Russian scientist Nikolai Bernstein (see Bernstein 1967) refers to moments, but in fact preferred to use segment accelerations as overall representations of segmental efforts. It is hard to argue that joint moments offer much more information. A 200 or 20 N·m joint moment is in itself fairly meaningless. It is only by comparing specific cases like the walking and running examples in figures 5.14 and 5.15 that we can develop a relative basis for the magnitudes of joint moments.

The problem with analyzing limbs segmentally is that it can promote the misconception that each joint is controlled independently. We have already stated that two-joint muscles are poorly treated with this method. In addition, the individual segments of an extremity interact. For example, we earlier discussed the fact that thigh moments affect the movement of the leg. Therefore, in terms of joint moments, we know that a hip moment will affect the thigh; however, we do not establish the resulting effect on the leg. Many studies have noted the importance of intersegmental coordination (see, for example, Putnam 1991); in fact, Bernstein in particular cited the utilization of segment interactions as the final step in learning to coordinate a movement. Clinicians are also beginning to recognize such effects in various populations, particularly in amputees. Systemic analyses such as the Lagrangian approach, outlined in chapter 10, may be preferable for these situations.

SUGGESTED READINGS

Bernstein, N. 1967. *Co-ordination and Regulation of Movements.* London: Pergamon Press.

Miller, D.I., and R.C. Nelson. 1973. *Biomechanics of Sport.* Philadelphia: Lea & Febiger.

Özkaya, N., and M. Nordin. 1991. *Fundamentals of Biomechanics.* New York: Van Nostrand Reinhold.

Plagenhoef, S. 1971. *Patterns of Human Motion: A Cinematographic Analysis.* Englewood Cliffs, NJ: Prentice Hall.

Winter, D.A. 1990. *Biomechanics and Motor Control of Human Movement.* 2nd ed. Toronto: John Wiley & Sons.

Zatsiorsky, V.M. 2001. *Kinetics of Human Motion.* Champaign, IL: Human Kinetics.

CHAPTER 6

Energy, Work, and Power

D. Gordon E. Robertson

Energy is a well-known physical quantity that, despite its notoriety, is not well understood. For instance, physicists have yet to identify any atomic or subatomic particle that corresponds to a basic unit, or quantum, of energy. One of the difficulties with understanding energy is that it takes many forms. Matter itself is one form of energy, which Einstein was able to quantify with his most famous equation, $E = mc^2$, but this energy is only manifest when the matter itself is torn apart. Other forms of energy include nuclear, electrical, thermal (heat), solar, light, chemical, and the one of greatest interest to biomechanists, mechanical.

In this chapter, we are mainly concerned with mechanical energy, and so we

- examine the concepts of energy, work, and the laws of thermodynamics;

- introduce the concepts of conservation of mechanical energy and a conservative force;

- outline methods for directly measuring mechanical work, called direct ergometry;

- outline methods for indirectly quantifying mechanical work, including the use of results from inverse dynamics; and

- examine the relationship between work and the cost of doing work, called mechanical efficiency.

ENERGY, WORK, AND THE LAWS OF THERMODYNAMICS

Energy can be defined as the ability to perform work; in other words the ability to affect the state of the matter. In a sense, energy is the motion of particles or the potential to create motion. For instance, heat, a ubiquitous form of energy, is the rate of vibration of molecules. The greater the vibration or agitation, the greater the heat or thermal energy. The quantity of heat in a substance is measured by its temperature. All matter vibrates to a certain extent; as this vibration is reduced, the matter's temperature is reduced and we say it is *cooler*. The lowest temperature, called *absolute zero*, corresponds to the complete absence of vibration, which, according to the third law of thermodynamics, can never be achieved, that is, no substance can be lowered to absolute zero (0 kelvin).

Thermodynamics is the field of study concerned with energy and its quantification, transmission, and transduction (change) from one form to another. It was once thought that energy was a fluid (caloric) that flowed from one object to another. We now understand that energy is a property of matter. According to the first law of thermodynamics, also called the law of conservation of energy, the quantity of energy in the universe is a constant. In simpler terms, we can say that in a closed system, the quantity of energy in the system is a fixed amount. A closed system is a volume that no energy can enter or leave. The energy within the system can change form, but as long as no energy enters or escapes the volume, the total amount inside is a fixed amount. Of course, it is not a simple matter to quantify all the energy sources within a system. If one can measure some of the sources and assume that the others do not change, then by monitoring changes in the known sources, any transformations of energy can be evaluated to determine the work done and the mechanical efficiencies of any machines within the volume.

The second law of thermodynamics, first elucidated by Rudolph Clausius in 1865, states that when energy is transformed from one form to another—for example, when electricity produces light, water power produces electricity, or biochemical energy produces a muscle contraction—some of the energy is wasted and can no longer be transformed into another usable form of energy. Clausius named this unusable energy entropy, to sound like energy. Entropy can be considered energy that can longer perform useful work.

Work can be defined as the changing of energy to another useful form of energy, also called *transduction*.

For example, when a fire heats up a pot of water or steam creates motion in a piston, or when electricity passing through a tungsten wire creates light, energy is transduced from one form to another. In every such transformation, some energy is produced that is not directly associated with the intended work. Much of this lost energy takes the form of heat. The heating of gears, springs, surfaces, and air is usually wasted energy. In some cases, this heat can be collected and reused; but some energy always escapes into the surrounding atmosphere and cannot be recovered. This heat becomes entropy. Entropy is forever increasing as the universe ages; in other words, the universe is gradually burning down or becoming increasingly chaotic. We will leave it to the physicists and philosophers to predict the eventual outcome. For the biomechanist, it is enough to understand that there are costs associated with the transformation of energy. This cost is manifested in the heating of machines, muscles, and the surrounding environment.

Figure 6.1 shows a simplified schematic of the flow of energy through the human body and identifies several areas where entropy occurs. For example, in the conversion of chemical energy to mechanical

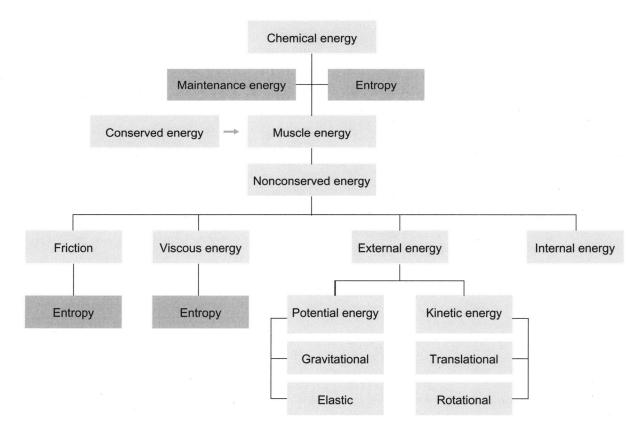

° **Figure 6.1**   Energy flow through the human body to the environment. Maintenance energy includes energy to all tissues, excluding the skeletal muscles. Entropy includes all energy that can no longer be recovered to perform useful work as a result of heating the environment or creating turbulence in fluids in the environment (air or water). Conservative energy is the energy that recycles by changing form within segments or exchanges between and among segments or other bodies.

energy, some energy is lost as heat that is dissipated by the skin or gases expired into the environment. Two other means of loss include mechanical friction and viscosity. Frictional losses occur whenever the body rubs against the environment or tissues within the body are subjected to internal friction. Viscous losses occur because of drag forces induced by moving through fluid media, such as water or air, or as a result of the viscoelastic properties of various tissues within the body.

Mechanical work is the work done when the total mechanical energy of a body changes. This principle, called the *work-energy relationship,* is based on Newton's second law. With the appropriate mathematical operations, Newton's second law can be transformed into the following relationship:

mechanical work = change in mechanical energy

$$W = \Delta E = E_O - E_N \qquad (6.1)$$

where the total mechanical energy of a body *(E)* is defined as the sum of the potential and kinetic energies of the body. These types of energies are defined later in this chapter. Figure 6.2 illustrates the relationship between work and energy. Notice that between points A and B, the mechanical energy of the body changes. This change in energy represents the work that was done on the body. If the duration of the work is known, the *average power* can be computed from:

$$\overline{P} = \Delta E / \Delta t = \text{work} / \text{duration} = W / \Delta t \qquad (6.2)$$

The *instantaneous power* at any particular instant in time can also be computed by taking the time derivative of the energy history. That is,

$$P = dE/dt \qquad (6.3)$$

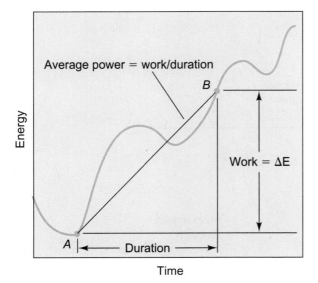

○ **Figure 6.2** Relationship between energy, work, and average power.

Note that the units of measure of work and energy must be equivalent. By international agreement, the units of work and energy are called joules. The *joule* is the work done by a 1-newton (N) force that moves an object through 1 m in the direction of the force. The joule is dimensionally equivalent to the units used for moments of force, newton meters (N·m). Physicists therefore gave work and energy units a different name—the *joule* (J)—in honor of James Prescott Joule. Joule established numerous relationships between the work and heat energy of electrical motors. The joule is the International System of Units (SI) unit of work or energy (mechanical, electrical, solar, and so on) and the newton meter is the SI unit for moment of force or torque. The SI unit for power is the *watt* (abbreviated *W),* which is defined as the rate of doing work at 1 J per second. The watt was named to honour James Watt, who designed efficient steam engines that helped power the Industrial Revolution.

When work is done by muscles, some of the energy produced may be used to move internal structures and some may be used to do work on the environment. The former is called *internal energy* the latter, *external energy* (see figure 6.1). Some of the work done may also be recycled by conservative forces. This pathway is also included in figure 6.1. Energy recycled in this way reduces the amount of muscle work and chemical energy required by the system. Work done on the environment can be done against friction or viscous forces (drag) or can be manifested in changes in the potential and/or kinetic energies of objects in the environment. For example, when an object is lifted to a height, its *gravitational potential energy* is increased. *Elastic potential energy* is increased by depressing a springlike object, such as a diving board, springboard, or pole vault pole. When a ball is thrown, the external energy done by the body appears in the form of *translational* and/or *rotational kinetic energy* of the ball. The translational kinetic energy appears as the ball's linear velocity, whereas the rotational kinetic energy manifests itself in the ball's spin.

The internal energy referred to in figure 6.1 appears as potential and kinetic energy, but instead of being used to perform work on external objects, it is used to move internal structures. These appear as movements of the upper and lower extremities and are considered to be a cost of performing the task. Some tasks, such as level walking, running, and cycling, require small amounts of external work to get the person up to speed, but then require mainly internal work to maintain the speed. Other tasks, such as lifting and vertical jumping, require small amounts of internal work, but relatively large amounts of external work. The following sections outline methods for computing these various quantities for a wide variety of human movements.

CONSERVATION OF MECHANICAL ENERGY

To conserve mechanical energy, special circumstances must occur. *Conservation of mechanical energy* occurs when all the forces and moments of force acting on a body or a single segment are conservative. That is, the resultant force acting on the body is a conservative force. Forces or moments of force are *conservative* when the work they do in moving a body from one point to another is independent of the path taken, that is, when the work done depends only on the location of the two points. Conservative forces include gravitational forces, the force of an ideal spring, elastic collisions, the tensile force of an ideal pendulum, and the normal force of a frictionless surface. Illustrations of these conservative forces appear in figure 6.3.

Another corollary that follows from the definition of conservative forces is that they do no work after acting through a closed path. For example, if a

weight is attached to a stretched spring and released and then the spring returns to its original stretched length without assistance from other forces, then no work is done and the spring force is considered to be conservative.

The work done by a nonconservative force is affected by the path that is taken when moving an object from one point to another. Nonconservative forces include frictional, viscous (e.g., fluid friction), and viscoelastic (e.g., muscle and ligament) forces and plastic deformations. Frictional forces are nonconservative because the amount of work they do increases with the length of the path taken from one point to another. The longer the path, the more work is required. The same is true for viscous forces. Viscous forces also increase the work done depending on how fast the path is taken—the faster the motion through a viscous medium, the greater the work. In general, the resistance offered by viscous forces increases with the square of the velocity. This is why the most efficiently run races are those in which a constant speed is maintained throughout the race.

Most conservative forces are ideals that cannot be realized physically. For example, an ideal spring returns all the energy that is imparted to it by compression or elongation. In reality, all springs obey the

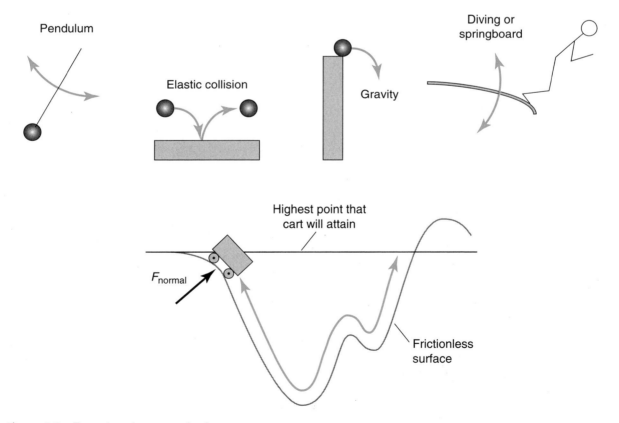

Figure 6.3   Examples of conservative forces.

second law of thermodynamics and lose some energy in the form of heat. Similarly, "pure" elastic collisions do not occur in nature. An *elastic collision* occurs when a body is dropped from a certain height and rebounds to the same height. In reality, all such collisions result in lower rebound heights and some heating and/or deformation of the surfaces. A *plastic collision* occurs when the body does not rebound at all.

Gravitational forces are truly conservative because, in the absence of nonconservative forces, any change in potential energy causes a corresponding change in kinetic energy. For example, if an object is allowed to fall, its kinetic energy will increase by the same amount as its potential energy decreases. Thus, the total energy of the body remains constant (that is, its energy is conserved).

The tensile forces of pendulums are also considered to be conservative and typically work in conjunction with gravitational force. Figure 6.4 illustrates a simple pendulum. At time A, the pendulum or body is motionless at a particular height. All of its energy is in the form of gravitational potential energy, but as the body is released, its height and potential energy decrease. Simultaneously, its translational kinetic energy increases in exactly the same amount that its potential energy decreases until the bottom of the swing is reached. At that point (midway between A and B), the body has no potential energy; instead, all its energy is in the form of kinetic energy. The kinetic

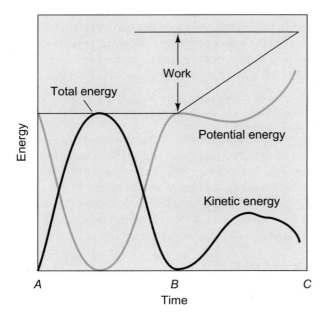

Figure 6.4 Energy of a pendulum. The pendulum is conservative from time A to time B. Between times B and C, work is done to raise the pendulum to a higher energy level. Amounts of translational and potential energies are also shown.

energy provides the necessary impetus to swing the body back to its original height at time B. From time B to C, a nonconservative force acts to raise the body to a new height and a new energy level—that is, work is done. If the body is then allowed to swing under the influence of only conservative forces (gravity and the tensile force of the pendulum string), the body will cycle back and forth at the new energy level, interchanging potential and kinetic energy at no cost to the system.

In humans, particularly during walking, some energy is conserved by permitting parts of the body to act like simple pendulums. There is, however, another way conservation can take place. It is possible to permit energy to transfer between adjacent segments, as is the case with compound pendulums. For this to occur, the segments must be connected with nearly frictionless joints, and the muscles that act across the joints must be passive or act to facilitate the transfer of energy. Elftman (1939a, 1939b), Winter and Robertson (1979), and others have shown that such situations occur during the swing phase of gait. They showed that energy originally in the leg and foot transfers to assist in the thigh's upward swing-through. Furthermore, this same energy transfers back to the leg and foot immediately before heel-strike, as the thigh reaches maximum flexion. Such mechanisms reduce the chemical energy costs needed for these movements. If these conservative mechanisms were not utilized, more chemical and mechanical energy would have to be produced to supply the necessary work to propel the segments. For example, during sprinting (Lemaire and Robertson 1989; Vardaxis and Hoshizaki 1989), there is insufficient time for the action of simple or compound pendulum motions to swing the leg through. These researchers showed that in some athletes, energy had to be supplied at the rate of 4000 J/s (watts) to swing the thigh upward and approximately 3600 W to drive it back down prior to touchdown on the track. Athletes cannot wait for conservative forces to supply the energy necessary to cycle the lower extremity during running and sprinting; instead, nonconservative forces are required. The measurement of the work done by nonconservative forces is called *ergometry*.

ERGOMETRY: DIRECT METHODS

Ergometry is literally the measurement of work. In practice, an ergometer is any system or technology that quantifies the work done during an activity. Many commercial ergometers are used by kinesiologists and physical educators, but the most common one is probably the bicycle ergometer. Other devices commonly found in research laboratories and exercise rooms include rowing machines, running treadmills, and exercise machines; the latter, however, are so poorly calibrated that they are unsuitable for accurate scientific measurements.

To qualify as an ergometer, a device must measure at least two things—force and displacement or moment of force and angular displacement. This is because work is defined as the product of force times the displacement or distance traveled. More accurately, work is defined as the scalar or dot product of force and displacement. Thus, the work done by the resultant force acting on a body is:

$$W = \vec{F} \cdot \vec{s} = F_x s_x + F_y s_y = Fs \cos \phi \qquad (6.4)$$

where W is work, $\vec{F}$ and (F_x, F_y) are the resultant forces, $\vec{s}$ and (s_x, s_y) are the displacement, and ϕ is the angle between the force and the displacement vectors. When a moment of force is used, the work done is quantified by:

$$W = M\theta \qquad (6.5)$$

where M is the resultant moment of force and θ is the angular displacement of the object. If a body has both a nonzero resultant force and a nonzero moment of force, then the work done is the sum of the two equations just set out. Note that to add the work done by the resultant force and resultant moment of force, the units of measure must be the same. The units for the work of a force appear to be newton meters, but these units are dimensionally equivalent to the joule. The work done by the resultant moment of force appears to be in units of newton meters radians, but because newton meters are equivalent to joules and radians are dimensionless, these units are dimensionally equivalent to the units of the work produced by a force, namely joules.

In many cases, the work done can be measured using forces or moments of force. For example, on a Monark bicycle ergometer, the work done can be measured by the moment of force produced by the flywheel braking system times the number of rotations of the flywheel (times 2π to convert flywheel rotations to radians). The moment of force must be in units of newton meters and is dependent upon the position of the load arm of the ergometer. Most ergometers are calibrated using the obsolete unit of a kilopond, which is equivalent to the weight (force) of a mass of 1 kg. To convert a kilopond load to its equivalent force in newtons, multiply the load by the gravitational acceleration value of 9.81 m/s². That is,

$$F \text{ (newtons)} = load \text{ (kiloponds)} \times 9.81 \text{ (newtons/kilopond)} \qquad (6.6)$$

To obtain the moment of force created by the load, multiply the force times the radius of the flywheel. This distance is 25.46 cm (the circumference is 160.0 cm) for a Monark bicycle ergometer. Therefore, the work done (W) using the moment of force approach is:

$$W = M\theta = Fr\theta \qquad (6.7)$$

where F is the load in newtons, r is the radius of the flywheel, and θ is the angular displacement of the flywheel. The force or load is actually a braking belt that creates a rotational friction equivalent to a linear frictional force acting tangentially to the flywheel. Thus,

$$\begin{aligned} W = F \text{ (N)} \times radius \text{ (m)} \times \\ rotations \text{ (revolutions)} \times \\ 2\pi \text{ (radians/revolution).} \end{aligned} \qquad (6.8)$$

To measure work for a Monark bicycle ergometer, the number of crank rotations usually is used instead of the number of flywheel revolutions. In a standard ergometer, every *crank* (not flywheel) revolution is equivalent to 6 m displacement of a point on the rim of the flywheel. Therefore,

$$\begin{aligned} W = load \text{ (kiloponds)} \times 9.81 \times \\ crank \ rotations \times 6 \text{ (m/rotation)} \end{aligned} \qquad (6.9)$$

In each case, the units of work are in joules.

A treadmill can also be used as an ergometer. A person running at a constant speed up an incline is considered to have done an amount of work equivalent to raising the body weight up to a certain height. The height is computed by multiplying the speed of the treadmill times the length of the run in seconds times the sine of the angle of the incline. That is,

$$\begin{aligned} W = body \ weight \text{ (N)} \times speed \text{ (m/s)} \times \\ duration \text{ (s)} \times \sin\gamma \end{aligned} \qquad (6.10)$$

where γ is the angle of incline of the treadmill. Notice that consideration is given not to the actual distance traveled, but rather only to the height the person would be raised if lifted to an equivalent height. It is also assumed that no work is done against the treadmill belt as a result of stretching or friction and that the speed is constant throughout the run.

EXAMPLE 6.1

Calculate the work done on a bicycle ergometer if the person pedals against a 2.50 kilopond workload for 20 s at the rate of 60 revolutions/s. Assume that each revolution is equivalent to 6 m of linear motion of the flywheel.

➤ *See answer 6.1 on page 280.*

ERGOMETRY: INDIRECT METHODS

In a sense, all methods of determining the mechanical energy of a system are indirect because there is no direct way to measure the flow of energy into the system like there is, for example, at an electrically metered house. When we refer to the indirect measurement of mechanical energy, we are referring to methods that quantify the motion of bodies, called their *kinetic energies,* and the positions of the bodies with respect to a fixed frame of reference, called their *potential energies.* The sum of the kinetic and potential energies of a body yields the body's *total mechanical energy.*

To simplify the analysis of some systems, the motion and position of the system can be reduced by considering only the system's center of gravity as a point mass. When analyzing a point mass, only two types of mechanical energy are present, *translational kinetic energy* and *gravitational potential energy.* A better model of the human body assumes that the body is a system of interconnected rigid bodies or segments. In this case, each segment includes an additional type of energy called *rotational kinetic energy.* It is also possible, although to date it has rarely been attempted, to model each segment as a deformable body, in which case *elastic potential energy* is also included in the computation of total mechanical energy.

Each of these energies can be quantified by knowing the kinematics of the body or its segments. In the following two sections, methods for determining the energy of point masses and systems of rigid bodies are presented. An alternative approach utilizes the work-energy relationship. This relationship equates the work done on a body to the change in mechanical energy of the body. Therefore, the changes in mechanical energy can be computed by determining the work done on the body by both internal and external forces. This approach is described in the section called Inverse Dynamics Methods.

POINT MASS METHODS

If a body can be assumed to behave as a single point mass, its energy can be quantified from the linear kinematics of its center of gravity. The most common means of obtaining linear kinematics is to digitize the coordinates of markers attached to the body and then perform finite difference calculus to obtain the linear velocities (as outlined in chapter 1). A simpler but less accurate approach is to estimate the location of the center of mass and determine its trajectory. Either way, the total energy of a point mass

is the sum of its gravitational potential energy plus its translational kinetic energy. That is,

gravitational potential energy:
$$E_{gpe} = mgy \qquad (6.11)$$

translational kinetic energy:
$$E_{tke} = 1/2\ mv^2 = 1/2\ m(v_x^2 + v_y^2) \qquad (6.12)$$

total mechanical energy:
$$E_{tme} = E_{gpe} + E_{tke} \qquad (6.13)$$

where *m* is the mass of the body, *g* is 9.81 m/s², *y* is the height of the body above the horizontal reference axis, and (v_x, v_y) is the velocity of the point with respect to the stationary reference axes. Several researchers have used this approach to quantify the work done during walking, running, and sprinting (Cavagna, Saibene, and Margaria 1963, 1964; Cavagna, Komarek, and Mazzoleni 1971).

EXAMPLE 6.2

Calculate the work done and the power produced when a 180.0 kg person starts from rest and achieves a speed of 6.00 m/s in 4.00 s. Assume that the person runs on a level surface and that the rotational energy is negligible.

➤ *See answer 6.2 on page 280.*

SEGMENTAL METHODS

The point mass method has been criticized for being inappropriate for human body analyses (Williams and Cavanagh 1983; Winter 1978) because it underestimates the true mechanical energy of the system. The more accurate method is to divide the body into segments and then sum the energies of each segment to determine the total body's energy. This method was first used by Wallace Fenn to determine the energetics of running (1929) and sprinting (1930).

To compute the total mechanical energy of a segment, usually only the gravitational potential energy and the translational and rotational kinetic energies are calculated. This is the correct formula for a rigid body, but if the body deforms, the elastic potential energy should also be included. In most biomechanical studies, it is not possible to compute the elastic potential energy because the amount of deformation is too small to measure and the deformation-force relationship is too expensive or difficult to obtain. Even if it were possible to determine these factors, the amount of energy stored in this way does not warrant the expense.

These equations define how the total mechanical energy (E_{tme}) of a segment is computed:

gravitational potential energy:
$$E_{gpe} = mgy \qquad (6.14)$$

translational kinetic energy:
$$E_{tke} = 1/2\ m\ v^2 = 1/2\ m(v_x^2 + v_y^2) \qquad (6.15)$$

rotational kinetic energy:
$$E_{rke} = 1/2\ I\omega^2 \qquad (6.16)$$

total mechanical energy:
$$E_{tme} = E_{gpe} + E_{tke} + E_{rke} \qquad (6.17)$$

where m is the mass of the segment, g is 9.81 m/s^2, y is the height of the segment above the reference axis, (v_x, v_y) is the velocity of the segment center of gravity, I is the segment's mass moment of inertia about its center of gravity, and ω is the angular velocity of the segment.

To obtain the total body's mechanical energy (E_{tb}), the individual segments' total mechanical energies are summed. That is,

total body mechanical energy:
$$E_{tb} = \sum_{s=1}^{S} E_{tme,s} \qquad (6.18)$$

where S is the number of segments in the human body model and $E_{tme,s}$ is the total mechanical energy of segments. The number of segments varies from study to study depending on the motion and whether bilateral symmetry is assumed. For example, Winter, Quanbury, and Reimer (1976) used three segments to investigate the work of walking by assuming that the head, arms, and trunk could be modeled as a single segment. Martindale and Robertson (1984) used six segments by assuming bilateral symmetry for rowing (on water and ergometer rowing). Twelve segments were employed by Williams and Cavanagh (1983) for the analysis of running and by Norman, Caldwell, and Komi (1985) for cross-country skiing. Sometimes a single extremity is analyzed, in which case only two or three segments are quantified (Caldwell and Forrester 1992).

EXAMPLE 6.3

Calculate the mechanical energy of an 18.0 kg thigh that has a linear velocity of 8.00 m/s and a rotational velocity of 20.0 rad/s and whose center of gravity is 1.20 m high. The thigh's moment of inertia is 0.50 kg·m^2.

➤　*See answer 6.3 on page 280.*

Although this method of calculating energy is relatively simple because only kinematic data and body segment parameters are needed, the method has several drawbacks for calculating the work done. In principle, calculating the external work done requires only a measurement of the changes in energy throughout the motion. In other words, the external work $(W_{external})$ is defined as:

$$W_{external} = \sum_{n=1}^{N} \Delta E_{tb,n} = E_{tb,N} - E_{tb} \qquad (6.19)$$

where N is the number of frames of the time-sampled motion, $E_{tb,n}$ is the total mechanical energy of the body (the sum of segmental total mechanical energies) at time frame n. Notice that you only need to measure the energy before and after the duration of the motion to obtain the total *external* work done on the body because all of the intermediate measures cancel out.

This measure is useful whenever the body increases its height or its linear or angular speed, but in situations in which speed is constant or locomotion is along a level surface, the external work done is zero. As long as the work is measured at the same point in the motion's cycle, there will be no change in height and no change in linear or angular velocity, and therefore no changes in any of the potential or kinetic mechanical energies. This is called the *zero-work paradox* (Aleshinsky 1986) because, although no external work is done (excluding work done against dry and fluid frictional forces), *internal work* is done to cycle the extremities throughout the motion. (More about this paradox will be described later in the section on Mechanical Efficiency.) It is also obvious that one can "locomote" across a particular distance in an efficient way or in any number of inefficient ways that clearly consume more chemical and mechanical energy to achieve. For example, one could walk normally over a 10 m distance or hop from side to side over the same distance. The hopping motion obviously wastes energy, but the external work done in both cases is the same if the person arrives at the same point with the same velocity.

Norman et al. (1976) proposed a method of calculating *pseudo-work* that was later modified by Winter (1979a) to quantify the internal work during locomotion. They assumed that any change in total body mechanical energy required that mechanical work be done and that monitoring these changes in mechanical energy would permit calculation of the total mechanical work and internal work done. In effect, the method calculated the internal work by computing the total work done and then subtracting the external work done. The total work done (W_{total}) was equal to the sum of the absolute values of the changes in total mechanical energy. That is,

$$W_{total} = \sum_{n=1}^{N} \left| \Delta E_{tb,n} \right| \qquad (6.20)$$

The internal work was then computed by subtracting the external work (defined previously) from the total work (above):

$$W_{internal} = W_{total} - W_{external}. \qquad (6.21)$$

Several researchers used these methods to investigate various locomotor movements, including overground walking (Winter 1979a), treadmill walking (Pierrynowski, Winter, and Norman 1980), load carriage (Pierrynowski, Norman, and Winter 1981), rowing (Martindale and Robertson 1984), and cross-country skiing (Norman, Caldwell, and Komi 1985; Norman and Komi 1987), to name a few.

Aleshinsky (1986) has criticized this approach because it inaccurately measures the total mechanical work and therefore the internal work. This approach, called the *absolute energy approach* or the *mechanical energy approach*, assumes that any simultaneous energy increase and decrease of the segments reduce the total mechanical energy expenditure. Williams and Cavanagh (1983) tried to correct this flaw by permitting transfers of energy only between adjacent segments, but Winter and Robertson (1979) had already disproved this concept by showing that some of the energy generated at the ankle transfers from the foot to the leg, to the thigh, and even to the trunk. Wells (1988) demonstrated that the work done and energy transferred within the body can be estimated using an algorithm that predicts the recruitment patterns of mono- and biarticular muscles. The algorithm partitions the net moments of force at each joint into either mono- or biarticular muscles based on whether activating a two-joint muscle could reduce the levels of muscle force.

In summary, the total work and internal work done, as calculated from mechanical energy changes, can be used to estimate the mechanical work. This method can be used when other, more accurate methods cannot be applied. A better method, supported by Elftman (1939a), Robertson and Winter (1980), Aleshinsky (1986), and Ingen Schenau and Cavanagh (1990), uses inverse dynamics to compute the net moments of force at each joint and then measures the work done by these sources.

From the Scientific Literature

Winter, D.A. (1976) Analysis of instantaneous energy of normal gait. *Journal of Biomechanics* **9:** 253-7.

This paper illustrates how quantifying the instantaneous energy patterns of linked-segment movements can identify conservation of mechanical energy. In this study, the segments of the lower extremity were studied in 2-D (the sagittal plane) during walking along a level surface. The potential, translational kinetic, rotational kinetic, and total energy patterns of the leg (shank), thigh, and torso were calculated, as were the total body energies. The total body energy was estimated by considering the trunk, head, and upper extremities one segment and assuming bilateral symmetry of the lower extremities so that the patterns of one lower extremity could be phase-shifted to estimate the other side.

The study reached several important conclusions about walking. For example, the results showed that at normal walking speeds, rotational kinetic energies can be ignored because of their small magnitudes. More importantly, Winter identified the relative importance of translational kinetic energy changes in comparison with the other forms of mechanical energy changes for the thigh (figure 6.5b) and particularly the leg segments (figure 6.5a). In contrast, the torso's energy variations for translational kinetic energy and potential energy were similar in magnitude.

Winter explained that the leg segment could not conserve energy because the kinetic and potential energies increased and decreased synchronously. Thus, no passive energy exchange like the one that occurs with pendular motion could take place between these two energy sources. Conversely, the thigh (figure 6.5b) and torso (figure 6.5c) segments exhibited periods during which their potential and translational kinetic energy patterns were asynchronous—one source increased while the other decreased and vice versa. This phenomenon reduces the muscular

(continued)

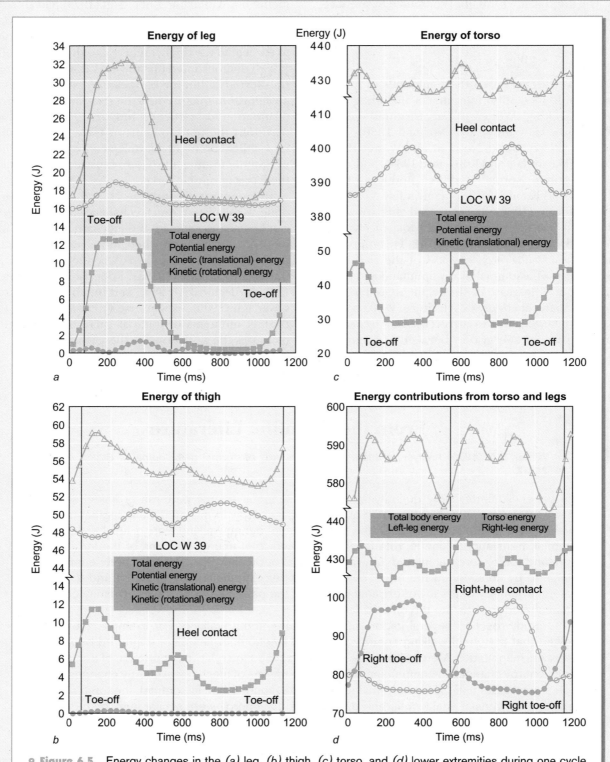

Figure 6.5 Energy changes in the *(a)* leg, *(b)* thigh, *(c)* torso, and *(d)* lower extremities during one cycle of walking.

Reprinted, by permission, from D.A. Winter, 1976, "Analysis of instantaneous energy of normal gait," *Journal of Biomechanics* 9: 253-257.

(continued)

work costs and shows that energy conservation occurred between the two segments. Winter also showed (figure 6.5d) that conservation of energy may occur between one lower extremity and the other; there were overlapping periods during which one extremity's total energy pattern decreased while the other's increased and vice versa. Although these data do not prove conservation occurs from one side to the other, they do support the possibility. The mechanical energy approach used in this study cannot distinguish if conservation occurs between and among segments, nor can it distinguish if muscles acting across each segment act concentrically or eccentrically to cause the respective increases or decreases in energy. Later, we present a more sophisticated method that uses inverse dynamics to determine the work done across adjacent segments.

INVERSE DYNAMICS METHODS

Because the work on a body can be measured by the changes in mechanical energy or the work done by the resultant forces and moments of force acting on the body, another approach to computing the external, internal, and total work done is also possible. Inverse dynamics permits the calculation of the net forces and moments of forces acting on individual segments of an n-link system of rigid bodies. This section details how the results from an inverse dynamics analysis can be used to compute total body work measures.

Procedures for computing net forces and moments of force were presented in chapter 5. If the body is assumed to be deformable and to have friction in its joints, then the net forces at these joints may do work against friction, and energy may be stored and released because of the elasticity of the tissues. These *sinks* and *sources* of energy are difficult to quantify and therefore are usually assumed to be negligible or are attributed to other sources, such as the net moments of force. A *sink* is a structure that can dissipate or store energy and may or may not return the energy. In the body, muscles are sinks when they contract eccentrically to reduce the mechanical energy of the system. Ligaments, bones, bursae, and cartilage also act as energy sinks. Energy *sources* are structures that supply or return stored energy. Muscles are the primary biological structures that supply energy to the body. Elastic structures are both sources and sinks. They receive energy during deformation (elongation or compression) and then return some of the energy after the deforming force is released (although some stored elastic energy is lost as heat).

Most researchers model the human body as a linked system of rigid bodies and assume that the joints are frictionless. These assumptions simplify the computation of mechanical work by eliminating the possibility of the net forces affecting the total work done by the body or its internal or external work. The only sources of work become the net moments of force acting at each joint and any external forces acting on the body. Aleshinsky (1986) presented a thorough explanation for this phenomenon. In the sections that follow, the equations for computing work based on the net forces and moments of force are presented, along with a discussion of how the two methods (mechanical energy and inverse dynamics) can be applied together to better understand how energy is supplied, transferred within the body, and dissipated. First, a segmental approach is taken to investigate the transmission and use of mechanical energy by the segments. Second, a joint analysis is described and the role of the moments of force at each joint and the transfers of energy by the net forces at each joint are defined. Third, methods for computing the total body power requirements are described. Last, the relationships between the two methods and the differences that are often observed are investigated.

Segmental Power Analysis

Segmental energy analysis requires only knowledge of segmental kinematics and inertial properties. Using inverse dynamics methods to obtain the work done by the moments of force at each joint requires additional information and a more complex analysis. The additional information includes the history of any external force that is in direct contact with the segment of interest. For example, to compute the work done at the ankle during stance, the GRF acting against the foot must be measured (Robertson and Winter 1980; Williams and Cavanagh 1983), so force platforms are necessary in gait laboratories. During the swing phase, however, a force platform is not needed because no external force (excluding gravity) acts on the foot. A number of studies of sprinting and running have been done in this way (Caldwell and Forrester 1992; Chapman et al. 1987).

The following equations describe how to compute the power delivered to a segment from the net forces and moments of force at the segment's connections with the rest of the body.

$$\text{force power: } P_F = \vec{F} \cdot \vec{v} = F_x v_x + F_y v_y \quad (6.22)$$

$$\text{moment power: } P_M = M_j \omega_j \quad (6.23)$$

total power delivered to segment:

$$P_S = \sum_{j=1}^{J} \left(P_{F_j} + P_{M_j} \right) \quad (6.24)$$

where J is the number of joints or other structures that are directly connected to the segment. For example, a foot segment has one attached segment (the leg) and the forearm has two (wrist and elbow), but the trunk could have two (both thighs), four (both thighs and both shoulders), or five (thighs, shoulders, and neck) connections, depending upon how it is modeled. Note that a foot in contact with a surface can gain or lose energy to the surface if there is relative motion (e.g., slippage, deformation, or motion such as with an elevator, escalator, diving board, or bicycle pedal). In such cases, the foot is considered to have two sources of power. For most locomotor studies, a single connection is the norm.

Another advantage of the inverse dynamics approach is that the powers delivered to each segment can be measured from the segment's rate of change of mechanical energy. This equality, derived from the work-energy relationship and sometimes called the *energy balance* or *power balance,* has been used to check the validity of assumptions about modeling segments as rigid bodies and synchronizing external forces with cinematographic or videographic data (Robertson and Winter 1980). It has also been used to indirectly determine the deformation energy of the foot during stance (Robertson, Hamill, and Winter 1997; Winter 1996). The following equations express these relationships:

rate of change of mechanical energy:

$$P_E = \frac{dE_{tme}}{dt} \approx \frac{\Delta E_{tme}}{\Delta t} \quad (6.25)$$

instantaneous power delivered to a segment:

$$P_S = \sum_{j=1}^{J} \left(P_{F_j} + P_{M_j} \right) \quad (6.26)$$

power balance: $P_E = P_S$ \quad (6.27)

where E_{tme} is the total mechanical energy of a segment, P_F and P_M are the powers delivered by the forces and moments of force, respectively, and J is the number of joints connected to the segment.

Joint Power Analysis

Joint power analysis is a simplified version of segmental power analysis. It refers to examining the flow of energy (power) across a joint that results from the net force and moment of force at the joint. The powers from net forces are the same as those used in segmental power analyses, but the moment-of-force powers require the relative angular velocities (ω_j) of the joints rather than the segmental angular velocities. In other words, the power provided by the net moment of force, called the moment power, is the product of the net moment of force times the difference between the angular velocities of the two segments that make up the joint ($\omega_p - \omega_d$).

$$\text{force power: } P_F = \vec{F} \cdot \vec{v} = F_x v_x + F_y v_y \quad (6.28)$$

$$\text{moment power: } P_M = M_j \omega_j = M_j \left(\omega_p - \omega_d \right) \quad (6.29)$$

This method is the one most often used for the analysis of a wide variety of human movements ranging from walking (Winter 1991), jogging (Winter and White 1983), race walking (White and Winter 1985), running (Elftman 1940), and sprinting (Lemaire and Robertson 1989) to jumping (Robertson and Fleming 1987), kicking (Robertson and Mosher 1985), skating (de Boer et al. 1987), and cycling (van Ingen Schenau et al. 1990).

In general, power histories are presented simultaneously with the net moment histories and the joint angular velocities or displacements. Figure 6.7 (page 139) shows how these data are typically presented. The graphs show the powers produced by the hip moment of force of an elite sprinter. Notice that the angular velocity curve at the top indicates when the joint is flexing or extending. That is, when the angular velocity is positive, the hip is flexing, and when it is negative, the hip is extending. Notice that the peak flexion velocity is nearly 15 rad/s, or approximately 860°/s, a speed significantly greater than what can be tested on isokinetic dynamometers such as the KinCom, Biodex, or Cybex, which are limited to velocities below 360°/s.

The net moment curve indicates the direction of the net moment of force. For example, a positive moment of force indicates a flexor moment, whereas a negative moment indicates an extensor moment. The sense of the moment of force (flexor or extensor) depends on which joint is being analyzed and which way the subject is facing. If the subject is facing sideways and to the right, a positive moment of force about the mediolateral axis is flexor for the hip, extensor for the knee, and dorsiflexor for the ankle. If the moment of force is negative, the hip moment is extensor, the knee moment is flexor, and the ankle moment is plantarflexor.

From the Scientific Literature

Robertson, D.G.E., and D.A. Winter. 1980. Mechanical energy generation, absorption and transfer amongst segments during walking. *Journal of Biomechanics* **13**:845-54.

This paper had two purposes—to partially validate the calculation of inverse dynamics and to show how segmental power analyses can be used to investigate the delivery of power to the segments of the lower extremity during walking. The first purpose was realized by applying the work-energy relationship—more specifically, its derivative. That is, the instantaneous power of a rigid body is equal to the power delivered to the body by forces and moments of force. Computing the instantaneous power of a body requires only knowledge of the body's kinematics, namely its linear velocity, angular velocity, and height above a reference (usually the ground). From these (presented earlier in this chapter), the translational kinetic energy, rotational kinetic energy, and gravitational potential energy are computed. Summing these to obtain the total mechanical energy and taking the time derivative yield the instantaneous power of the body. The same can be done for segments of the body. These powers are not affected by GRF measurements or synchronizing motion data with external force data and therefore are relatively more accurate than the powers of the forces and moments of force determined by inverse dynamics methods (see chapter 5).

In theory, the instantaneous power of a body must be equal to the sum of the powers delivered to the body from all forces and moments of forces that act on the body. Because the force and moment powers can be contaminated by errors resulting from improper measurement or synchronization of external forces (e.g., GRFs), the authors felt that agreement between the two measures of a segment's power requirements provided evidence that the inverse dynamics calculations were accurate.

They tested this principle for the three segments of the lower extremity during walking and found excellent agreement for all but the foot segment during the periods of early weight acceptance and late push-off. They concluded that the poor agreement probably resulted from poor spatial or temporal synchronization of the GRFs with the motion kinematics. In a more recent paper by Siegel, Kepple, and Caldwell (1996), the authors suggest that the cause of these discrepancies may be the assumption that the foot is a rigid body. They showed that the errors between the instantaneous powers and the sum of the powers from the forces and moments were significantly reduced by modeling the foot as a deformable body and using three-dimensional (3-D) equations of motion. Winter (1996) also supported this concept, whereas Robertson, Hamill, and Winter (1997) proposed that modeling the foot as two segments by dividing it at the metatarsal-phalangeal joints would improve the power discrepancies.

The second purpose of the paper was to show how these two measurements of power can be used to understand how a segment's energy is supplied by the net forces and moments of forces that act on the segment. Figure 6.6 shows the various ways a moment of force acting across a joint can transfer energy between, supply energy to, or dissipate energy from adjacent segments. The authors showed that each moment of force can simultaneously transfer and generate or dissipate (absorb) energy, depending on the adjacent segments' and joint's angular velocities. They also demonstrated the importance of energy transfers between segments as a result of the moments of force acting across the joints. These transfers had magnitudes comparable to the magnitudes of the powers generated during concentric contractions or dissipated during eccentric contractions. These latter roles of the muscle (doing positive and negative work) are usually considered to be the only functions of the moments of force and the muscles. This research showed that an equally important role of muscles and moments of force is to transfer energy from segment to segment and thereby to conserve energy and reduce the physiological costs of motion.

(continued)

Decription of movement	Type of contraction	Directions of segmental original velocities	Muscle function	Amount, type, and direction of power
Both segments rotating in opposite directions				
(a) Joint angle decreasing	Concentric	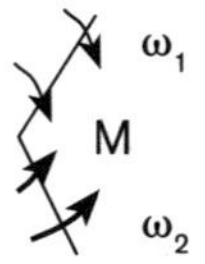 ω_1 M ω_2	Mechanical energy generation	$M\omega_1$ generated to segment 1. $M\omega_2$ generated to segment 2.
(b) Joint angle increasing	Eccentric	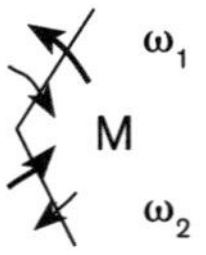 ω_1 M ω_2	Mechanical energy absorption	$M\omega_1$ absorbed from segment 1. $M\omega_2$ absorbed from segment 2.
Both segments rotating in same direction				
(a) Joint angle decreasing (e.g., $\omega_1 > \omega_2$)	Concentric	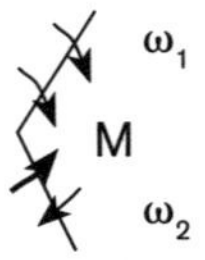 ω_1 M ω_2	Mechanical energy generation and transfer	$M(\omega_1 - \omega_2)$ generated to segment 1. $M\omega_2$ transferred to segment 1 from 2.
(b) Joint angle increasing (e.g., $\omega_2 > \omega_1$)	Eccentric	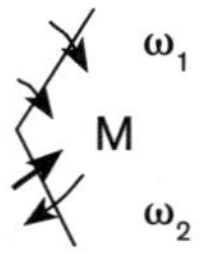 ω_1 M ω_2	Mechanical energy absorption and transfer	$M(\omega_2 - \omega_1)$ absorbed from segment 2. $M\omega_1$ transferred to segment 1 from 2.
(c) Joint angle constant ($\omega_1 = \omega_2$)	Isometric (dynamic)	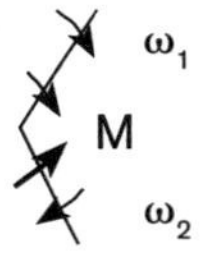 ω_1 M ω_2	Mechanical energy transfer	$M\omega_2$ transferred from segment 2 to 1.
One segment fixed (e.g., segment 1)				
(a) Joint angle decreasing ($\omega_1 = O, \omega_2 > O$)	Concentric	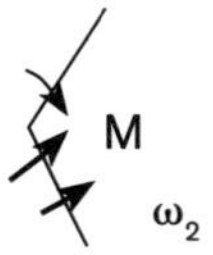 M ω_2	Mechanical energy generation	$M\omega_2$ generated to segment 2.
(b) Joint angle increasing ($\omega_1 = O, \omega_2 > O$)	Eccentric	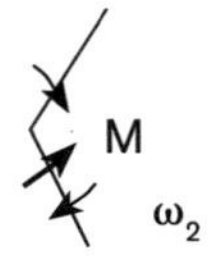 M ω_2	Mechanical energy absorption	$M\omega_2$ absorbed from segment 2.
(c) Joint angle constant ($\omega_1 = \omega_2 = O$)	Isometric (static)	 M	No mechanical energy function	Zero.

○ **Figure 6.6** Possible transfers, generation, and absorption of energy by the moment of force acting across a joint.

Reprinted, by permission, from D.G.E. Robertson and D.A. Winter, 1980, "Mechanical energy generation, absorption and transfer amongst segments during walking," *Journal of Biomechanics* 13: 845-854.

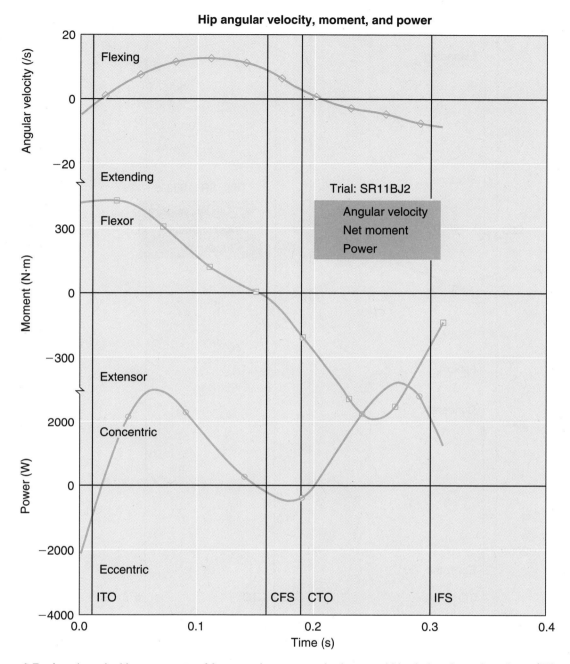

Hip angular velocity, moment, and power

(continued)

° **Figure 6.7** Angular velocities, moments of force, and powers at the knee and hip during the swing phase (ITO to IFS) of an elite male sprinter running at 12 m/s. The data displayed are from the ipsilateral side. ITO = ipsilateral, toe-off; IFS = ipsilateral, foot-strike; CTO = contralateral, toe-off; CFS = contralateral, foot-strike.

The power curve, which is the product of the other two curves, shows when positive and negative work are being done. Positive work, also called concentric work, is done when the power history is positive, for example, between 0.01 and 0.16 s of the hip power. By looking at the moment curve above it, one can readily determine which moment of force (extensor or flexor) created the power. The area under the power curve determines the amount of work done. Conversely, when the power history is negative, for example between 0.01 and 0.13 of the knee power, eccentric or negative work is being done by the associated moment of force. Many authors use power histories to determine the sequence of events during a movement cycle (see Robertson and Winter 1980; Winter 1983c).

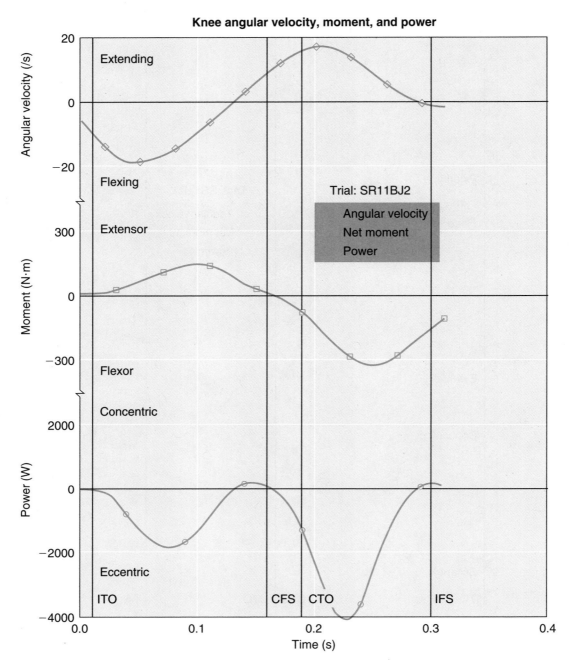

° **Figure 6.7**   *(continued)*

Total Body Work and Power

Computing total body work—equivalent to the external work done by or on the body—requires the summation of the work done by all of the body's moments of force. The work done by a moment of force, as outlined earlier, is the time integral of the product of the moment of force times the angular velocity of the joint it crosses. In equation form:

$$W_{tb} = \int_0^T \sum_{j=1}^J P_j \, dt \approx \sum_{n=1}^N \sum_{j=1}^J P_{n,j} \Delta t \qquad (6.30)$$

where J is the number of joints in the body, N is the number of time intervals, and $P_{n,j}$ is the moment power produced ($= M_{n,j} \, \omega_{n,j}$) by the jth net moment of force at time n. This relationship assumes that no external structure (elevator, escalator, diving board, car, bicycle) does work on the body and that there are no losses of energy at the joints from friction or compression or from deformation of the skeletal system. These latter situations create forces that consume power and reduce the total body's mechanical energy. If the skeletal system is assumed to be rigid and the joints are assumed to be frictionless, only the moments of force can increase or decrease the total mechanical energy.

To determine the total body's instantaneous power, you simply add up the moment powers of all the joints at any instant in time (n). That is,

total body power:

$$P_{tb} = \sum_{j=1}^{J} \int_{0}^{T} M_j\,\omega_j\,dt \approx \sum_{j=1}^{J} \sum_{n=1}^{N} M_{j,n}\,\omega_{j,n}\,\Delta t \qquad (6.31)$$

Cappozzo et al. (1975, 1976) were among the first researchers to attempt to quantify the total body's mechanical energy and power in this way. They also validated their measure against the change in energy as measured by segmental energy changes. That is,

$$W_{tb} = \Delta E_{tme} \qquad (6.32)$$

Using relatively crude data-smoothing techniques, they showed that the two measures agreed. Differences were attributed to errors in the data-collection procedures and the assumption that segments were rigid bodies.

RELATIONSHIP BETWEEN METHODS

The segmental and inverse dynamics methods are different ways of obtaining the work and power produced and distributed within the body. This relationship exists because the work-energy relationship demonstrates that the change in energy of a body or segment results from the work done on the body or segment. That is, $W = \Delta E$. Equivalently, the power of a body or segment is equal to the rate of change of the body's or segment's total energy. That is,

$$P = \Delta E / \Delta t \qquad (6.33)$$

The two sides of these equations can be determined using the two methods just outlined. Table 6.1 shows an expanded version of these equations for a simple four-segment model of the body. Note that the powers (P) have been replaced by each segment's associated force ($P_F = \vec{F} \cdot \vec{v}$) and moment ($P_M = M\omega$) powers. This model assumes that the subject performs a bilaterally symmetric motion and that the head, arms, and trunk are combined into a single segment called *trunk*.

Table 6.1   Segment and Total Powers of a Four-Segment Model of the Body

Joint	Ankle	Knee	Hip	Total
Foot	$+F_{ankle} \cdot v_{ankle} + M_{ankle}\omega_{foot}$			$= \dfrac{\Delta E_{foot}}{\Delta t}$
Leg	$-F_{ankle} \cdot v_{ankle} - M_{ankle}\omega_{leg}$	$+F_{knee} \cdot v_{knee} + M_{knee}\omega_{leg}$		$= \dfrac{\Delta E_{leg}}{\Delta t}$
Thigh		$-F_{knee} \cdot v_{knee} - M_{ankle}\omega_{thigh}$	$+F_{hip} \cdot v_{hip} + M_{hip}\omega_{thigh}$	$= \dfrac{\Delta E_{thigh}}{\Delta t}$
Trunk			$-F_{hip} \cdot v_{hip} - M_{hip}\omega_{trunk}$	$= \dfrac{\Delta E_{trunk}}{\Delta t}$
Total	$M_{ankle}(\omega_{foot} - \omega_{leg})$ or $M_{ankle}\omega_{ankle}$	$+ M_{knee}(\omega_{leg} - \omega_{thigh})$ or $+ M_{knee}\omega_{knee}$	$+M_{hip}(\omega_{thigh} - \omega_{trunk})$ or $+ M_{hip}\omega_{hip}$	$= \dfrac{\Delta E_{Total}}{\Delta t}$

F_{ankle} = net ankle force as applied to foot; F_{knee} = net knee force as applied to leg; F_{hip} = net hip force as applied to thigh; v_{ankle} = velocity of ankle; v_{knee} = velocity of knee; v_{hip} = velocity of hip; M_{ankle} = net ankle moment as applied to foot; M_{knee} = net knee moment as applied to leg; M_{hip} = net hip moment as applied to thigh; ω_{foot} = angular velocity of foot; ω_{leg} = angular velocity of leg; ω_{thigh} = angular velocity of thigh; ω_{trunk} = angular velocity of trunk; ΔE_{foot} = change in total mechanical energy of foot; ΔE_{leg} = change in total mechanical energy of leg; ΔE_{thigh} = change in total mechanical energy of thigh; ΔE_{trunk} = change in total mechanical energy of trunk; ΔE_{Total} = change in total mechanical energy of total body; Δt = duration of motion; ω_{ankle} = angular velocity of ankle; ω_{knee} = angular velocity of knee; and ω_{hip} = angular velocity of hip.

Notice that all of the $\vec{F} \cdot \vec{v}$ terms cancel out and only the $M\omega$ terms are preserved when all the segmental powers are summed. This is because it is assumed that no energy is dissipated across the joints from frictional losses or compression of the joint articulating surfaces, and thus the joint forces do no work on the segments. They do transfer energy from segment to segment, but do not influence the body's overall mechanical energy. Only the moments of force across the joints increase or decrease the body's energy levels. The final equation defines the rates (that is, the powers) at which the moments affect the body, and it corresponds to the method of calculating the total body power outlined earlier.

MECHANICAL EFFICIENCY

Traditionally, mechanical efficiency (ME) is defined as either the work done by a system divided by the energy cost of running the system (times 100%) or the power output over the power input. That is,

$$ME = 100\% \times W_{output} / W_{input}$$
$$= 100\% \times P_{output} / P_{input} \qquad (6.34)$$

For mechanical or electrical systems, measuring the input and output work or power is relatively easy. For example, an engine's input cost is measured by the amount of fuel consumed or electricity used. The output work or power is more difficult to determine, but, depending on what the engine is used for, it is measured by the useful work done. Of course, no machine can achieve 100% efficiency. Some energy is wasted as frictional heat or viscous damping or mechanical wear on the system parts. Typical mechanical systems rarely achieve MEs of greater than 30%, and electrical systems rarely of greater than 40%.

For a biological system, ME is defined as the mechanical work done over the physiological cost times 100% (Cavagna and Kaneko 1977; Williams 1985; Zarrugh 1981). In general, the mechanical work done is considered to be the *external work* done by the body on its environment. This quantity was defined earlier. More recently, however, the work done has been defined as the *total mechanical work* done by the body—in other words, the external work plus the internal work done. These terms were also defined earlier in this chapter. The reason that this redefinition has occurred is the *zero-work paradox* (Aleshinsky 1986).

The zero-work paradox occurs when a person or machine ambulates (walks, runs, paddles, crawls, or otherwise moves) at a constant average speed along a level surface. The input cost can be measured by calculating the energy cost of the activity. In the case of humans, the energy cost is measured from the utilization of biological fuels, such as adenosine triphosphate (ATP) or creatine phosphate (CP), or indirectly by the amount of oxygen consumed by the activity. In these situations, however, the mechanical work done is zero because there is no change in the mechanical energy of the body. Because the body moves along a level surface, there is no change in gravitational potential energy and because the body has a constant speed, there is no change in kinetic energy. (The cost of friction is considered to be negligible.)

Clearly, the person (or machine) can traverse a distance efficiently or inefficiently. For example, she could walk at a self-selected pace in a straight line from point A to B, or she could meander between the two points, hop from one foot to another, or walk with a severely disturbed pattern as a result of a poorly constructed prosthesis or a neurological disorder. In all cases, if the person starts and arrives at the same speed, no mechanical work can be said to have been done if only the *external work* is measured.

Of course, the physiological cost reflects any abnormal, inefficient movements during a locomotor task; thus, the most efficient gait patterns have the smallest costs, and it is likely that the self-selected walking gait in the example just discussed was the most efficient gait. But the physiological measurements cannot tell the researcher where the inefficiencies lie. To solve this problem, various researchers have included the internal work done in the numerator so that ME is defined as

$$ME = 100\% \times (\text{external work} + \text{internal work})$$
$$/ (\text{physiological cost})$$

The numerator (external plus internal) is also called the total mechanical work and is best measured by summing the integrals of the absolute powers produced by the net moments of force, but it may also be estimated by summing the absolute values of the changes in total body mechanical energy. These measures, in addition to evaluating the efficiency of the locomotor pattern, also aid in identifying where mechanical energy is produced and dissipated, and they potentially can identify where inefficiencies exist.

Although calculating the mechanical energy costs in this way is imperfect, at least valuable insights into how people perform an activity can be gained. However, another source of difficulty with quantifying ME in biological systems is determining exactly how to calculate the physiological costs. Because it is very difficult to quantify the exact amount of biological fuel required to perform an activity, physiologists and biomechanists have used indirect calorimetry to estimate the energetic costs. It is assumed that the oxygen consumed is equivalent to the energetic costs. This works well as long as submaximal activities are investigated, but activities that exceed the *anaerobic threshold* result in the accumulation of lactic acid and an *oxygen debt*. The oxygen debt is the amount of oxygen that needs to be consumed to restore the body to its equilibrium state. This cost must be included as part of the physiological cost of the movement, but it is not included if the experimenter stops measuring oxygen cost at the end of the activity. To correctly assess the physiological cost, the researcher should quantify the oxygen cost until the debt is repaid. The difficulty lies in determining exactly when this occurs and the length of time it takes to restore equilibrium. Whereas the debt is created quickly, recovery is quite slow, and for vigorous activities it may take several hours. This makes the data-collection expensive and limits the number of subjects that can be tested per day.

Another difficulty that must be considered is whether the whole oxygen cost should be used to quantify the physiological cost of an activity. Part of the oxygen cost is maintenance energy—the energy required to keep the nonmuscular tissues functioning (see figure 6.1). This energy can be measured from the basal metabolic rate (BMR), which is the metabolic cost of maintaining vital functions. These requirements are necessary to scientifically and reliably quantify BMR:

- The subject should have fasted for between 14 and 18 hours.
- The subject should be lying supine, quietly awake, following a restful night of sleep.
- The subject should not have exerted himself within the past 3 hours.
- The subject's body temperature should be within the normal range and the ambient temperature should be thermoneutral.

Clearly, this measure is difficult and expensive to obtain. Some researchers have proposed that a more appropriate measure is the oxygen cost of the person's resting state immediately before the activity. For most activities, this is the cost of standing. Thus, to compute the physiological cost of walking, one would measure the oxygen cost of standing, then subtract this amount from the oxygen cost during walking. Although this may be appropriate for some projects (because it increases sensitivity), it is not typical of evaluations of mechanical systems. It is up to the researcher to determine the best approach and to report precisely how efficiency was computed.

From the Scientific Literature

Ingen Schenau, G.J. van, and P.R. Cavanagh. 1990. Power equations in endurance sports. *Journal of Biomechanics* 23:865-81.

This survey article "attempts to clarify the formulation of power equations applicable to a variety of endurance activities." The authors outline some of the approaches taken to investigate the energetic costs of locomotor activities. Their concern is not only with the biomechanical costs of motion, but also with the considerations necessary for measuring the physiological costs. The authors point out that there still is no completely accepted way of "accurate[ly] accounting [for] the relationship between metabolic power input and the mechanical power output," but they do present the best currently available techniques. The paper reviews the research on the energetics of running, cycling, speed skating, swimming, and rowing.

This is an excellent overview of how equations based on Newtonian mechanics can be used to derive the mechanical power output of human locomotor activities. By lumping together all the external forces acting on the body, the authors derived an expression for the power production during locomotion. The expression equates the summation of the moment powers for all the joints (on the left side of the equation) to the rate of change

(continued)

of segmental mechanical energy minus the power that is delivered to the environment through the external forces ($\vec{F}_{external}$). That is,

$$\sum M_j \omega_j = \frac{d\Sigma E_{tme}}{dt} - \sum \left(\vec{F}_{external} \times \vec{v}_{external} \right) \tag{6.35}$$

where M_j is the net moment of force at each joint, ω_j is the joint angular velocity, E_{tme} is the segment total mechanical energy, $\vec{F}_{external}$ is any external force acting on the body, and $\vec{v}_{external}$ is the velocity of the point of application of each external force. To the right-hand side of the equation could be added any powers resulting from external moments of force ($M_{external}\,\omega_{external}$), such as when a cyclist grips handlebars, but these types of forces are rarely encountered in biomechanics.

SUMMARY

This chapter detailed how to compute or measure mechanical energy, work, and power of planar human motions. Methods for computing whole-body work, the work done on a single segment, and the work done by the moments of force at each joint were presented. One of the most useful tools in the biomechanist's toolkit was outlined—joint power analysis, which tells the researcher where mechanical energy is consumed, where it is transmitted within the body, and where it is produced. This analysis requires that an inverse dynamics analysis be done, but it adds important information about how the moments of force at each joint contribute to the energetics of the musculoskeletal system. Chapter 7 presents additional information that deals with 3-D motion. Fortunately, it is not difficult to extend the principles outlined here to 3-D motions.

SUGGESTED READINGS

Alexander, R.M., and G. Goldspink. 1977. *Mechanics and Energetics of Animal Locomotion*. London: Chapman & Hall.

Cappozzo, A., F. Figura, M. Marchetti, and A. Pedotti. 1976. The interplay of muscular and external forces in human ambulation. *Journal of Biomechanics* 9:35-43.

Ingen Schenau, G.J. van, & Cavanagh, P.R. (1990) Power equations in endurance sports. *Journal of Biomechanics*, 23: 865-881.

Winter, D.A. 1987. *The Biomechanics and Motor Control of Human Gait*. Waterloo, ON: Waterloo Biomechanics.

Winter, D.A. 1990. *Biomechanics and Motor Control of Human Movement*. 2nd ed. Toronto: John Wiley & Sons.

CHAPTER 7

Three-Dimensional Kinetics

Joseph Hamill and W. Scott Selbie

The term *kinetics* was defined earlier in this book in reference to a two-dimensional (2-D) calculation: It is the study of the forces and torques that cause motion of a body. Specifically, for human movement, biomechanists attempt to determine the forces that result from muscle contractions and the torques that are produced, which together bring about the movement of the segments and thus of the total body. Vaughan (1996) suggested that coordinated movement results from the activation of many muscles and that it is the tension in these muscles that causes the kinematics that we observe. Therefore, the study of the kinetics of human movement is important because it allows researchers to gain insight into basic mechanisms of movement. The concepts and methods of analysis in three dimensions (3-D) are not that much different from those in 2-D—we still use the principles of the second law of motion, superposition, and the method of sections. However, there are several considerations that must be addressed, just as there were when shifting from 2-D to 3-D kinematic analysis. It should be noted that, just as with 3-D kinematic analysis, multiple planar views of 2-D forces and moments should not be considered a 3-D analysis. In this chapter, we once again use human locomotion—specifically, lower-extremity motion—to illustrate how a 3-D kinetic analysis is undertaken.

This chapter, like chapter 3 on 3-D kinematics, is an introduction to the methodology used in the calculation of 3-D kinetics. There are a number of approaches that can be used, but in this chapter, we demonstrate only one. We hope that after learning this method, students will be able to extend their knowledge to other methods used in the literature. Therefore, we are not heavily burdened with theory in this chapter. Once again, we take a how-to approach. We will be greatly aided by the operations that we used in the 3-D kinematics chapter, and several of these concepts are extended in this chapter. The purposes of this chapter are

- to present one method of calculating 3-D joint moments and powers,
- to present a step-by-step approach to a 3-D kinetic calculation,
- to illustrate the types of data required to accomplish a 3-D kinetic calculation, and
- to present different marker sets that are used in the literature.

In biomechanics, we can either input the joint forces and torques to predict the movement of the system or compute joint moments and forces from a combination of measured external forces, segment kinematics, and segment inertial characteristics. The first technique is referred to as a *forward dynamics* approach, whereas the latter is an *inverse dynamics* approach. In this chapter, we deal with only the inverse dynamics approach, which uses one of two techniques to conduct kinetic analyses: the *Newton-Euler* (e.g., Bresler and Frankel 1950) and

Lagrangian (e.g., Whittlesey and Hamill 1996) approaches. Both of these methods derive a set of equations of motion. In this chapter, however, we discuss only the Newton-Euler approach, because it is by far the most often used method in 3-D kinetic analyses of human motion.

The major limitations of the 3-D kinetic analysis described in this chapter are the same as those in a 2-D analysis. The segments are assumed to be rigid. Although this is an obvious idealization, it can be justified to some degree for the leg segment, but not for the foot or thigh segments. The thigh has a great deal of muscle tissue surrounding the femur, and some researchers use a wobbling mass model to represent the segment (Pain and Challis 2001). The foot consists of multiple segments rather than a single rigid body. In the inverse dynamics calculation, both thigh and foot are generally modeled as rigid bodies, but the Newton-Euler inverse dynamics method does not require the joints to be constrained (Crowninshield and Brand 1981). This is not true when using other methods, including the Lagrangian formulation.

The second and most important problem is the *indeterminacy* of the solution. Figure 7.1 is a schematic diagram of the foot segment. All structures (with the exception of the soft tissue) and the forces

that they exert as well as the ground reaction force (GRF) are illustrated. Of these, only the GRF is measured. In a 2-D kinetic analysis, we have three equations of motion (relating to the three degrees of freedom), whereas for a 3-D kinetic analysis of a single rigid body, we have six equations of motion (relating to the six degrees of freedom, three translational and three rotational). However, in this analysis, we still have many more unknowns than we have equations. The result is that we have an *indeterminate solution*. In order to solve this problem, we can apply *d'Alembert's principle*, which states that the external forces acting on a rigid body are equivalent to the effective forces of the various particles forming the body. Therefore, we can consider that all bone and muscle forces are reduced to a single vector resultant force and torque.

LABORATORY SETUP

Many of the laboratories that conduct 3-D analyses of human motion employ multiple cameras and one or more force platforms. A multicamera system is necessary to faithfully reconstruct the markers placed on the subject. For each situation, the minimum number of cameras can be determined. However, most clinical gait laboratories operate with at least six cameras. Only one force platform is necessary to collect unilateral data, but a multiple force platform setup allows the researcher to collect bilateral data. All of the commercial data-collection systems employ software that synchronizes the force and the kinematics data. A clinical gait laboratory setup is illustrated in figure 7.2.

DATA REQUIRED FOR THREE-DIMENSIONAL ANALYSIS

The types of data that are necessary and thus the possible sources of error in the calculation of 3-D joint moments and forces are the same as those for the 2-D calculation. Regardless of the calculation approach used, the equations of motion include segment inertial parameters consisting of the segment mass, center of mass, and moment of inertia; kinematic data consisting of velocities and accelerations of joint positions; and forces and moments, including the GRF (figure 7.3). These data are utilized in the equations of motion. The unknowns in these equations are the joint forces and moments, whereas the known quantities are the GRF, the segmental kinematics, and the segment inertial parameters.

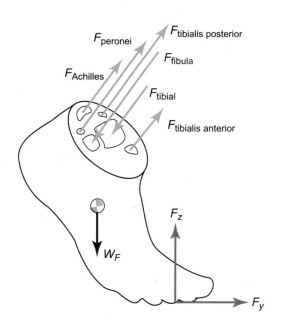

○ **Figure 7.1** A free-body diagram of the foot with all external forces identified.

Reprinted, by permission, from C.L. Vaughan, 1996, "Are joint torques the holy grail of human gait analysis?" *Human Movement Science* 15: 423-443.

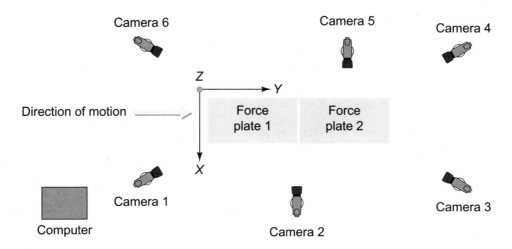

Figure 7.2 Typical camera experimental setup for a bilateral 3-D kinetic analysis of a walking stride.

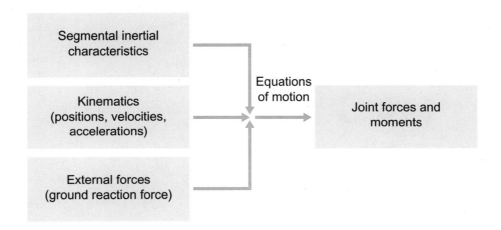

Figure 7.3 Flowchart for the inverse dynamics approach.

ANTHROPOMETRY

Anthropometry is described in more detail in chapter 3 of this book. Segmental anthropometry, which measures segment masses, segment lengths, segment center-of-mass locations, and segment moments of inertia about the principal axes, can be determined by a number of different means. They can be derived from cadavers (as in Chandler et al. 1975; Clauser, McConville, and Young 1969; Dempster 1955), from direct measurement (as in Brooks and Jacobs 1975; Zatsiorsky and Seluyanov 1985), from regression equations (as in Vaughan, Davis, and O'Connor 1992), or from mathematical modeling (as in Hanavan 1964; Hatze 1980). None of these approaches is more correct than the others for determining these parameters. The choice should be based on the population being studied and the level of accuracy sought by the researcher.

A critical factor in a 3-D kinetic analysis is locating the joint center. In certain instances, joint centers can be calculated from the markers strategically placed on the subject during a standing calibration trial. This technique is described later in this chapter. However, other researchers have used such techniques as radiographic (Deluzio et al. 1999) and anatomical measurements (Andriacchi et al. 1980). Once the joint center and the center of mass of each segment have been established, we use these data to define the segment. These parameters include the linear velocity and acceleration of the center of mass of each segment, the angular velocity and acceleration of each segment, and the moment arms for the moment calculations.

From the Scientific Literature

Kirkwood, R.N., E.G. Culham, and P. Costigan. 1999. Radiographic and non-invasive determination of the hip joint center location: Effect on hip joint moments. *Clinical Biomechanics* **14:227-35.**

The purpose of this study was to determine which of several noninvasive methods most accurately locates the hip joint center. The hip joint center was determined using X-rays and four other methods that utilized measured distance between bony landmarks. The most accurate noninvasive method was to take the midpoint of a line between the pubic symphysis and the anterior superior iliac spine and move the point inferiorly by 2 cm. This method located the hip joint 0.7 cm medially and 0.8 cm superiorly to the actual hip joint location as determined by X-ray (figure 7.4). The 95% confidence interval of the maximum difference in the joint moments calculated using the best noninvasive method and the radiographic method suggested that the true differences in the frontal and sagittal planes could be zero, whereas in the transverse plane the true difference was very small.

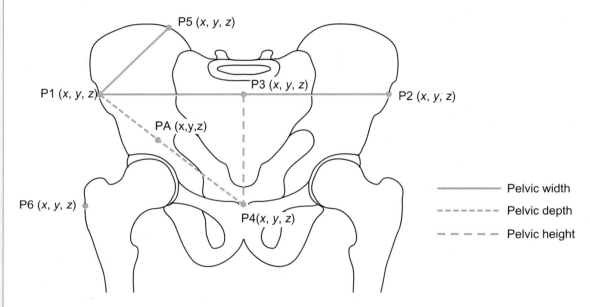

° **Figure 7.4**   A schematic of the pelvis with the parameters used in the four methods to calculate the hip joint parameters. PA represents the midpoint of the line segment P1 to P4. Using the best method, the hip joint was located 1.5 to 2.0 cm distal to PA.

Reprinted, by permission, from R.N. Kirkwood, E.G. Culham, P. Costigan, 1999, "Radiographic and non-invasive determination of the hip joint center location: Effect on hip joint movements," *Clinical Biomechanics* 14: 227-235.

KINEMATIC DATA

The second type of data required is the kinematic data describing the segment accelerations (both linear and angular) and the moment arms for the proximal and distal forces. These data are obtained by extending the techniques and marker set used in the previous chapter on 3-D kinematics. A number of marker sets in the biomechanics literature have been used in the calculation of 3-D kinetics. In this chapter, we describe the use of only one marker set—the one used in the previous chapter on 3-D kinematics—in the hope that readers can extend the concepts to any marker set (figure 7.5).

Several extra markers are placed on the subject to define the lower-extremity joint centers. These extra markers are used only during the standing calibration trial and can be removed when the actual data

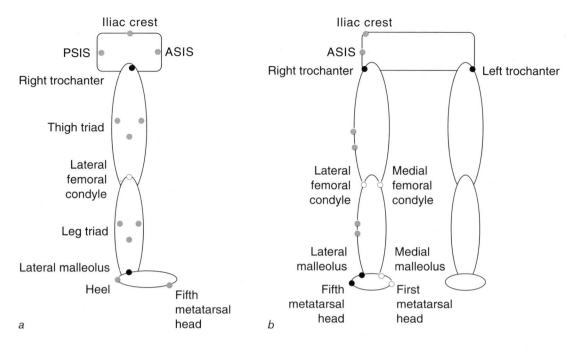

Tracking markers

Tracking and calibration markers

Calibration markers

Figure 7.5   *(a)* Sagittal and *(b)* frontal views of marker setup for a calibration frame. Note that some markers are used for tracking only, some only for the calibration, and some for both. The joint centers are defined for the hip by markers on the right and left greater trochanters, for the knee by markers placed on the lateral and medial femoral epicondyles, and for the ankle on the lateral and medial malleoli. PSIS = posterior superior iliac spine; ASIS = anterior superior iliac spine.

collection begins. Therefore, we refer to *tracking* markers that remain on the subject both in the calibration and data-collection processes and to *calibration* markers that are removed after the calibration is completed. Some markers perform both a tracking and calibration function.

The need for these extra markers will become apparent later. In 2-D, we used a marker on the lateral side of the segment in the sagittal plane to represent the joint center, knowing that it was displaced laterally from the real joint center. Because we can calculate forces only along the horizontal and vertical axes and moments only about a mediolateral axis through the joint center, this is a satisfactory assumption. However, for 3-D, we must know the internal location of the joint center. These extra markers help us to determine the joint centers of the proximal and distal joints that make up a segment. For example, the markers on the lateral and medial epicondyles of the knee help us to determine the knee joint center and those on the lateral and medial malleoli, the ankle joint center.

These extra markers also help us to develop a *local* or *segment coordinate system* (LCS) that is effectively attached to the bone of the segment. We consider the origin of the LCS reference frame to be at the center of mass of the segment. However, the origin could be placed anywhere along the line between the proximal and distal joint centers.

EXTERNAL FORCES

The third type of data involves the direct measurement of external forces. In human locomotion, the most prominent external force is the GRF that generally is measured via a force platform. It is of great importance that we synchronize the force platform data with the kinematic data both temporally and spatially. Most advanced motion-capture systems allow for simultaneous collection of kinematic and force platform data at compatible sampling rates. Therefore, the only data-collection problem that we must concern ourselves with is spatial synchronization. This is accomplished by establishing the origin of the global coordinate system (GCS) relative to the origin of the force platform. One way to do this is to make the origin of the GCS coincident with the

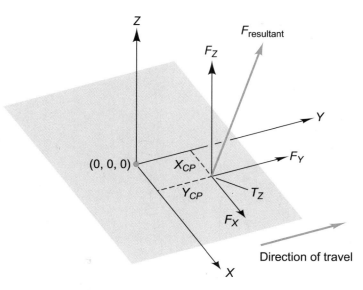

○ **Figure 7.6** Schematic of force platform showing the GRF vector ($F_{resultant}$), the components of the vector, and the center of pressure (X_{CP}, Y_{CP}). For data collection and spatial synchronization of force platform and kinematic data, the origin of the GCS (X,Y,Z) is placed at the origin of the force platform.

origin of the force platform (figure 7.6). However, it does not matter where the origin of the GCS is placed as long as its relationship to the force platform is known. For example, some gait laboratories establish the origin of the GCS at one of the corners of the force platform. However, the end result must be that all measurements of the kinematics of the system and the center of pressure (COP) from the force platform are in the GCS. This can be accomplished with the transformation techniques discussed in chapter 2.

When collecting GRF data, most gait laboratories make several assumptions that cannot be made in other situations. We must recognize that in the collection of GRF data, six signals are acquired from the force platform. These six signals are in the form of voltages that must be scaled to represent three orthogonal forces (F_z, vertical; F_y, anteroposterior; and F_x, mediolateral) and three moments (M_z, vertical; M_y anteroposterior; and M_x, mediolateral) about each of the axes. However, nine parameters are required for a 3-D kinetic analysis. These parameters include the three force components, the XYZ components of the COP vector, and the free torques about each of the axes. The three force components are three of the signals collected. However, the components of the COP and the three free moments either must be calculated from the six scaled signals that were collected or assumptions must be made concerning these parameters. The X and Y compo-

nents of the COP can be calculated using the following formulae:

$$X_{CP} = -\frac{M_y + F_x \times d_z}{F_z} \tag{7.1}$$

$$Y_{CP} = -\frac{M_x - F_y \times d_z}{F_z} \tag{7.2}$$

where M_x and M_y are the measured moments about the x and y force platform axes, respectively; F_x, F_y, and F_z are the mediolateral, anteroposterior, and vertical GRF components; and d_z refers to the distance that the origin of the force platform is beneath its surface. For the third component of the COP, we can assume that Z = 0, although most force platforms have their origin just below the platform surface. Because we are considering the origin of the force platform to be coincident with the GCS, the calculated values of the COP are in the GCS system.

For the three free moments, we can assume that the shear torques, T_x and T_y, are equal to zero. The free moment is calculated as:

$$T_z = M_z + (F_x \times Y) - (F_y \times X) \tag{7.3}$$

where M_z is the measured moment about the vertical axis of the force platform, F_x and F_y are the shear components of the GRF, and X and Y are the components of the COP.

The assumptions made for three of the necessary parameters must be evaluated by each gait laboratory to ensure that they are true. If, for example, a handle were placed on the force platform or on a nearby obstacle and an upward force was applied against it, then these assumptions would not necessarily be true. For cases such as these, other measurement devices must be used to quantify one or more of the nine parameters.

SOURCES OF ERROR IN THREE-DIMENSIONAL CALCULATIONS

The combination of GRF, kinematic, and anthropometric data leaves many possible sources of error in the calculation of joint forces and moments. Of these, the GRF data are probably the most accurate and reliable. Although the segmental inertial parameters vary widely based on the technique used to derive them, they do not greatly contribute to the end result. The data derived from the kinematics, however, are a different story. The kinematic information needed to calculate joint torques is highly dependent on marker placement and the associated skin movement. For example, the calculation of joint centers from these markers affects the location of the center of mass that, in turn, affects the calculation of moment arms. The moment arms are multipliers of the joint forces and the GRF and ultimately have a profound effect on the end result.

In terms of the joint torques themselves, the dominance of the GRF in the calculation indicates that distal joint torques and forces are less variable than are those of the more proximal joints. Also, torques about the mediolateral axis (i.e., in the sagittal plane) are more reliable than torques about the secondary plane axes.

THREE-DIMENSIONAL KINETICS CALCULATIONS

The remainder of this chapter describes one particular method of calculating 3-D joint force and moments. There are other methods that accomplish the same task, and students can apply what they learn here in understanding the steps in other methods. For clarity, all parameters in the LCS of the segment are designated with lowercase symbols, whereas those in the GCS are designated with uppercase symbols.

OVERVIEW OF THE METHOD

Two separate steps must be taken to obtain the necessary information for a 3-D kinetic analysis. First, the calibration trial is recorded, with all of the tracking and calibration markers on the subject. The calibration markers used to define the individual segments can be removed after the calibration trial data have been collected. For the trial, the subject stands in the field of view of the cameras so that her feet are pointed in the positive Y direction. This trial is used to create the LCS of each segment and to locate the proximal and distal joint centers and the segment center of mass. In addition, transformation matrices are formed to determine the locations of the calibration markers from the locations of the tracking markers.

In the next step, the data trials are collected. In these trials, only the tracking markers on the subject and the GRFs are monitored. The analysis uses the segment information determined in the calibration trial along with the GRFs, segment kinematics, and anthropometric data to calculate the joint kinetics.

CALCULATIONS USING THE CALIBRATION DATA

A number of steps that use this calibration trial data must be completed before you can proceed to the joint kinetics calculations. These procedures include (1) calculating the joint centers, (2) determining the segment center of mass, (3) calculating the local bone-embedded coordinate system (LCS), and (4) calculating a transformation matrix. In keeping with the protocol established in chapter 2, all uppercase letters represent parameters in the GCS and lowercase letters represent parameters in an LCS.

Calculation of the Joint Centers

Figure 7.7a represents an idealized leg segment with the proximal markers $\vec{L}_1$ and $\vec{M}_1$ on the right and left femoral epicondyles and the distal markers $\vec{L}_2$ and $\vec{M}_2$ on the lateral and medial malleoli. These are calibration markers, some of which are removed after the calibration frame is obtained. A number of studies have described methods for determining joint centers, particularly the hip joint center (Andriacchi et al. 1980; Deluzio et al. 1999; Kirkwood, Culham, and Costigan 1999). In this chapter, we describe one rather simple method of determining joint centers. The virtual joint center of the knee ($\vec{V}_1$) is defined as a point that is some percentage (r) of the distance from $\vec{L}_1$ to $\vec{M}_1$. The ankle joint ($\vec{V}_2$) is similarly defined using $\vec{L}_2$ and $\vec{M}_2$. Generally, then, the joint centers are defined as

$$\vec{V}_1 = \vec{L}_1 + r_l\left(\vec{M}_1 - \vec{L}_1\right) \qquad (7.4)$$

where $\vec{L}_1$ is the lateral proximal calibration marker, $\vec{M}_1$ is the lateral distal calibration marker, and r is the proportion of the distance from the lateral to the medial markers. For the hip joint center, r is 25% of the distance between the right and left greater trochanter markers. For the knee and ankle joints, r is 50% of the length between the markers (figure 7.7b). Note that this calculation is done with the coordinates of the markers in the GCS in the calibration position.

It can be assumed that the long axis of the segment is defined by a line between the centers of the proximal and distal joints of the segment. It is further assumed that the center of mass lies on the long axis of the segment. The precise distance of the center of mass from the proximal joint center is obtained from a published anthropometric data set. As stated previously, there are a number of methods for obtaining this information for each segment. For this chapter, we will estimate the location of the segment center of mass (h_l) as 0.433 of the length of the leg segment as

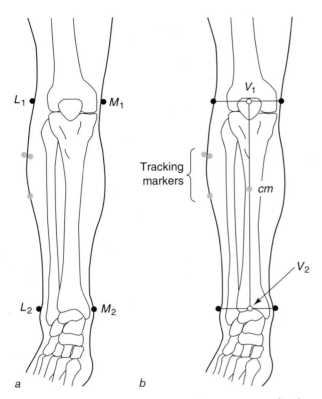

Figure 7.7 *(a)* Using the calibration frame markers $\vec{L}_1$, $\vec{L}_2$, $\vec{M}_1$, and $\vec{M}_2$, *(b)* virtual joint centers are calculated from the lateral and medial markers. A line between the virtual joint centers describes the z-axis orientation of the segment. A proportion of the distance along the z-axis orientation line gives the location of the segment's center of mass.

measured from the proximal joint center (Dempster 1955). Thus, the center of mass $(\vec{O})$ can be calculated in GCS coordinates as

$$\vec{O} = \vec{V}_1 + h_I\left(\vec{V}_2 - \vec{V}_1\right) \qquad (7.5)$$

where $h_I = 0.433$ and $\vec{V}_1$ and $\vec{V}_2$ are the virtual joint centers (figure 7.7b). This formula can be used with other body segments by changing h_I appropriately.

Defining the Local Coordinate System of the Segment

The LCS of the segment on the right limb is created so that the z-axis is directed from distal to proximal, the y-axis from posterior to anterior, and the x-axis from medial to lateral. The origin of the LCS is located along the line connecting the segment's ends (i.e., the z-axis) at the segment's center of mass. For the left limb, we maintain the same orientation of the segment LCS by multiplying the mediolateral forces and moments by –1.

The $\vec{k}'$ unit vector is collinear with the vector from the distal joint center to the proximal joint center.

The unit vector along the long axis of the segment toward the proximal joint center is defined as

$$\vec{k}' = \frac{\vec{V}_1 - \vec{O}}{\left|\vec{V}_1 - \vec{O}\right|} \qquad (7.6)$$

where $\vec{V}_1$ is the proximal joint center vector of the segment and $\vec{O}$ is the center of mass.

Next, the $\vec{i}'$ unit vector of the segment can be defined (figure 7.8a). The plane of the four markers—$\vec{L}_1$, $\vec{M}_1$, $\vec{L}_2$, $\vec{M}_2$—is defined as the mediolateral (z-x) plane. Generally, a least-squares plane is fitted to the four targets so that the sum of squares distance between the targets and the frontal plane is minimized. The $\vec{i}'$ unit vector can then be defined as

$$\vec{i}' = \frac{\vec{L}_1 - \vec{V}_1}{\left|\vec{L}_1 - \vec{V}_1\right|} \qquad (7.7)$$

where $\vec{L}_1$ is the proximal lateral marker vector and $\vec{V}_1$ is the proximal joint center.

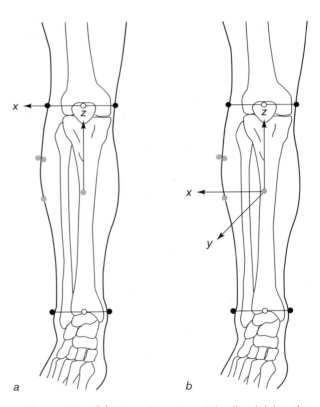

Figure 7.8 *(a)* The unit vector of the line joining the center of mass and the proximal joint center forms the z-axis of the segment. The lateral and medial markers define the x-y plane, from which the x-axis is defined. *(b)* The y-axis of the embedded system is defined as the cross product of the z- and x-axes.

Finally, the $\vec{j'}$ unit vector is defined (figure 7.8b) as the line perpendicular to the y-z plane by taking the cross product of the two already-calculated unit vectors, $\vec{k'}$ and $\vec{i'}$:

$$\vec{j'} = \vec{k'} \times \vec{i'} \qquad (7.8)$$

Calculation of the Transformation Matrix

We can generate a transformation matrix ($T_{global2local}$) that transforms the markers from $\vec{L}_1$, $\vec{L}_2$, $\vec{V}_1$, and $\vec{V}_2$ in the GCS to $\vec{l}_1$, $\vec{l}_2$, $\vec{v}_1$, and $\vec{v}_2$ in the LCS. This transformation matrix is a 4×4 matrix that combines both the position and orientation vectors. Thus, the position and orientation of the LCS can be described with reference to the global reference frame. The technique for determining this transformation matrix is described in Spoor and Veldpaus (1980) and Veldpaus, Woltring, and Dortmans (1988) and is not discussed in this chapter. Note that if $T_{global2local}$ describes the transformation from global coordinates to local coordinates, then its transpose describes the transformation of local coordinates to global coordinates, designated as $T_{local2global}$. The $T_{global2local}$ matrix is a combined rotation and translation transformation matrix that takes the form

$$\left[T_{global2local}\right] = \begin{bmatrix} 1 & 0 & 0 & 0 \\ D_{21} & R_{22} & R_{23} & R_{24} \\ D_{31} & R_{32} & R_{33} & R_{34} \\ D_{41} & R_{42} & R_{43} & R_{44} \end{bmatrix} \qquad (7.9)$$

where the first column (D_{21} through D_{41}) represents the position vector and the remaining columns represent the rotation matrix (R_{22} through R_{44}).

We can now create a similar transformation matrix to transform the triad of tracking markers placed on each segment from the GCS ($\vec{P}_1$, $\vec{P}_2$, $\vec{P}_3$) to the LCS ($\vec{p}_1$, $\vec{p}_2$, $\vec{p}_3$). First, we must calculate the triad of markers in the LCS. This can be accomplished as follows:

$$\vec{p}_1 = \vec{P}_1 - \vec{O} \qquad (7.10)$$

$$\vec{p}_2 = \vec{P}_2 - \vec{O} \qquad (7.11)$$

$$\vec{p}_3 = \vec{P}_3 - \vec{O} \qquad (7.12)$$

where P_1, P_2, and P_3 are the tracking markers in the GCS and O is the origin of the LCS. From these values, we can generate a transformation matrix ($T_{global2local}$) that transforms from $\vec{P}_1$, $\vec{P}_2$, $\vec{P}_3$ in the GCS to $\vec{p}_1$, $\vec{p}_2$, $\vec{p}_3$ in the LCS, again by using the technique described by Spoor and Veldpaus (1980).

To summarize, as part of the calibration process, these calculations do three things: (1) determine the virtual joint center of each segment from the calibration markers, (2) enable us to determine for each segment an LCS that is embedded in the segment, and (3) enable us to use the tracking markers and joint center positions to calculate transformation matrices between the LCS and GCS for each segment that will be used in the actual trial.

CALCULATIONS USING THE TRIAL DATA

Once again, several steps must be taken to lead us toward our goal of calculating 3-D joint kinetics. These steps include the calculation of (1) segment kinematics, (2) joint kinematics, (3) joint forces and moments, and (4) joint power.

Segment Kinematics

For each segment in each frame of the data trial, the cluster of markers on the segment is digitized and the coordinates are represented in the GCS. These markers ($\vec{P}_1$, $\vec{P}_2$, $\vec{P}_3$) are transformed into the segment's LCS.

$$\vec{p}_i = T_{global2local} \vec{P}_i \qquad (7.13)$$

Then, the lateral joint markers ($\vec{L}_1$ and $\vec{L}_2$) and the virtual joint centers ($\vec{V}_1$ and $\vec{V}_2$) are transformed into the GCS using the inverse of the transformation matrix above ($T_{local2global}$) (see figure 7.9).

Several techniques are often used to calculate segment kinematics. The necessary parameters are the linear velocities and accelerations of the center of mass and the distal and proximal joint centers in addition to the angular velocity and acceleration of the segment. It should be pointed out that the angular velocity and acceleration of the segment cannot be determined through differentiation of the Euler angle representation of the segment.

We next describe a technique that employs a least-squares estimation of the *instantaneous screw axis* (ISA), which is also referred to as the *instantaneous helical axis* (Sommer 1992). The procedure is complicated, and our description is cursory. For greater detail, readers should refer to Sommer (1992). The first step in the procedure is to determine the centroid of the three tracking markers that make up a segment. By differentiating the equation that represents the general movement of the centroid, the position and direction of the ISA can be determined.

The input to the calculation of the ISA consists of four matrices. Three of these describe the position, linear velocity, and linear acceleration of the cluster of three tracking markers on a segment. These matrices are formed as

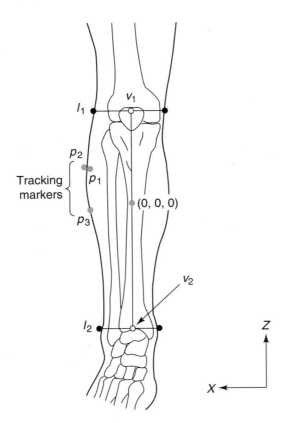

o **Figure 7.9**   Relationship of the tracking markers (p_1, p_2, and p_3) to the calibration markers l_1 and l_2 and the virtual joint centers v_1 and v_2 in the local reference frame.

$$[P] = \begin{bmatrix} P1_x & P2_x & P3_x \\ P1_y & P2_y & P3_y \\ P1_z & P2_z & P3_z \end{bmatrix} \quad (7.14)$$

$$[V] = \begin{bmatrix} \dot{P1}_x & \dot{P2}_x & \dot{P3}_x \\ \dot{P1}_y & \dot{P2}_y & \dot{P3}_y \\ \dot{P1}_z & \dot{P2}_z & \dot{P3}_z \end{bmatrix} \quad (7.15)$$

$$[A] = \begin{bmatrix} \ddot{P1}_x & \ddot{P2}_x & \ddot{P3}_x \\ \ddot{P1}_y & \ddot{P2}_y & \ddot{P3}_y \\ \ddot{P1}_z & \ddot{P2}_z & \ddot{P3}_z \end{bmatrix} \quad (7.16)$$

The fourth matrix is a weighting matrix for the marker measurements. This weighting matrix is often used to assign a reliability value to the different tracking markers when the data are particularly noisy. However, in most cases the identity matrix (3 × 3) is used and thus no weighting is given to any particular markers.

Given these matrices, the output from the algorithm described by Sommer (1992) consists of the

segment angular velocity vector [$\vec{\omega}$] and the segment angular acceleration vector [$\vec{\alpha}$]. Other parameters generated from this procedure allow us to calculate the linear velocity and acceleration of the proximal and distal virtual joint centers of a segment and the segment center of mass linear velocity ($\vec{v}_{CM}$) and acceleration ($\vec{a}_{CM}$).

Kinetics

Note that vectors in the LCS of the segment are designated with lowercase symbols and those in the GCS, with uppercase symbols. In order to calculate the kinetics of the foot segment, it is important to note that all of the necessary parameters—including the GRFs and torques, the COP, the force on the segment resulting from gravity, the segment center of mass accelerations, the proximal and distal moment arms, and the proximal and distal joint center locations—must be transformed into the LCS:

$$GCS \quad \Rightarrow \quad LCS_{foot}$$

$$(F_{GRF\_X}, F_{GRF\_Y}, F_{GRF\_Z}) \Rightarrow (f_{GRF\_X}, f_{GRF\_Y}, f_{GRF\_Z})$$

$$(0, 0, T_z) \Rightarrow (0, 0, t_z)$$

$$(X, Y, 0) \Rightarrow (x, y, 0)$$

$$(0, 0, -MG) \Rightarrow (mg_x, mg_y, -mg_z)$$

$$(\ddot{X}_{CM}, \ddot{Y}_{CM}, \ddot{Z}_{CM}) \Rightarrow (\ddot{x}_{CM}, \ddot{y}_{CM}, \ddot{z}_{CM})$$

$$(D_{1x}, D_{1y}, D_{1z}) \Rightarrow (d_{1x}, d_{1y}, d_{1z})$$

$$(D_{2x}, D_{2y}, D_{2z}) \Rightarrow (d_{2x}, d_{2y}, d_{2z})$$

$$(V_{1x}, V_{1y}, V_{1z}) \Rightarrow (v_{1x}, v_{1y}, v_{1z})$$

$$(V_{2x}, V_{2y}, V_{2z}) \Rightarrow (v_{2x}, v_{2y}, v_{2z})$$

As was the case in the 2-D inverse dynamics calculation, the only unknown in the equations of motion are the forces and moments at the proximal joint of the segment. Also as in the 2-D computation, we calculate the translational dynamics first (i.e., solve for the joint reaction forces), then calculate the rotational dynamics (i.e., solve for the joint torques).

To closely examine the 3-D inverse dynamics method, we will use the foot segment as an example and calculate the joint forces and moments acting at the ankle (figure 7.10). The following procedures are carried out for each instant in time (i.e., each frame of data). The first step for each segment is to calculate the translational dynamics, meaning the ankle joint reaction forces. The equation for the translational dynamics of the foot in vector form is:

$$j\vec{r}f_{Ankle} = m\vec{a}_{CM} - m\vec{g} - \vec{f}_{GRF} \quad (7.17)$$

where $j\vec{r}f_{Ankle}$ is the vector describing the ankle joint reaction force, m is the segment mass, $\vec{a}_{CM}$ is the linear acceleration vector of the center of mass, $m\vec{g}$

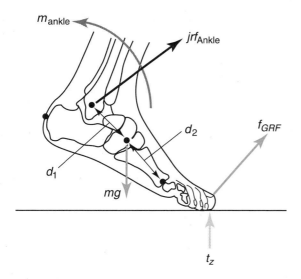

Figure 7.10   The FBD of the foot segment. For calculation, all parameters must be represented in the LCS of the foot.

is the gravity force vector (note that the components of the gravity vector are 0, 0, –mg in the GCS), and $\vec{f}_{GRF}$ is the GRF vector. The expanded form of this equation is

$$jrf_{Ankle\_x} = m\ddot{x}_{CM} - f_{GRF\_x} - mg_{\_x} \tag{7.18}$$

$$jrf_{Ankle\_y} = m\ddot{y}_{CM} - f_{GRF\_y} - mg_{\_y} \tag{7.19}$$

$$jrf_{Ankle\_z} = m\ddot{z}_{CM} - f_{GRF\_z} - mg_{\_z}, \tag{7.20}$$

where jrf_{Ankle_x}, jrf_{Ankle_y}, and jrf_{Ankle_z} are the components of the ankle joint reaction force; g is the acceleration resulting from gravity; m is the segment mass; $\ddot{x}_{CM}$, $\ddot{y}_{CM}$, and $\ddot{z}_{CM}$ are the components of the center of mass acceleration; f_{GRF_x}, f_{GRF_y}, and f_{GRF_z} are the orthogonal components of the GRF; and $mg_{\_x}$, $mg_{\_y}$, and $mg_{\_z}$ are the components of the weight of the segment in the LCS.

The next step in the procedure is to calculate the rotational dynamics, that is, the ankle joint moment. Taking moments about the segment center of mass, the vector form of the rotational dynamics equation of the ankle is

$$j\vec{m}_{Ankle} = I\alpha - \left(\vec{d}_1 \times j\vec{r}f_{Ankle}\right) - \left(\vec{d}_2 \times \vec{f}_{GRF}\right) - \vec{t} \tag{7.21}$$

where $j\vec{m}_{Ankle}$ is the vector describing the ankle joint moment, I is the moment of inertia matrix, α is the angular acceleration matrix, $\vec{d}_1 \times j\vec{r}f_{Ankle}$ is the vector describing the moment resulting from the joint reaction force, $\vec{d}_2 \times f_{GRF}$ is the vector describing the moment resulting from the GRF, and $\vec{t}$ is the ground reaction torque vector.

To present the rotational dynamics equation in expanded form, we must first calculate the vector describing the moment about the x, y, and z axes

of the proximal joint center and, for the foot, the GRFs. For the foot segment, we calculate the moment vector resulting from the joint reaction forces at the ankle, $j\vec{m}_{JRF}$, by taking the product of the joint forces and the respective distances from the ankle center of mass to the line of action of the joint reaction forces. This is calculated using the cross product

$$\vec{m}_{JRF} = \vec{d}_1 \times j\vec{r} f_{Ankle} \tag{7.22}$$

where $\vec{d}_1$ is the vector of the xyz distance between the center of mass and the proximal joint center and $j\vec{r}f_{Ankle}$ is the vector of the xyz components of the proximal joint reaction force. Similarly, the moment vector resulting from the GRF, $\vec{m}_{GRF}$, is calculated as the cross product of the distance between the center of mass and COP vectors and the GRF vector.

$$\vec{m}_{GRF} = \left[\vec{d}_2\right] \times \left[\vec{f}_{GRF}\right] \tag{7.23}$$

where $[\vec{d}_2]$ is the vector of the xyz distance between the center of mass and the COP and $[\vec{f}_{GRF}]$ is the vector of the xyz components of the GRF.

The moment at the proximal (ankle) joint of the foot segment, taken about the center of mass of the segment, can then be calculated as

$$m_{Ankle\_X} = I_{XX}\alpha_X + (I_{ZZ} - I_{YY})\omega_{ZZ}\omega_{YY} \\ - m_{GRF\_X} - m_{jrf\_X} - t_x \tag{7.24}$$

$$m_{Ankle\_Y} = I_{YY}\alpha_Y + (I_{XX} - I_{ZZ})\omega_{XX}\omega_{ZZ} \\ - m_{GRF\_Y} - m_{jrf\_Y} - t_y \tag{7.25}$$

$$m_{Ankle\_Z} = I_{ZZ}\alpha_Z + (I_{YY} - I_{XX})\omega_{YY}\omega_{XX} \\ - m_{GRF\_Z} - m_{jrf\_Z} - t_z \tag{7.26}$$

where I_{XX}, I_{YY} and I_{ZZ} represent the components of the moment of inertia vector; α_{XX}, α_{YY} and α_{ZZ} represent the components of the segment angular acceleration; ω_{XX}, ω_{YY} and ω_{ZZ} represent the components of the segment angular velocity; m_{GRF_X}, m_{GRF_Y} and m_{GRF_Z} are the components of the moment resulting from the GRF; m_{jrf_X}, m_{jrf_Y} and m_{jrf_Z} are the components of the moment resulting from the joint reaction force; and t_x, t_y and t_z are the ground reaction torques. In gait, we usually assume that t_x and t_y equal zero.

In a 3-D inverse dynamics procedure, the calculations go from distal to proximal (just as they did in 2-D) and the values calculated for the more distal segment are used in the calculations for the next, more proximal segment. Note that in the 3-D procedure, all force and moment calculations are conducted in the LCS of the segment in question. Therefore, once the calculations for a segment have been completed, the force and moment vectors must be transformed into the GCS as follows, using the appropriate transformation matrix:

$$\left[J\vec{R}\,F_{Ankle} \right] = \left[T_{local2global} \right] \left[j\,\vec{r}\,f_{Ankle} \right] \qquad (7.27)$$

$$\left[J\,\vec{M}_{Ankle} \right] = \left[T_{local2global} \right] \left[j\vec{m}_{Ankle} \right] \qquad (7.28)$$

Figure 7.11 illustrates a free-body diagram (FBD) for the leg segment. For the leg, all the necessary parameters including the ankle joint reaction forces and moments from the foot segment calculation must be transformed into the local leg coordinate system.

Once this is done, calculation of the proximal forces and moments at the proximal joint (the knee) can proceed. When these calculations are complete, the knee joint reaction force and knee moment are transformed into the GCS. For the thigh computations, the knee joint reaction force and knee moment are transformed into the LCS of the thigh for the calculation of the hip joint reaction force and hip moment. Subsequently, the hip joint reaction force and hip torque are transformed into the GCS. The equations for the leg and thigh segments are analogous to those for the foot segment with the exception that the GRF components are replaced by the analogous components of the distal joint reaction force. In vector form these equations are

$$j\,\vec{r}\,f_{Prox} = m\,\vec{a}_{CM} - m\,\vec{g} - j\,\vec{r}\,f_{Distal} \qquad (7.29)$$

$$j\,\vec{m}_{Proximal} = I\alpha - \left(\vec{d}_1 \times \vec{f}_{jrf\_Prox} \right)$$
$$- \left(\vec{d}_2 \times \vec{f}_{jrf\_Distal} \right) - j\,\vec{m}_{Distal} \qquad (7.30)$$

For the leg and thigh, $\vec{d}_1$ is the vector of the xyz distance between the segment center of mass and the proximal joint center, whereas $\vec{d}_2$ is the vector of the xyz distance between the center of mass and the distal joint center.

Figure 7.12 illustrates the mean (±1 standard deviation) 3-D lower-extremity moments for a normal population (Meglan and Todd 1994). In these graphs, the moments are presented in the distal segment coordinate system (i.e., the ankle moments are in the foot LCS).

Joint Angles

The calculation of joint angles was described in chapter 2. Cardan angles can be calculated using an Xyz sequence of rotations, as was also explained in that chapter.

Joint Power

In the section on 2-D kinetics, joint power was defined as the product of the joint moment and the joint angular velocity. In 3-D kinetics it is similarly defined, although now we will calculate the power about all three axes. The joint power vector can be calculated using the joint moment vector and the segment angular velocity vector in the GCS:

$$\vec{P} = J\vec{M} \times \left(\vec{\omega}_{Proximal} - \vec{\omega}_{Distal} \right) \qquad (7.31)$$

where $J\vec{M}$ is the vector representing the XYZ components of the joint torque and $\vec{\omega}_{Proximal}$ and $\vec{\omega}_{Distal}$ are the vectors representing the XYZ proximal and distal segment angular velocities, respectively.

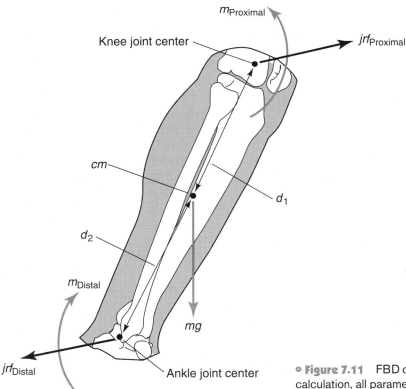

° **Figure 7.11**   FBD of the leg segment. For the inverse dynamics calculation, all parameters must be represented in the LCS of the leg for knee forces. The FBD of the thigh would be similar.

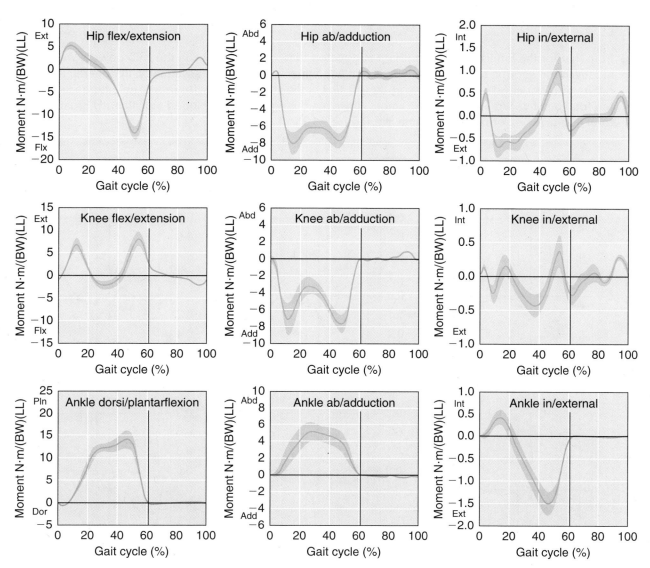

○ **Figure 7.12** Joint moments of the ankle, knee, and hip during a stride cycle of walking based on normal population data. Moments are scaled by body weight and leg length. The blue line represents the mean value, whereas the blue zone indicates ±1 standard deviation. The black line indicates the beginning of the nonsupport phase of the stride.

Reprinted from Meglan, D., Todd, F. (1994). Kinetics of Human Locomotion. In *Human Walking*. J. Rose and J. G. Gamble (eds.). pp. 73-99. Baltimore, MD: Williams and Wilkins.

From the Scientific Literature

Allard, P., R. Lachance, R. Aissaoui, and M. Duhaime. 1996. Simultaneous bilateral able-bodied gait. *Human Movement Science* **15:327-46.**

The purpose of this paper was to report on lower-extremity muscle power and mechanical energy during gait over two consecutive stride cycles. The walking speed of the subjects was 1.30 m/s. The relative duration of the stance phase for the right limb was not significantly different

(continued)

from that for the left limb. These authors suggested that the differences in the limb powers were mostly in the sagittal plane and reflected gait adjustments rather than asymmetry. In general, the differences occurred during the absorption portion of the stance phase. There was no difference in the total positive work between the limbs. However, the right limb developed significantly greater total negative work than the left limb did (figure 7.13).

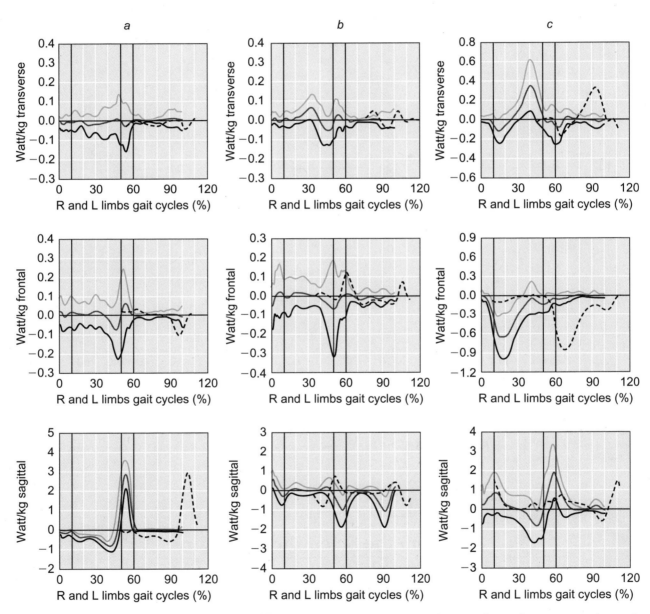

Figure 7.13 *(a)* Mean ankle; *(b)* knee; and *(c)* hip power in the transverse (top), frontal (middle), and sagittal (bottom) plane with standard deviations (blue line = +1SD and black line = –1SD) for the right limb (gray line). Only the mean values are given for the left limb (dashed line). The right limb stance phase occurs from 0 to 61%. The swing phase of the left limb occurs from 11 to 61%, and the subsequent left limb toe-off occurs at 112%.

7.13a Reprinted, by permission, from P. Allard et al., 1996, "Simultaneous bilateral able-bodied gait," *Human Movement Science* 15: 327-346.

PRESENTATION OF THE DATA

The joint forces and moments that are calculated from an inverse dynamics procedure may be 3-D vectors defined in the GCS, as suggested by Winter, Eng, and Ishac (1995). Their rationale was that, in gait, the individual's line of progression is generally aligned along one of the planes of the global reference system, and thus the sagittal plane of the lower extremity and the line of progression will be essentially the same. However, it is difficult to interpret these forces and moments, particularly if the individual's local segment coordinate system is not aligned with the GCS. Therefore, we must present the joint forces and moments in a coordinate system that can be readily understood regardless of the subject's orientation to the GCS. The methods that are generally employed include transforming the forces and moments to (1) the coordinate system of the more proximal segment that constitutes the joint, (2) the coordinate system of the more distal segment that constitutes the joint, or (3) a body-fixed anatomical reference frame. However, no international standard has been adopted, and so readers must take note of the coordinate system in which the moments are presented.

There is no readily apparent reason for presenting the forces and moments in the LCS of either the proximal or distal segment, and some researchers favor one or the other. The point is that the joint forces and moments are represented in a coordinate system that is unique to the subject and is independent of the subject's position in the GCS. One drawback to representing the joint forces and moments in either system is that it is assumed that the basic orientations of the segments in all subjects are the same.

The most common body-fixed anatomical reference frame, proposed by Grood and Suntay (1983), is the *joint coordinate system* (JCS). It was discussed previously, when we presented methods for calculating 3-D joint angles. This coordinate system has the flexion/extension axis fixed in the distal segment, the axial rotation axis fixed in the proximal segment, and the adduction/abduction axis perpendicular to the other two axes. It is readily apparent that these axes are not necessarily orthogonal and thus that the transformed forces and moments are not necessarily orthogonal. It should be noted, however, that the JCS transformation is the most commonly reported and clinically accepted method of presenting 3-D kinetic data.

It is also necessary to discuss the scaling of 3-D joint kinetics. Scaling is used to attempt to remove between-subject differences in the joint moments. In a clinical setting in which an individual's data are being collected and a report is generated on that subject, there is no need to scale the data, and in fact, it is probably inappropriate to do so. However, in a research study in which data are collected on multiple subjects in different conditions, it probably is appropriate to scale the data. The most-reported methods of scaling joint kinetic data are scaling to body mass, as in N·m/kg (Vaughan 1996; Winter, Eng, and Ishac 1995), or to body weight (BW) and leg length (LL), as in N·m/(BW)(LL) (Meglan and Todd 1994).

A final point in presenting the results of an inverse dynamics calculation concerns the interpretation of the force and moments. The gait analysis literature includes three documented approaches to describing joint moments. The first method presents the net torque generated by muscles crossing a joint, which is called the *internal moment*. The internal moment is the product of the inverse dynamics calculation. The second interpretation is based on the GRF being the dominant input to the body. For example, in this interpretation, the extensor muscles of the legs during the stance phase of gait counteract the GRF. This is referred to as the *external moment,* and it is opposite in sign to the internal moment. The third interpretation is based on the sum of all extensor moments at the hip, knee, and ankle. In this approach, all extensor moments are assigned a positive value. As a consequence, the hip and ankle moments have different sign conventions with respect to the laboratory. Regardless of which interpretation has been used, it is important that biomechanists be aware of the method.

SUMMARY

There are several different methods for computing 3-D kinetics, although only one, the Newton-Euler procedure, is discussed in this chapter. The assumptions for a 3-D kinetic analysis are exactly the same as those for a 2-D analysis, and thus the limitations are also the same. A 3-D analysis involves merging three types of data: (1) external force data (usually the GRF), (2) the 3-D coordinates of markers describing the individual segments, and (3) the anthropometric data of the individual segments. Error associated with each of these types of data must be taken into consideration.

A number of different marker sets can be used, but there should be a minimum of three noncollinear markers per segment. The determination of 3-D kinetics involves collecting a calibration trial, in which the subject has both tracking and calibration markers attached. The tracking markers remain on the subject during the actual trials, but the calibration markers are removed. The calibration markers allow for the calculation of the virtual joint centers and the LCS of the segment. In addition, a transformation matrix is determined between the tracking markers on the segment and the calibration positions that will define the segment. This matrix is used to determine the calibration marker positions after those markers have been removed.

When the trial data are collected, the inverse dynamics procedure is initiated from the most distal segment—for example, the foot—to the most proximal segment, the thigh. For each segment, all information necessary for calculating the proximal joint moments is transformed into the LCS of that segment. For example, to calculate the ankle moments, the data are transformed into the LCS of the foot. When the proximal moments are calculated, the forces and moments are then transformed back into the GCS for use in the calculations for the next, more proximal segment. This procedure continues until the most proximal moments and forces (in our example, the hip moments and force) are calculated. Joint power is calculated as the product of the net moment and the joint angular velocity in the GCS.

There are many different methods of presenting 3-D kinetic data. These methods include presenting the data in either the GCS or the LCS of either the proximal or distal segment. A further method of presenting the kinetic data involves transforming the data into the JCS. It should be noted, however, that this coordinate system is not orthogonal.

Kinetic data are often presented unscaled or scaled to either body mass or to body weight and leg length. Whether the data are scaled or not depends on what the researcher is representing in the study.

In gait analysis, three documented approaches are used to describe the calculated joint moments: the internal moment, the external moment, and the support moment. It is critical that readers are aware of the approach being described.

SUGGESTED READINGS

Allard, P., A. Capozzo, A. Lundberg, and C. Vaughan. 1998. *Three-Dimensional Analysis of Human Locomotion.* Chicester, U.K.: John Wiley & Sons.

Nigg, B.M., and W. Herzog. 1994. *Biomechanics of the Musculo-skeletal System.* Toronto: John Wiley & Sons.

Vaughan, C.L., B.L. Davis, and J.C. O'Connor. 1992. *Dynamics of human gait.* Champaign, IL: Human Kinetics.

Zatsiorsky, V.M. 2002. *Kinetics of Human Motion.* Champaign, IL: Human Kinetics.

Part III

Additional Techniques

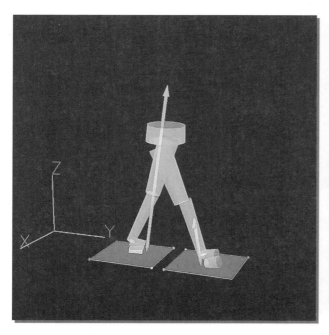

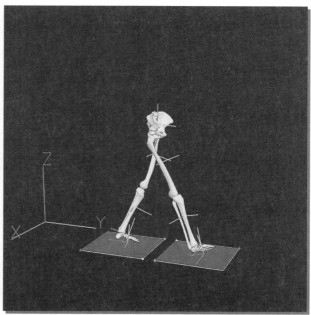

CHAPTER 8

Electromyographic Kinesiology

Gary Kamen

The use of electromyography—the study of muscle electrical activity—can be quite valuable in providing information about the control and execution of voluntary (and reflexive) movements. This chapter provides an overview of the use of electromyography (EMG) in biomechanics. In this chapter, we

- determine the basis of the muscle fiber action potential and how it propagates along the muscle fiber,
- determine the characteristics of the electromyographic signal,
- understand the basic features of EMG electrodes,
- examine some of the technical issues that alter the characteristics of the EMG signal,
- determine what variables are usually implemented to describe the EMG signal, and
- review some examples to illustrate how EMG has been used to understand human movement.

PHYSIOLOGY OF THE ELECTROMYOGRAPHIC SIGNAL

Although researchers must understand several technical features to process the EMG signal, the signal itself has a physiological origin. The first section of this chapter briefly explains the physiological concepts that underlie the EMG signal's origin and how it propagates along muscle fibers.

MUSCLE FIBER ACTION POTENTIAL

To produce muscular force, muscle fibers must receive an impulse from a motoneuron. Once a motoneuron is activated by the central nervous system (CNS), an electrical impulse propagates down the motoneuron to each motor endplate. At this specialized synapse, ionic events occur that culminate in the generation of a muscle fiber *action potential* (AP).

Resting Membrane Potential

Even at rest, muscle is an excitable tissue from which electrical activity can be recorded. Normally, the inside of the muscle fiber has an electrical potential of about –90 millivolts (mV). This voltage gradient results from the presence of different concentrations of sodium (Na^+), potassium (K^+), and chloride (Cl^-) ions across the sarcolemma. The resting membrane potential is about 9 to 15 mV more positive in slow-twitch fibers, apparently because of their greater Na^+ permeability and higher intracellular Na^+ activity than those of fast-twitch fibers (Hammelsbeck and Rathmayer 1989; Wallinga-De Jonge et al. 1985). Moreover, the resting membrane potential is not necessarily fixed; it can be altered, for example, by exercise training (Moss et al. 1983).

Action Potentials

The AP is the neural messenger responsible for activating every segment of the muscle fiber so that each sarcomere contributes to the generation of muscular force. The process begins with a change in the muscle fiber membrane's permeability to Na⁺. Because these Na⁺ ions are in relatively greater concentration outside the muscle fiber, any change in permeability results in an influx of Na⁺ across the membrane. Eventually, sufficient Na⁺ ions enter the cell to reverse the polarity of the membrane potential so that the inside of the muscle fiber becomes positive (by about 30 mV) with respect to the surrounding extracellular medium. As the membrane's potential polarity reverses, the permeability of the membrane to K⁺ changes, causing K⁺ to exit the cell. It is largely this efflux of K⁺ that repolarizes the cell and restores the resting membrane potential.

Propagation of the Muscle Fiber Action Potential

To ensure complete electromechanical activation of the muscle fiber, an AP produced in one small segment adjacent to the neuromuscular junction must spread to adjoining sections. By a passive process, the AP propagates along each adjacent section of the muscle fiber and in both directions from the neuromuscular junction so that the entire muscle fiber is electrically activated. As the AP spreads, the membrane potential at each subsequent muscle fiber section changes from negative to positive and back to negative as each adjacent muscle fiber area is successively activated. The deeper portions of the muscle fiber also need to be electrically activated and so, by means of the transverse tubule system, the AP propagates to the deeper sections as well. The axons of the motoneurons of some muscle fibers are slightly longer than those of others, so it

may take slightly longer to begin activation in those fibers (figure 8.1).

Muscle Fiber Conduction Velocity

The varying rates of transmission of the propagated AP determine many of the characteristics of the EMG. APs moving at slow rates, for example, contribute low-frequency components to the surface EMG. Thus, understanding the factors that determine the rate at which this propagation occurs is important. Because AP generation is an ionic process, the velocity at which the AP is conducted along the muscle fiber is dependent on the rate at which these ions can be exchanged. The passive membrane permeability characteristics determine part of this exchange rate, along with the active metabolic mechanism for pumping Na⁺ back out of the fiber.

Determinants of Muscle Fiber Conduction Velocity

Differences in conduction velocity among different muscle fibers can be attributed to both histochemical and architectural features of the muscle fiber. The amplitude of the muscle fiber AP tends to be larger in fast-twitch fibers. Moreover, the shapes of the APs of fast- and slow-twitch fibers differ, causing the fast-twitch fiber AP to occur more quickly (including depolarization and repolarization) than the corresponding slow-twitch AP. Consequently, fast-twitch fibers have faster conduction velocities than slow-twitch muscle fibers. Larger-diameter muscle fibers also produce larger APs than do smaller fibers (Andreassen and Arendt-Nielsen 1987), partly

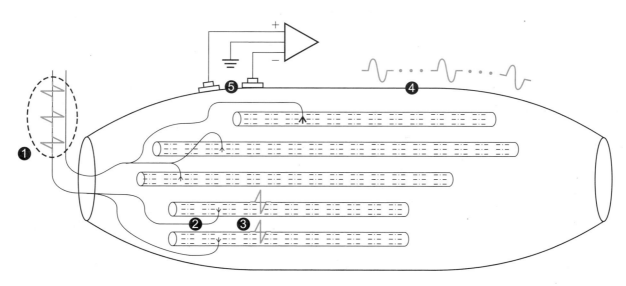

○ **Figure 8.1**   The motoneuron AP initiates the process of muscle fiber excitation (1). The AP arrives at all of the motor endplates innervated by the motoneuron (2). By electrochemical processes, a muscle fiber AP is initiated and propagates along the length of the muscle fiber (3). The sum of all muscle fiber potentials activated by one motoneuron produces a motor unit AP (4), which can be recorded at the skin surface with amplifiers used specifically for biological signals (5).

because of the greater activity of Na⁺. Atrophied fibers have distinctly slower conduction velocities (Buchthal and Rosenfalck 1958). Increases in the length of muscle fibers tend to decrease conduction velocity, and this may result from other architectural changes that occur in the fiber (Dumitru and King 1999).

MOTOR UNIT ACTION POTENTIAL

Each motoneuron typically innervates several hundred muscle fibers, although the number varies in different muscles from around 10 to as many as several thousand. This characteristic—termed the *innervation ratio*—is computed by determining the number of muscle fibers per motoneuron. The muscle fibers within a single motor unit tend to be distributed throughout the muscle, although some muscles may have a more focused distribution (Windhorst, Hamm, and Stuart 1989). The individual unit of motor action is the *motor unit*—one motoneuron and all of the muscle fibers innervated by that motoneuron.

The motor unit action potential (MUAP) represents the summated electrical activity of all muscle fibers activated within the motor unit. The amplitude of the MUAP is partly determined by the innervation ratio. In addition, motor units with more (or larger) muscle fibers have a larger MUAP. However, there also are some temporal dispersion issues that define the shape of the MUAP (figure 8.2).

Motor Unit Activation

The production of muscular force is controlled by the action of numerous muscles acting across a joint. In a single muscle, muscular force is initiated by activating an increasing number of motor units in a process termed *recruitment*. To produce almost any muscular action, smaller motor units are recruited first, and successively larger motor units are recruited as the force requirement increases. The nervous system also controls how frequently motor units are activated, the quantification of which is termed the motor unit *discharge* or *firing rate*. As the motor unit fires at faster rates, it produces an increasing amount of muscular force.

Muscular force can be altered in other ways, as well. For example, at the onset of a muscular contraction, when considerable effort may be required to overcome the inertia of the limb to be moved, motor units may fire in two short latency bursts before beginning a regular firing rate. The twin bursts are *doublets* that have the potential to produce larger forces than might be expected from the addition of two motor unit force twitches (Clamann and Schelhorn 1988). Two or more motor units frequently fire simultaneously in a process called *synchronization*. The exact role of motor unit synchronization in muscular force production is not clear, although it seems to occur more often than would be expected with chance alone.

In performing an action like wrist flexion, we tend to activate other muscles *(synergists)* that perform similar actions. Wrist flexion might be performed by activating the flexor digitorum profundus and superficialis, flexor carpi ulnaris and radialis, palmaris longus, and other wrist flexors. Later in this chapter, we will show that electrode sensors placed over a skin surface serve as detectors of EMG

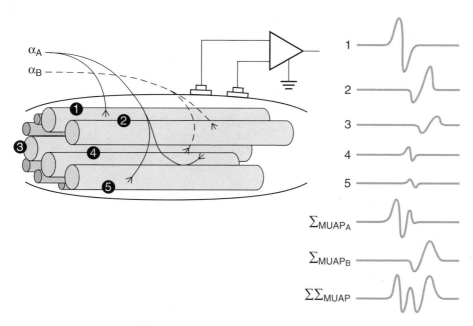

○ Figure 8.2 The contribution that each fiber's AP makes to the EMG signal depends in large part on the depth of the fiber; note that fiber 5 contributes a smaller AP than fiber 1. The temporal characteristics of the signal also depend on the electrode–motor endplate distance, as well as the terminal lengths and diameters of the motoneurons. Two motor units are shown here, with the amplitude of each motor unit represented as the algebraic sum of the individual muscle fiber APs (Σ_{MUAP}). The overall signal is the algebraic sum of all motor units ($\Sigma\Sigma_{MUAP}$).

activity over a potentially large volume. If neighboring muscles perform similar actions, then that activity is recorded by the surface electrodes.

Characteristics of the Electromyographic Signal

In the mildest muscle contraction, a single motor unit may be activated. This is recorded at the surface as the MUAP, followed by electrical silence until the unit's next firing. As the desired force increases, other motor units may be recruited and fire at ever-increasing frequency. At any point in time, the EMG signal is a composite electrical sum of all of the active motor units. A large peak in the EMG signal might be the result of the activation of two or more motor units separated by a short interval. Note that the signal has both positive and negative components. When the signal crosses the baseline, a positive phase of one MUAP is likely balanced by the negative phase of other MUAPs.

The amplitude of the surface-recorded EMG signal varies with the task, the specific muscle group under study, and many other features. Naturally, EMG amplitude increases as the intensity of the muscular contraction increases. However, the relationship between EMG amplitude and force frequently is nonlinear. Moreover, cocontraction activity from antagonists may require compensatory activity from the agonist muscle group from which EMG recordings are being made. Thus, one cannot assume that increases in EMG activity are indicative of parallel increases in force (Redfern 1992; Solomonow, Baratta et al. 1990). Issues regarding the relationship between EMG amplitude and force are discussed later in this chapter.

RECORDING AND ACQUIRING THE ELECTROMYOGRAPHIC SIGNAL

EMG activity can be recorded using either a monopolar or a bipolar recording arrangement (figure 8.3). In monopolar recordings, one electrode is placed directly over the muscle and a second electrode is placed at an electrically neutral site, such as a bony prominence. In general, monopolar signals yield lower-frequency responses and less selectivity than bipolar recordings. Although monopolar recordings are frequently used during static contractions (Ohashi 1995, 1997) and in a variety of clinical investigations involving needle electrodes (Dumitru, King, and Nandedkar

1997), the monopolar recording is inherently less stable and would be a poor choice for measuring nonisometric contractions. Monopolar recordings are appropriate for the assessment of H- and T-reflexes and muscle M-waves, however (Mineva, Dushanova, and Gerilovsky 1993).

Bipolar (or *single differential*) recordings are considerably more common. In a bipolar recording arrangement, two electrodes are placed in the muscle or on the skin overlying the muscle and a third neutral, or ground, electrode is placed at an electrically neutral site. This configuration utilizes a differential amplifier that records the electrical *difference* between the two recording electrodes. Thus, any signal that is common to the two inputs is greatly attenuated. The feature that allows the amplifier to attenuate these common signals is called *common-mode rejection,* and the extent to which signals common to both inputs are attenuated is described by the *common-mode rejection ratio* (CMRR). These CMRRs are expressed in either a linear or logarithmic scale. A very good commercial amplifier might have a CMRR of 100 decibels (dB). One can convert CMRR in a dB scale using the formula

$$CMRR_{(dB)} = 20 \log_{10} CMRR_{(linear)} \qquad (8.1)$$

the CMRR is equivalent to 100,000:1.

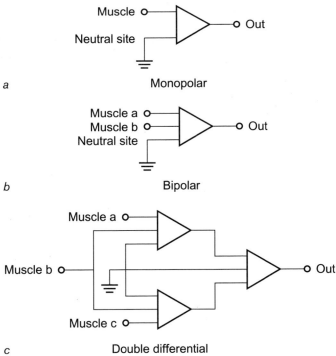

a

b

c

○ **Figure 8.3** EMG signals can be acquired using either a *(a)* monopolar or *(b)* bipolar configuration. Some applications, however, may require specialized amplifiers, such as the *(c)* double differential recording technique.

In a typical laboratory or field environment, there may be considerable radio frequency (RF) and line activity from electrical outlets, lights, or other line signals. These signals are typically at 50 or 60 Hz (depending on the recording location), and so they are directly in the frequency range represented by the EMG signal. There may also be ambient RF signals in the atmosphere at other frequencies. Because the differential amplifier reduces the signals appearing in-phase at both amplifier inputs, their influence

is greatly reduced. There are instances in which both monopolar and bipolar recordings may be valuable, such as in investigating the geometry of muscle fibers during changes in muscle length (Gerilovsky, Tsvetinov, and Trenkova 1989).

Care must be used in setting the amplifier gain. If the amplitude of the EMG signal exceeds the amplifier's gain, the amplifier is saturated and distortion occurs in the form of *clipping* (figure 8.4). Clipping occurs when the amplitude of the output

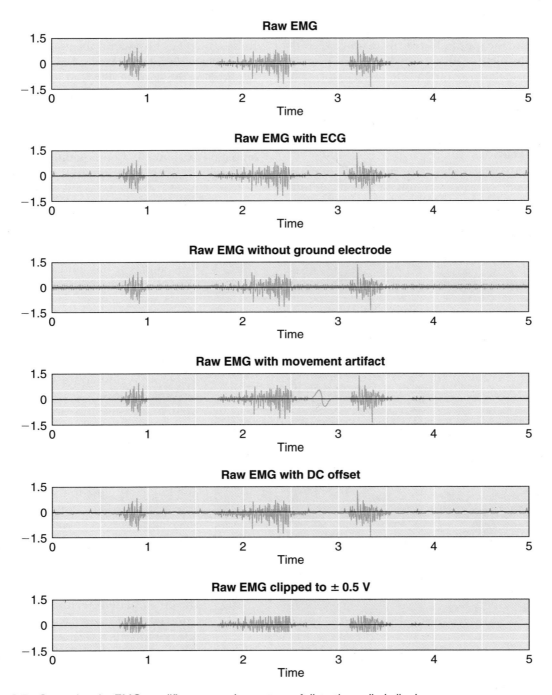

○ **Figure 8.4**   Saturating the EMG amplifier can produce a type of distortion called clipping.

exceeds the bounds of the power supply. On the other hand, if the gain is set too low, the resolution of the signal after analog-to-digital (A/D) conversion will be low. Ideally, the gain should be set so that the amplitude of the signal is matched to the range of the A/D converter.

ELECTRODE CHARACTERISTICS AND GEOMETRY

A large variety of EMG electrodes are available (figure 8.5). The choice of electrode depends on the motor task to be explored, the nature of the research question, and the specific muscle from which recordings are to be made. This section discusses the use of EMG electrodes placed on the skin surface, as well as the use of indwelling fine-wire and needle electrode designs. A few other electrode designs currently in research use are also noted.

Surface Electrodes

The first EMG electrodes were simple conducting surfaces made of various kinds of metal, includ-

ing silver, gold, stainless steel, and even tin. These plate-type electrodes are commonly used for clinical applications such as assessing sensory and motor nerve conduction velocity, F-waves, M-waves, and H-reflexes (Oh 1993). However, some technical problems may occur with these surface electrodes. There is normally a 30 mV potential between the inside and outside skin layers. When the skin is stretched, the potential decreases to about 25 mV, and the resulting 5 mV change is recorded as motion artifact. Obviously, motion artifact produced by skin transients can be a problem in recording situations in which appreciable movement is expected. However, these artifacts can frequently be minimized by using silver-silver chloride electrodes (Webster 1984) and by abrading the skin lightly to reduce the normal 50 kΩ impedance across the skin surface. These surface-applied electrodes serve as detectors of EMG activity in the underlying muscles, but they can also be excellent antennae for other RF activity in the ambient environment. Consequently, interelectrode impedance has to be minimized to minimize RF activity. After the electrodes are applied to the skin, an impedance meter can be used for this purpose. Prior to applying the electrodes, impedance can be minimized by carefully preparing the skin, removing dead skin cells and skin oils, and increasing local skin blood flow. Minimizing cable distances, using shielded cables, and braiding individual electrode cables together also help to minimize RF interference.

One technique that successfully reduces these artifacts is locating the first-stage amplifier as close to the electrode site as possible. For example, friction and movement in the cable insulation can produce artifacts. These artifacts can be reduced by using a unity gain operational amplifier at each electrode. Frequently called *active* electrodes, these devices position the amplifier very close to the surface sensor (see figure 8.5). The signals from the first-stage amplifiers frequently have a higher signal-to-noise ratio (SNR) and thus are "cleaner" signals (Hagemann, Luhede, and Luczak 1985). Several authors have described the low-cost construction of on-site preamplified electrodes (Hagemann, Luhede, and Luczak 1985; Johnson et al. 1977; Nishimura, Tomita, and Horiuchi 1992). Disposable electrodes are also commercially available for applications that require them.

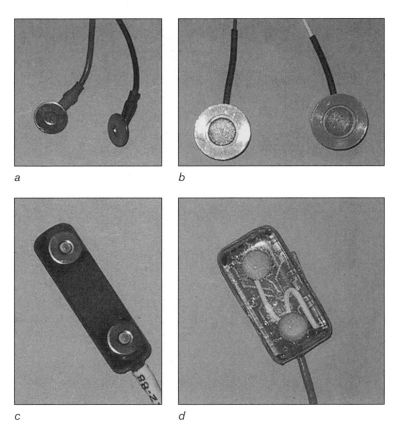

a   b

c   d

◦ Figure 8.5 Some of the many surface electrodes commercially available: *(a)* plate electrodes, *(b)* silver-silver chloride cup electrodes, *(c)* fixed interelectrode distance plate electrodes, and *(d)* preamplified electrodes.

Limitations of Surface Electrodes

Surface electrodes have limited use in recording activity from deeper muscles or deeper portions of large muscles. The estimated effective recording area of surface electrodes ranges between 10 and 20 mm from the skin surface (Barkhaus and Nandedkar 1994; Fuglevand et al. 1992). It is also difficult to use them to record signals from small muscles, because it can be difficult to discern whether the signal is arising from the underlying muscle or an adjacent muscle. This *crosstalk* is a serious problem that is discussed later in the chapter. There is some evidence that deeper motor units may be smaller than their superficial counterparts, so surface EMG recordings must be interpreted in the context of a possible bias toward recording from larger and more glycolytic motor units (Lexell et al. 1983).

Fine-Wire Electrodes

An alternative procedure that facilitates recording from deeper or smaller muscles is the use of fine-wire electrodes (Basmajian and Stecko 1962; Burgar, Valero-Cuevas, and Hentz 1997). This configuration generally consists of two fine-diameter insulated wires (about 75 μm diameter) that are threaded through a hollow needle cannula. The tips of the wires constitute the recording surfaces, and these are either cut flush or the last millimeter or so of insulation is removed from the wire. The greater the amount of insulation that is removed, the greater is the recording volume. The distal centimeter of these wires is bent backward so that the wires and needle can be inserted into the muscle. After the needle is inserted, it is carefully removed, leaving just the wires as the recording electrodes. The cannula can either be removed or left off to the side during the experiment, and the ends of the wires are then connected to an amplifier.

Needle Electrodes

Many types of needle electrodes are commercially available, and they are frequently used to monitor the activity of one or more individual motor units rather than the EMG activity of the composite muscle. Concentric electrodes consist of a small wire placed in the middle of a hollow cannula. The wire is set in place with epoxy. The cannula is then cut at an acute angle (about 15°), leaving a shiny wire surface that is then referenced to the cannula for a bipolar recording.

Other Electrode Designs

Many other types of EMG electrodes have been designed for different purposes. Longitudinal array electrodes are used to record the features of propagating muscle fiber APs (Masuda, Miyano, and Sadoyama 1985; Merletti, Farina, and Granata 1999). Multiple 2-D array electrodes consisting of nine or more electrode surfaces arranged in a grid are used to investigate the architectural features of muscle (Thusneyapan and Zahalak 1989).

Recording individual muscle fiber APs during high-force contractions requires specialized electrodes. There have been some excellent attempts to record individual APs using surface multielectrode arrays (Rau, Disselhorst-Klug, and Silny 1997), but again, a surface detector is biased toward muscle fibers closer to the skin surface. APs from individual muscle fibers are usually recorded from wire electrodes (as just described), multiple wire electrodes (Hannerz 1974; Shiavi 1974), or needle electrodes (Kamen et al. 1995; Sanders, Stålberg, and Nandedkar 1996).

Electrode Geometry and Electrode Placement

The importance of correctly placing surface electrodes over the muscle cannot be overemphasized. Certainly, the electrodes must be placed in a position from which APs from the underlying muscle fibers can be recorded. In general, this means placing the electrodes away from highly tendinous areas. The motor point (the area where the nerve enters the muscle) is *not* a good location for surface electrodes. For studies requiring several measurement sessions, the endplate zone has the potential to produce the most variable EMG signals. Because MUAPs propagate in both directions from the neuromuscular junction, signals recorded over the motor point are frequently subjected to algebraic subtraction from a differential amplifier, resulting in the cancellation of EMG signals common to both electrodes. Standard locations for electrodes can be found in various sources (LeVeau and Andersson 1992; Zipp 1982). However, the orientation of the electrodes with respect to the muscle fibers is also important. If the electrode is not placed parallel to the muscle fibers, the amplitude of the signal may be reduced by as much as 50% (Vigreux, Cnockaert, and Pertuzon 1979). The frequency content of the EMG signal is also affected by off-parallel electrode placement. Consequently, although the muscle fiber pennation angle can be difficult to determine, every effort should be made to orient the electrodes parallel to the muscle fibers. Anatomical atlases are available for assistance (Cram, Kasman, and Holtz 1998).

Interelectrode Geometry

Although the evidence is equivocal, the amplitude of the EMG signal may be affected by the interelectrode

From the Scientific Literature

Masuda, T., H. Miyano, and T. Sadoyama. 1992. The position of innervation zones in the biceps brachii investigated by surface electromyography. *IEEE Transactions on Biomedical Engineering* **32:36-42.**

EMG techniques are sometimes useful in revealing morphological features in muscle. In this paper, Masuda and colleagues describe the construction of a linear surface electrode array that is placed along the surface of the biceps brachii muscle. Muscle fiber APs are detected by the multiple electrodes and followed and tracked as they propagate along the muscle. When the array is placed in the vicinity of the innervation zone (the so-called *motor point*), the AP polarity reverses, reliably and accurately marking the location of the motor point. The authors describe a computer program that automatically calculates this position using regression techniques. In this way, surface EMG techniques are quite useful to muscle morphologists.

distance—the distance between electrode pairs. One researcher found that an interelectrode distance of about 60 mm produces the greatest EMG amplitude when surface electrodes with a 7 mm diameter recording area are used (Vigreux, Cnockaert, and Pertuzon 1979). However, more recently, Jonas, Bischoff, and Conrad (1999) failed to find any difference in compound muscle AP amplitude using different surface electrode types, different recording areas, and different interelectrode distances. EMG frequency characteristics are also amenable to change with electrode spacing, with higher spectral frequencies obtained from electrodes placed closer together (Bilodeau et al. 1990; Moritani and Muro 1987). In placing EMG electrodes, it is best to keep constant as many conditions as possible between recording sessions and among subjects to minimize variability from any of these factors.

ELECTROMYOGRAPHIC SIGNAL PROCESSING

Advances in both analog and digital electronics and in signal-processing techniques, as well as continued advances in our understanding of the EMG signal, have changed the way we process and analyze the signal. This section focuses on the most common techniques available for acquiring and filtering the signal. We also consider some of the technical problems inherent in processing the EMG signal.

Application of Analog and Digital Filtering

Knowledge of the inherent frequency characteristics and prevailing sources of RF noise and interference

of any biological signal allows some decisions to be made regarding the kinds of filtering techniques to be used. Analysis of the EMG signal frequency spectrum indicates that little (if any) of the signal is contained at frequencies below about 10 Hz or above 1 kHz. Indeed, for the surface EMG signal, the upper frequency limits of the signal are even more bandwidth-limited, with the highest frequency components found at about 400 Hz. Indwelling signals (recorded within muscles) contain higher-frequency content (Gerleman and Cook 1992). Consequently, it is frequently recommended that the surface EMG signal be acquired using a high-pass filter set at about 10 Hz, with an anti-aliasing low-pass filter set at about 1 kHz. However, opinions vary on the exact band-pass characteristics to be used. Software can be used after recording to apply further digital signal processing.

Analog-to-Digital Data Acquisition

Knowledge of the inherent frequency characteristics of the EMG signal is also important in choosing a sampling rate. Certainly, the Nyquist limit must be considered, requiring as it does a sampling rate that is at least twice that of the highest-frequency component in the signal. For the surface EMG signal, this generally requires a sampling rate of at least 1,000 samples/s, and sometimes higher. A low-pass, anti-aliasing filter set at the Nyquist limit (the highest frequency in the signal) should then be implemented to prevent signals above this frequency from distorting the true signal.

Many manufacturers offer 50 to 60 Hz notch filters to attenuate RF activity from lights or other equipment. However, there is considerable EMG activity at these line frequencies. Consequently, although the signal may look "cleaner," a significant portion of the EMG signal is also eliminated with these notch filters. If necessary, techniques other than the use of notch filters should be used to eliminate line-frequency interference. This may require changing the surrounding equipment environment, using alternative electrodes, improving the grounding configuration, or improving the electrode-skin interface.

INFLUENCES ON THE ELECTROMYOGRAPHIC SIGNAL

The large number of influences on the EMG signal are both physiological and technical in origin. Knowledge of the myriad sources that can contribute to the surface-recorded EMG signal is necessary to ensure correct interpretation.

Electrodes

As described earlier, electrode characteristics that can affect the frequency and amplitude content of the EMG signal include the type of electrode (surface, metal plate, silver-silver chloride, indwelling, and so on), electrode size, and the interelectrode distance. The electrode configuration (monopolar vs. bipolar) is also a determinant.

The characteristics of the underlying tissue also affect the EMG signal. A poor electrode-skin interface increases electrode impedance, contributing to a poorer SNR. A good skin-electrode interface is particularly important for nonamplified electrodes. The volume of subcutaneous fat can also affect the recording of the EMG signal. In general, lower levels of subcutaneous fat are associated with higher SNR.

Blood Flow and Tissue Influences

The tissue between the muscle and surface electrodes has a dramatic low-pass filtering effect. Consequently, the high-frequency characteristics of individual muscle fiber APs, particularly those from deeper fibers, are attenuated appreciably at the surface. This decay occurs exponentially with distance, so at even small distances from the muscle fiber, the higher-frequency characteristics of the AP are sharply attenuated. Even 100 μm away from the muscle fiber, about 80% of the signal strength is lost (Andreassen and Rosenfalck 1978). Low-pass filtering may also be induced by increased muscle blood flow, which increases dramatically during contraction. Thus, it is possible that EMG signal characteristics can be altered by factors unrelated to muscle electrical activity. Because the EMG signal can be affected by

changes in Na^+, situations that involve appreciable fatigue, dehydration, or interruption in muscle blood flow may also affect the EMG.

Muscle Length

Our knowledge of changes in muscle length during dynamic contraction has been extended considerably by the ultrasound studies conducted by Kawakami and colleagues (Ito et al. 1998). The rate of propagation of the muscle fiber AP changes as the muscle changes in length. This is demonstrated as a decrease in muscle fiber conduction velocity with muscle lengthening (Morimoto 1986). Also, the amplitude of the AP declines with increasing length (Gerilovsky, Tsvetinov, and Trenkova 1986; Hashimoto et al. 1994), likely as a result of morphological changes in the neuromuscular junction or the sarcolemmal membrane (Kim, Druz, and Sharp 1985). Moreover, the frequency characteristics are affected, with increasing muscle length shifting the spectrum toward the lower frequencies (Okada 1987). These changes in the characteristics of the EMG signal are one reason that it is difficult to interpret the signal during dynamic contractions.

Muscle Depth

Because the AP decays so rapidly as the distance from the recording electrodes increases, deeper muscle fibers are considerably more difficult to record from the surface than more superficial muscles are. As noted earlier, surface electrodes record signals only from these more superficial muscle fibers. Moreover, there is some evidence that the fiber type may vary with muscle depth. The deeper muscle fibers seem to have more slow-twitch characteristics, and so they are recruited at lower forces, whereas the more superficial fibers may be more fast-twitch and recruited at somewhat higher muscle forces. Consequently, the surface EMG may be biased in favor of recording from the more superficial, fast-twitch muscle fibers, which also generate the larger APs.

ANALYZING AND INTERPRETING THE ELECTROMYOGRAPHIC SIGNAL

The two important characteristics of the EMG signal are amplitude and frequency. Amplitude is an indicator of the magnitude of muscle activity, produced predominantly by increases in the number of active motor units and the frequency of activation, or the firing rate. The frequency of the signal is also affected by these factors. By activating more motor units, the number of spikes and turns in the surface EMG signal increases. Changes in firing rate also change EMG frequency characteristics. However, as discussed earlier, a number of other factors, both technical and physiological, also affect both the amplitude and frequency of the EMG signal. In this section, we discuss the major variables used to analyze the EMG and provide some information regarding their interpretation.

ELECTROMYOGRAPHIC AMPLITUDE

The major variables used to define EMG amplitude include peak-to-peak (p-p) amplitude, average rectified amplitude, root mean square (RMS) amplitude, linear envelope, and integrated EMG. These variables are discussed below.

Peak-to-Peak Amplitude

The peak-to-peak amplitude is one of the simplest ways to describe the magnitude of the EMG signal. This variable is particularly useful when the signal is highly synchronous—composed of multiple simultaneously firing motor units. For example, when a peripheral motor nerve is stimulated, most or all of the motoneurons are activated simultaneously to produce a synchronous signal called the M-wave (figure 8.6). When the intensity of the stimulation is increased sufficiently, all motoneurons are activated, resulting in the maximal EMG activity that the muscle is capable of producing. This maximal-amplitude M-wave can be described by calculating the negative-peak-to-positive-peak amplitude (*p-p amplitude*).

S   M   H

° **Figure 8.6**   The M-wave (M) is the synchronous electrical activity of all muscle fibers following an electrical stimulus (S). The H-reflex (H) is produced by the activation of Ia afferents.

Another example is the H-reflex, which is evoked by delivering a low-intensity electrical stimulus to the peripheral motor nerve in an effort to activate only the primary muscle spindle afferents. The orthodromic AP from these Ia afferents synapses on α motoneurons, resulting in motoneuron discharge. The resultant signal is similar in appearance to the M-wave, although smaller in magnitude, because it is rare for all of the motoneurons to be activated by this technique. The amplitude of the H-reflex can also be described using p-p amplitude.

Average Rectified Amplitude

The normal EMG (also called the *interference pattern*) is an alternating current (AC) signal, varying in both the positive and negative voltage directions. Unless there is some voltage offset in the signal acquisition/ amplification system, the mean of the signal is zero. Consequently, the mean value is *not* a valid indicator of EMG amplitude.

To compute a representative averaged amplitude measure over a period of time, the signal must first be rectified. Rectification involves converting the negative voltages to positive values (i.e., absolute values). After doing this, the average of the values is a nonzero amplitude measure termed the *average rectified amplitude* (figure 8.7).

Root Mean Square Amplitude

One alternative that does not require rectification is to compute the RMS amplitude as follows:

$$RMS\{EMG\,(t)\} = \left(\frac{1}{T}\int_{t}^{t+T} EMG^2(t)dt\right)^{1/2} \quad (8.2)$$

where *EMG* is the value of the EMG signal at each moment of time *(t)*, and *T* represents the duration of the analyzed signal. Because RMS amplitude incorporates the squared values of the original EMG signal, it does not require full-wave rectification.

Linear Envelope

Because the EMG is a time-varying signal with a zero mean, the value of the signal at any instant is not an indicator of the overall magnitude of EMG activity. However, an estimate of the "volume" of activity can be obtained using a variable called the *linear envelope*, which is computed by passing a low-pass filter through the full-wave rectified signal (figure 8.7). The linear envelope, then, is a type of moving average indicator of EMG magnitude. The exact selection of a frequency to be used for the cutoff is somewhat arbitrary, and the appropriate cutoff frequency depends on the application. Cutoff frequencies of 3 to 50 Hz have been suggested. Shorter-duration activities benefit by choosing a higher cutoff frequency, but generally a frequency of 10 Hz gives satisfactory results. However, note that the resolution of the high-frequency characteristics of the signal is attenuated. Consequently, the resolution available in computing onsets and offsets is reduced when the linear envelope is used for this purpose.

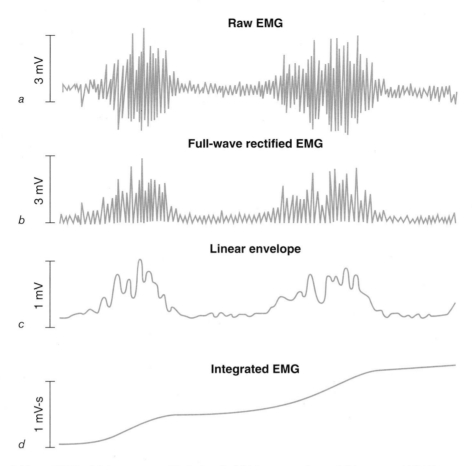

Figure 8.7 *(a)* Raw EMG. *(b)* Average rectified signal. *(c)* Linear envelope. *(d)* Integrated EMG.

8.7a Reprinted, by permission, from G.F. Harris and J.J. Wertsch, 1994, "Procedures for gait analysis," *Archives of Physical Medicine and Rehabilitation* 75: 216-225.

Integrated Electromyography

An integrator is an electronic device (or computer algorithm) that sums and totals activity over a period of time so that the total accumulated activity can be computed for a chosen time period. If the device is not reset, the totals continue to accumulate. Consequently, at a preset time, the output of the integrator is reset to zero and integration begins again. The term *integrated EMG* has a strict definition and is frequently misused and mistaken for average rectified EMG amplitude or RMS amplitude.

ELECTROMYOGRAPHIC FREQUENCY CHARACTERISTICS

After amplitude analysis, the next most common analytical method involves the frequency characteristics of the EMG signal. This can be accomplished by defining so-called *turning points* and zero crossings or identifying the median or mean frequency, as well as by other techniques that are discussed later.

Turning Points and Zero Crossings

One of the simplest ways to describe the frequency characteristics of the EMG signal is by counting spike peaks. Each time the signal changes direction, a new turning point is created. The number of turning points in peaks per unit time in the EMG is one estimate of the frequency content of the signal. Similarly, the number of times the signal crosses the zero baseline can be counted. This *number of zero crossings* variable is also a valid estimate of frequency content. Turning points and zero crossings are frequently used clinically to describe potential neuromuscular pathologies (Hayward 1983; Ronager, Christensen, and Fuglsang-Frederiksen 1989). Moreover, the number of zero crossings is well correlated with other frequency variables, like those obtained from spectral analysis (Inbar et al. 1986).

Mean and Median Frequency

Spectral analysis techniques are often used to describe the EMG frequency characteristics. In

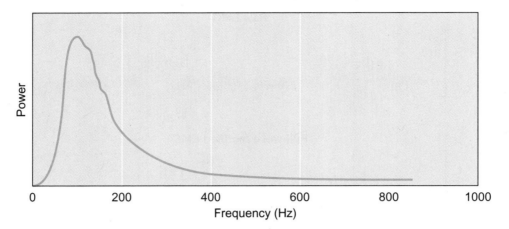

Figure 8.8 Typical frequency spectrum obtained from the surface EMG.

general, the surface-recorded EMG frequency spectrum is positively skewed, with a mean of approximately 120 Hz and a median value of about 100 Hz (figure 8.8). These frequency variables are often indicative of changes occurring in muscle fiber conduction characteristics, and thus they may be better interpreted as markers of peripheral muscular changes than as markers of neural or central drive.

A word of caution: EMG frequency characteristics are misinterpreted and overinterpreted all too often. For example:

- An increase in frequency does *not* necessarily indicate that more fast-twitch motor units are active. It may indicate a higher firing rate for slow-twitch motor units, activation of muscle fibers with higher conduction velocities, decreased motor unit synchronization, additional activation of synergist muscles, or other possibilities.

- Similarly, a decrease in frequency does *not* necessarily indicate an increase in motor unit synchronization. It could indicate a decrease in the total number of active motor units, a decrease in motor unit firing rate, a slowing of conduction velocity, or a change in the intramuscular milieu.

- The analysis of EMG spectral frequency characteristics during dynamic contractions is particularly difficult. To compute spectral frequency content, it is assumed that the signal is stationary—that is, that the frequency content does not change over the analysis interval. During isometric contractions, the stationarity assumption is reasonably well met, particularly for short time intervals. However, the EMG signal obtained during dynamic contractions generally violates

the stationarity assumption. The extent to which the stationarity of the signal is violated depends on the task. In rapid cycling, for example, there might be considerable violation of the stationarity assumption. One solution is to consider the analysis of short epochs, during which the signal would be *quasi-stationary* or *wide-sense stationary* (Hannaford and Lehman 1986). Other solutions currently under investigation involve alternative algorithms, such as the Choi-Williams distribution (Knaflitz and Bonato 1999) and wavelet analysis (see Karlsson, Yu, and Akay 2000). Consequently, analyzing and interpreting the frequency characteristics of the EMG signal during dynamic activity require particular caution.

OTHER ELECTROMYOGRAPHIC ANALYSIS TECHNIQUES

Amplitude and frequency analyses are the most common ways of interpreting the EMG signal, but many other techniques are also used. If only the start and end of muscle activity are needed, onset/offset analysis is suitable. This method plus several others, including the use of polar and phase plots, are presented below.

Onset-Offset Analysis

Frequently, electromyographers are interested in determining when muscle electrical activity begins and terminates. One criterion for determining onsets and offsets is to ensure that the high-frequency components of the signal have *not* been filtered or otherwise attenuated to any appreciable extent. Filtering can delay the identification of the onset-offset time, with the delay varying depending on the high-frequency content at the time of analysis

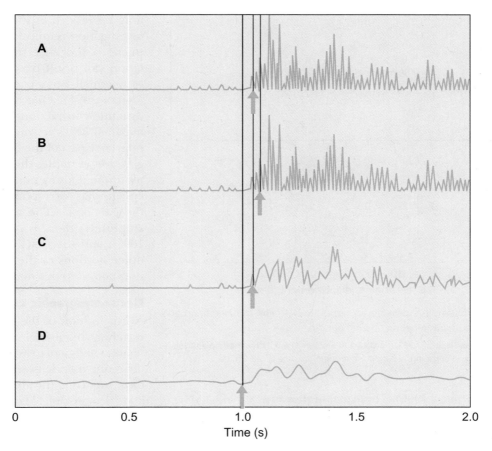

○ **Figure 8.9** Low-pass filtering can have a considerable impact on the identification of EMG onset time. *(a)* Full-wave rectified raw EMG. Note that the onset identification becomes increasingly inaccurate as the signal is increasingly smoothed using low-pass filters from *(b)* 500 Hz to *(c)* 50 Hz and finally to *(d)* 10 Hz.

Reprinted, by permission, from P.W. Hodges and B.H. Bui, 1996, "A comparison of computer-based methods for the determination of onset of muscle contraction using electromyography," *Electroencephalography and Clinical Neurophysiology* 101: 511-519.

(figure 8.9). Many of the algorithms that have been suggested for determining when the signal begins and ends are subjective and produce figures that require reader interpretation. Other methods are more objective, utilizing a threshold EMG activity or a change in the rate of EMG activation (for a review, see Hodges and Bui 1996). Recently, Li and Caldwell (1999) introduced a novel procedure to identify onsets and offsets with respect to kinematic events by using cross-correlation analysis.

Polar Plots/Phase-Plane Diagrams

At times, it is desirable to illustrate EMG changes with respect to a performance measure, such as changes in muscular force or joint displacement. Phase-plane diagrams (figure 8.10) attempt to link the kinematic characteristics of a movement with the resultant EMG activity of both agonist and antagonist muscles (Carrière and Beuter 1990). Polar plots are another alternative. For example, Dewald et al. (1995) asked healthy

and hemiparetic subjects to perform an isometric task that required combinations of elbow flexion-extension, shoulder abduction-adduction, and forearm pronation-supination. Their results illustrated that more EMG activity was observed in the contralateral than in the impaired side in the patients. Many other examples also illustrate the use of polar plots for EMG analysis (Buchanan et al. 1986; Buchanan, Rovai, and Rymer 1989; Chen et al. 1997).

Other Analysis Techniques

Many other analysis techniques have been used to describe and interpret the EMG signal, including wavelet analysis (Karlsson, Yu, and Akay 1999) and autoregressive models (Sherif, Gregor, and Lyman 1981), analysis of cepstral coefficients (Kang et al. 1995), neural network classification (Liu, Herzog, and Savelberg 1999), period-amplitude analysis (Betts and Smith 1979), recurrence quantification analysis (Filligoi and Felici 1999), and fractal analysis

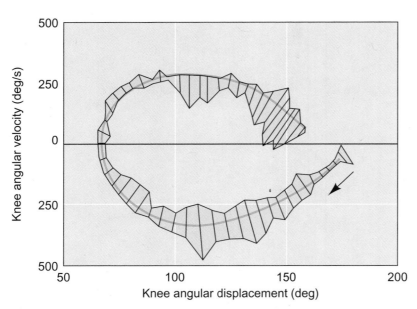

○ Figure 8.10   Phase plots provide an opportunity to link the kinematic and electromyographic characteristics of a movement.

Reprinted, by permission, from L. Carrière and A. Beuter, 1990, "Phase plane analysis of biarticular muscles in stepping," *Human Movement Science* 9: 23-25.

(Gitter and Czerniecki 1995). These techniques and many others being developed have merit and should be considered with regard to the analytical requirements of the specific application.

NORMALIZATION OF THE ELECTROMYOGRAPHIC SIGNAL

EMG data from different subjects, muscles, and days can be compared. In light of the characteristics that determine EMG magnitude, discussed earlier, it is likely that both technical and physiological factors will contribute to variations in EMG magnitude among these conditions. Investigators interested in changes in the EMG signal over the course of some treatment condition should normalize the signal and report changes in that normalized EMG signal; otherwise, inappropriate interpretations can result (Lehman and McGill 1999).

Many normalization techniques are available. Most frequently, subjects are asked to perform a maximal isometric contraction and the magnitude of the EMG signal of interest is normalized to the value obtained during this maximal contraction (e.g., Mathiassen, Winkel, and Hägg 1995). Another approach is to use the maximal M-wave amplitude. The muscle is electrically stimulated to produce the largest response (the maximal M-wave) and the p-p or area of the M-wave is used to normalize the EMG data.

The technique of using EMG amplitude from maximal isometric contractions for normalization may be less than ideal when the investigation requires dynamic contractions. Indeed, errors in interpretation can result from normalizing to maximal isometric contractions (Mirka 1991). One alternative for dynamic contractions is to use the maximal EMG activity during some reference part of the contraction. In gait, for example, the cycle can be partitioned into a number of logical epochs that correspond to phases of the gait cycle. The maximal EMG amplitude, then, is represented as 100%, and the EMG amplitudes in other portions of the gait cycle are normalized to this maximal value.

Electromyographic Reliability

Normalization of the EMG signal is not always required. In a simple investigation designed to assess EMG activity in one muscle group over several days, reporting the absolute value of the EMG magnitude is acceptable. Many studies have demonstrated that absolute value reporting is valid and reliable (Finucane et al. 1998; Gollhofer et al. 1990) and can be more meaningful than a relative score derived using normalization methods. EMG amplitude measurements obtained from submaximal isometric contractions appear to be more reliable than those from maximal contractions (Yang and Winter 1983), suggesting that researchers should also consider the EMG amplitude from submaximal contractions when performing normalization.

APPLICATIONS FOR ELECTROMYOGRAPHIC TECHNIQUES

As discussed in the previous section, the selection of electrodes, amplifiers, and filters, the A/D data-acquisition requirements, and the subsequent analysis procedures depend on the nature of the research question. In this section, several research areas requiring the use of EMG analysis are cited as examples of the kinds of procedures that might be conducted using EMG.

MUSCLE FORCE-ELECTROMYOGRAPHY RELATIONSHIPS

Electromyography can be used to determine the extent of muscular force based on the amplitude of the EMG signal. When a person throws a ball, for example, what external forces are produced, and what forces are placed on the muscles involved? In controlling prosthetic limbs, how much electrical activation should be used to achieve the desired level of force? Such research questions require knowledge of the relationship between EMG amplitude and muscular force.

Under isometric conditions, the relationship between muscular force is frequently linear: Incremental changes in muscular force are produced by linearly related incremental changes in EMG amplitude (Bouisset and Maton 1972; Jacobs and van Ingen Schenau 1992; Milner-Brown, Stein, and Lee 1975). These increases in EMG amplitude are presumably produced by a combination of motor unit recruitment and increases in motor unit firing rate. However, there are many exceptions to this linear EMG-force relationship. For example, a curvilinear relationship is frequently observed (figure 8.11): Small increments in force at the low- and high-force end of the scale may be accompanied by large increases in EMG amplitude (Clamann and Broecker 1979). Because the EMG-force relationship can be affected by the techniques used to process the EMG signal (Siegler et al. 1985), it is important to report the details of the technique used to acquire and process the EMG activity.

Our knowledge of the EMG-force relationship is built predominantly on experiments using isometric contractions. In producing a rapid elbow extension movement, Aoki, Nagasaki, and Nakamura (1986) found that the EMG amplitude during the first 100 ms is linearly related to kinematic features such as peak velocity and acceleration. However, in cycling, the EMG-force relationship may be linear in one plantarflexor—the soleus—and nonlinear in another, such as the gastrocnemius (Duchateau, Le Bozec, and Hainaut 1986).

During repeated contractions involving changes in muscle temperature or fatigue, the EMG-force relationship can be altered (see Dowling 1997). These observations make the assessment of the EMG-force relationship a viable research tool for assessing both central and peripheral contributions to muscular force. For example, the EMG-force relationship can be affected by pathologies such as hemiparesis (Tang and Rymer 1981).

The relationship between muscular force and EMG frequency characteristics is usually nonlinear (figure 8.12). In general, the mean and median power frequency increase rapidly with increases in muscular force to about 20 to 30% of maximal voluntary contraction (MVC) (Hagberg and Ericson 1982).

Thus, analysis of the EMG-force relationship requires an assessment of the neuromuscular and movement features. Issues to be considered include the type of muscle contraction (isometric vs. dynamic), the size of the muscles involved, the potential role of various agonists and antagonists, and the extent to which the EMG recording is representative of muscle electrical activation. Note that the vast majority of the extant literature has considered only the relationship between EMG activity and *external* force production. Although there have been many investigations of *in situ* forces in both human and animal models (Gregor and Abelew 1994; Gregor, Komi, and Jarvinen 1987; Landjerit,

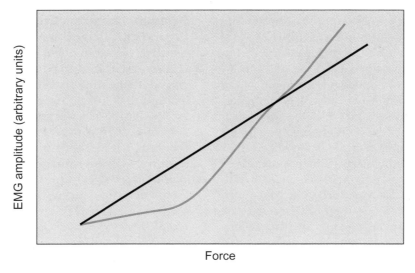

○ Figure 8.11 A linear relationship between EMG amplitude and external muscular force is often observed (black line). However, there are many exceptions in which there is a curvilinear relationship between the two variables.

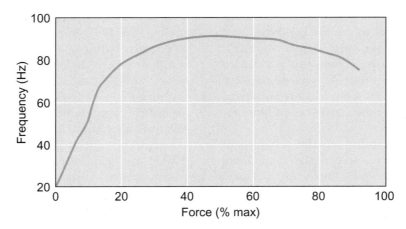

° Figure 8.12   The relationship between EMG mean power frequency and muscular force.

Maton, and Peres 1988), we lack a full understanding of the relationship between these internal forces and muscle electrical activation.

KINESIOLOGICAL ELECTROMYOGRAPHY

Kinesiological electromyographers often want to know if a particular muscle or muscle group is used during some physical activity, and if so, how much. The discussion at the beginning of this chapter revealed that there are a number of issues to be resolved, including the analysis of the EMG during dynamic contractions, crosstalk, and the difficulty of recording from deeper muscles.

Gait

The human gait is a frequent subject for EMG investigation. The order and magnitude of activation among the muscles of the thigh are of interest, and determining the activation order may require an appropriate algorithm and analysis technique for onset analysis. In such a case, the raw EMG signal may be useful in determining EMG onset times. EMG amplitude analysis using the linear envelope may require one of the normalization techniques. One advantage of the gait cycle is that it is repetitive and cyclical, lending itself to analysis over a number of different gait cycles.

In the assessment of pathological gait, we may want to determine if there are distinctive normal or abnormal gait patterns. EMG analysis can be used to discern a subtle gait disorder, such as in the pattern analysis Shiavi et al. (1992) used to describe some of the features accompanying anterior cruciate ligament injury. Other decision-making tools used by electromyographers to detect abnormal EMG patterns include neural network models, applied cluster analysis, and other EMG "expert systems" (Pattichis et al. 1999).

Another common gait problem that lends itself to clinical EMG analysis is *foot drop*. Failure to sufficiently activate the tibialis anterior (TA) prior to the heel-strike leads to this foot "slapping." In such cases, surface EMG records reveal that there is insufficient or no EMG activity in the TA (Kameyama et al. 1990). Thus, EMG analysis can be used to identify or confirm the origin of abnormal gait behavior.

Crosstalk

Because muscle is a volume conductor, electrical signals are transmitted indiscriminately through it, regardless of the origin of the signal. Thus, EMG activity produced in one plantarflexor muscle can readily be detected by electrodes placed over an adjacent plantarflexor. This EMG crosstalk can be considerable (Morrenhof and Abbink 1985) and is likely frequently underestimated. Crosstalk is affected by a number of factors.

Electrode Characteristics

Crosstalk can be greatly minimized by using more selective electrodes. Wire electrodes that have a smaller recording area than surface electrodes do considerably minimize crosstalk. A rehabilitation technique involving proprioceptive neuromuscular facilitation has been used to demonstrate the risks of EMG crosstalk. Earlier studies had shown that muscle EMG activity increases when a muscle is stretched during antagonist muscle contraction. When subjects performed plantar flexion followed by dorsiflexion, surface electrodes recorded the soleus muscle EMG activity during the dorsiflexion phase. However, no EMG activity was observed using wire electrodes (Etnyre and Abraham 1988). What had been thought to be cocontraction of agonist and antagonist muscles turned out to be crosstalk produced by the antagonist dorsiflexor muscles.

Subject Characteristics

Crosstalk increases with the size of the subcutaneous fat layer (Solomonow et al. 1994). Thus, crosstalk may be more frequently observed in muscles surrounded by large layers of subcutaneous fat (e.g., gluteus maximus) and in women and infants.

It should be noted that this crosstalk issue is controversial. Some investigators view the extent of crosstalk from adjacent muscles to be quite small (Winter, Fuglevand, and Archer 1994). Thus, whenever crosstalk may be a problem, the investigator must determine the extent to which it exists.

Identifying Crosstalk

Crosstalk can be identified using several techniques. The motor nerve innervating the antagonist muscle can be stimulated and the EMG inspected visually for evidence of crosstalk activity (Koh and Grabiner 1992). Cross-correlation analysis of two EMG signals (in two different muscles) can be conducted to determine if the two signals are strongly related at some time lag (Etnyre and Abraham 1988).

Minimizing Crosstalk

EMG activity from three surface electrodes placed in a series can be recorded using two conventional EMG amplifiers. The signal from these two amplifiers can then serve as input to a third differential amplifier, resulting in a *double-differential signal* (see figure 8.3). This double-differential recording technique may attenuate signals from distant sources (Koh and Grabiner 1992). Wire electrodes having a small area also can be used to minimize the area of muscle activity being recorded from each pair of electrodes.

ELECTROMYOGRAPHY IN ERGONOMICS

Repetitive stress injuries (RSIs) have become a topic of increasing concern for biomechanists, ergonomists, and other rehabilitation-science professionals. Injuries to the carpal tunnel have been traced to many types of light-manufacturing tasks (soldering, hammering, assembly tasking, and others). More recently, the use of computer keyboards and positioning devices (such as computer mice) that place excessive pressure on the carpal tunnel in the wrist has led to new designs and the need to identify individuals at risk for RSIs. Carpal tunnel problems are often seen in wheelchair users and require modi-

fication of the wheelchair or changes in propulsive techniques (Veeger et al. 1998). Manual laborers exposed to excessive upper-limb vibration or hammering frequently present with symptoms of carpal tunnel syndrome (Brismar and Ekenvall 1992). Biomechanists can use many EMG tools to identify tasks that threaten the integrity of the carpal tunnel.

Sensory and Motor Conduction Velocity

When excessive pressure is placed on any mixed nerve, the ability of both sensory and motor nerve fibers to conduct APs may be threatened. Sensory and motor conduction velocities in the median nerve can be readily assessed using noninvasive EMG techniques. To measure sensory conduction velocity, low-intensity stimuli are applied to the median nerve at the wrist, about 3 cm proximal to the distal wrist crease. These square-wave stimuli are used to antidromically elicit APs in the sensory nerves innervated by the median nerve. Ring electrodes placed around the second or third digit record these APs. Because the signals can be quite small (typically 20 to 50 μV), the average response of multiple trials must be computed using a technique called *spike-triggered averaging* (STA; figure 8.13). In this technique, multiple stimuli are delivered and each response is averaged with all of the previous responses to obtain an averaged signal across many trials. STA assumes that the noise present in each trial is random. Therefore, over a number of stimuli, the noise averages to zero, enabling the low-amplitude signal to be studied with improved resolution. When the stimuli are

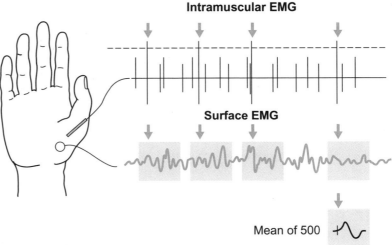

○ Figure 8.13 The STA technique is used to obtain an electrophysiological representation of motor unit size.

Reprinted, by permission, from R.G. Lee, P. Ashby, D.G. White, and A.J. Aguayo, 1975, "Analysis of motor conduction velocity in the human median nerve by computer simulation of compound muscle action potentials," *Electroencephalography and Clinical Neurophysiology* 39: 225-237.

applied, the time between stimulus onset and nerve AP (sensory nerve latency) is determined. As well, the distance between the stimulation site and the recording site is measured so that the nerve conduction velocity (in units of m/s) can be computed. In addition to sensory nerve pathologies, motor nerve conduction velocity can also be deleteriously affected by compression of the median nerve at the wrist. Similar stimulation techniques can be used to activate the motor neurons in the mixed median nerve to identify the presence of a conduction delay.

If these stimuli fail to produce an AP, then a complete conduction block exists between the stimulation and recording sites. Many researchers have published "normal" latency and conduction velocity values to help biomechanists discern the presence of peripheral nerve pathology (see DeLisa and Mackenzie 1982). These conduction measures are important indicators of pathology in electronic-assembly personnel (Feldman et al. 1987).

Other Ergonomic Applications

Surface EMG can also be useful for optimizing the use of the wrist in tasks requiring manual manipulation. For example, in determining the appropriate stiffness for a computer keyboard, Gerard et al. (1999) observed that subjective comfort was greatest for keyboards that had a key activation force of 0.83 N. EMG techniques were used to verify the subjective reports. Other studies comparing subjective reports to surface EMG activity have suggested that fewer upper-extremity disorders might be produced by using a centrally located trackball rather than a mouse positioned on either side of the keyboard (Harvey and Peper 1997). Procedures for counting the number of viable motor units in muscle have been used to determine the severity of carpal tunnel types of disorders (Cuturic and Palliyath 2000). EMG biofeedback techniques can be useful in rehabilitating individuals recovering from carpal tunnel syndrome and other occupational disorders (Basmajian 1989; Reynolds 1994). Similar EMG techniques have been applied in diagnosing back pain disorders (Ambroz et al. 2000; Ikegawa et al. 2000; Lariviere, Gagnon, and Loisel 2000).

ELECTROPHYSIOLOGICAL ASSESSMENT OF MUSCLE FATIGUE

Fatigue frequently accompanies short-term, high-intensity motor activity. Even low-intensity activity can be fatiguing if conducted for prolonged intervals. Consequently, the identification of fatigue is important in the workplace environment. Because fatigue can result from either peripheral (muscular)

or central (neural) mechanisms, the exact site of the fatigue can be determined with EMG techniques. EMG analysis may be a useful tool in redesigning workplace techniques to minimize fatigue.

Electromyographic Activity During Prolonged Isometric Contractions

At the onset of maximal isometric contractions, all motor units are active to the maximal extent. During a sustained maximal contraction, EMG amplitude decreases in parallel with the decline in muscular force. To what extent is this decline a result of either peripheral or central mechanisms? One way of identifying the site of fatigue is to measure the response of muscle to electrical stimulation, thereby eliminating the influence of the CNS. When a high-intensity electrical stimulus is applied to a mixed nerve, the motoneurons are orthodromically activated, resulting in muscle fiber contraction. This EMG response of muscle to electrical stimulation (the M-wave) is frequently used as a measure of maximal muscle electrical activity. The size of the M-wave varies with the muscle size, the number of muscle fibers, and a number of other characteristics.

With prolonged muscle activation, the amplitude of the M-wave usually declines, particularly during long-duration, low-intensity contractions. In some muscle fibers, the AP may cease to propagate along the full length of the fiber, which would explain part of the M-wave decline during fatigue (Bellemare and Garzaniti 1988).

During submaximal isometric contractions, EMG amplitude initially increases (see Krogh-Lund 1993), likely because of the increase in the number of motor units needed to sustain the contraction as contractile failure ensues. Also, the firing rate of newly recruited motor units increases (Maton and Gamet 1989). The CNS's ability to recruit new motor units and increase the motor unit firing rate during the fatiguing task points out the CNS's ability to adapt to changing conditions.

Changes in the frequency characteristics of the EMG signal during fatigue are considerably more complicated. The median frequency of about 100 Hz declines during a fatiguing exercise task, as does the number of zero crossings (Hägg 1981), indicating a greater predominance of lower-frequency activity. There are many sources of this frequency decline, which demonstrates why surface EMG data alone may not be sufficient to identify the mechanism underlying the fatigue phenomenon.

First, muscle fiber conduction velocity declines during fatiguing muscular contractions (Eberstein and Beattie 1985). The decline in muscle fiber conduction velocity results from a number of metabolic factors, including changes in Na⁺ and intramuscular pH (Juel 1988). However, many changes in neural control also occur with fatigue, and these cannot be identified on the basis of the surface EMG alone. Although the motor unit firing rate decreases with prolonged muscle contractions (Bigland-Ritchie et al. 1983), both simulation and experimental studies have concluded that firing rate changes exert little influence on the EMG power spectrum (Hermens et al. 1992; Pan, Zhang, and Parker 1989; Solomonow, Baten et al. 1990). Some researchers have suggested that motor units may be activated in bursts in a pattern called *synchronization,* which could have some advantages. For example, the relationship between motor unit firing rate and muscular force is nonlinear, and a few brief bursts could result in large increases in muscular force (Clamann and Schelhorn 1988). It is possible that synchronization would decrease the EMG power spectrum toward lower frequencies. However, there is as yet insufficient evidence that increased motor unit synchronization occurs with fatigue.

In long-duration tasks, subjects may be able to rotate the required activity among different synergists. In animal experiments, activating muscle compartments sequentially rather than synchronously decreases fatigue (Thomsen and Veltink 1997). During a 1-hour (h) experiment requiring 5% of muscular effort, Sjøgaard et al. (1986) found that different magnitudes of activity occurred in the human quadriceps muscles over the 1 h period, although this observation requires additional examination. Clearly, recording among numerous muscle synergists requires the use of rather selective recording techniques, such as intramuscular wire electrodes.

The idea of discerning with EMG analysis the relative contributions of slow-twitch and fast-twitch muscle fibers during a fatiguing contraction is intriguing. However, because of the myriad factors discussed, it is not possible to state with certainty that the contribution of slow-twitch or fast-twitch fibers is lesser or greater with fatigue.

Changes in EMG activity also follow dynamic contractions. For example, individuals who operate industrial sewing machines exhibit changes in EMG frequency characteristics during the course of a workday, suggesting the onset of fatigue (Jensen et al. 1993). EMG analysis was also used to determine that carrying excessive loads in one hand can result in fatigue and possible occupational injury (Kilbom, Hägg, and Kall 1992).

In similar studies, EMG analysis performed during fatiguing contractions has been useful in understanding post-polio syndrome (Cywinska-Wasilewska, Ober, and Koczocik-Przedpelska 1998; Sandberg, Hansson, and Stalberg 1999), metabolic diseases (Mills and Edwards 1984), chronic fatigue syndrome (Connolly et al. 1993), and other disorders. Using EMG analysis, Milner-Brown and Miller (1990) documented important therapeutic benefits from a pharmacologic treatment in patients with myotonic dystrophy.

SUMMARY

The EMG signal is physiologically rooted in the ionic charges at the muscle fiber membrane. Neural activation produces transient changes in these ionic concentrations, allowing the muscle fiber AP to propagate along the sarcolemma. Many electrodes, amplifiers, and analysis techniques are available for detecting, processing, and describing the signal. Users must be aware of a number of technical issues that can interfere with the characteristics of the EMG signal. Examples of EMG use in areas such as physical rehabilitation, clinical medicine, dentistry, gait analysis, biofeedback control, motor control, ergonomics, and fatigue analysis demonstrate the wide variety of research questions amenable to study using these techniques.

SUGGESTED READINGS

Loeb, J.E., and C. Gans. 1986. *Electromyography for Experimentalists.* Chicago: University of Chicago Press.

Luttmann, A. 1996. Physiological basis and concepts of electromyography. In *Electromyography in Ergonomics,* ed. S. Kumar and A. Mital, pp. 51-95. London: Taylor & Francis.

U.S. Department of Health and Human Services. 1992. *Selected Topics in Surface Electromyography for Use in the Occupational Setting: Expert Perspectives.* Washington, DC: National Institute for Occupational Safety and Health.

CHAPTER 9

Muscle Modeling

Graham E. Caldwell

In chapter 8, we discussed how the control signals used by the nervous system to elicit muscle activity can be monitored using electromyography (EMG). These EMG signals allow us to examine how the nervous system effects purposeful movement, yet they tell only a portion of the story of how human motion is produced. In this chapter, we

- introduce the *Hill muscle model*, named after the illustrious British Nobel laureate A.V. Hill,

- examine how muscles respond to the nervous system's signals to produce the forces that result in skeletal motion,

- consider the basic mechanical properties of muscle tissue that dictate how much force a muscle will produce for a given level of nervous stimulation,

- examine the dynamic interaction between Hill muscle model components, and

- discuss how such models can be used to represent specific individual muscles in *musculoskeletal models* that represent the anatomical and geometrical characteristics of human body segments and joints.

THE HILL MUSCLE MODEL

Muscle is unique because it can convert signals from the nervous system into force, which in turn can cause movement of the skeletal system. It is unfortunate that one of the first terms we learn is "muscle contraction," from which we infer that the muscle shortens in response to nervous input. In fact, the muscle may shorten, lengthen, or remain at a constant length, depending on other internal and external forces acting on the skeleton. One fact is incontrovertible under any loading situation: In response to a nervous signal, the muscle produces force. It would be expedient if the quantitative relation between input (nervous signal) and output (muscle force) were linear (i.e., *x* units of nervous stimulation yield *y* units of muscular force under any conditions). However, in the first half of the 20th century it became obvious to scientists studying muscle that this was not the case. Muscular tissue could produce different amounts of force for a given level of nervous input, depending on the exact circumstances under study. Perhaps the best known of the early scientists is A.V. Hill, who developed a simple but powerful conceptual model of muscle function. Since the 1950s, many important aspects of muscular function have been uncovered, some of which cannot be explained by or even contradict the predictions of the Hill model. For a thorough understanding of muscle mechanics, it is important to study the experimental findings and theoretical considerations of many, such as the sliding filament theory, the Huxley (1957) model, the Huxley and Simmons (1971) crossbridge model, the Morgan (1990) intersarcomere dynamics model, and Zahalak's (1981) distribution moment model. However, the Hill model is appropriate for describing muscle mechanics for the purposes of modeling and understanding most voluntary human movements. Therefore, we will explore the Hill model in some depth, although we encourage readers to investigate these other models to gain a greater understanding of all aspects of muscle function.

The basic Hill model (figure 9.1) consists of the *contractile component* (CC), the *series elastic component* (SEC), and the *parallel elastic component* (PEC). Each

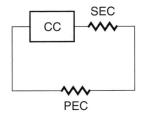

Figure 9.1   The three-component Hill muscle model consists of a contractile component (CC), series elastic component (SEC), and parallel elastic component (PEC).

component has mechanical characteristics that explain specific phenomena seen in experimental studies. It is important to realize that the model represents muscle behavior rather than structure. This means that anatomical correlates to the individual model components do *not* exist, although in some cases we can see how certain muscle structures are responsible for the need for a particular model component. First, we describe the model components separately and then examine how dynamic interactions between the components aid our understanding of muscular force production.

CONTRACTILE COMPONENT

In the Hill model, the CC is the "active" element that turns nervous signals into force. The magnitude of the CC force produced depends on its mechanical characteristics, which can be expressed in four separate relationships: stimulation-activation (SA), force-activation (FA), force-velocity (FV), and force-length (FL).

Stimulation-Activation

The first of the CC's mechanical properties concerns how the nervous system signal (stimulation) is related to the muscle's intrinsic force capability or potential (activation). Physiologically, this property reflects the excitation-contraction coupling process, in which α motor neuron action potentials (APs) trigger motor unit action potentials (MUAPs) that travel along muscle fibers. These MUAPs are carried inward through the transverse tubule system to the sarcoplasmic reticulum, where they cause the release of calcium ions into individual sarcomeres. This portion of the excitation-contraction coupling sequence can be considered the stimulation, because it is independent of the actual force production mechanism in the sarcomere at the level of the crossbridges, which link the thick and thin filaments containing the contractile proteins myosin and actin, respectively. The actin-myosin complex responds to the calcium ion

influx by changing from its resting state (no crossbridge attachment and no force potential) to an activated state in which force production can occur. This is the activation part of the stimulation-activation process. Note that the stimulation represents the input to the process and the activation represents the response, or output.

How much activation is produced for a given input of stimulation? This simple question cannot be easily answered, because it is difficult to quantify either stimulation or activation. As described in chapter 8, the level of stimulation is altered through the processes of motor unit recruitment and rate coding. How does one quantify the "amount" of stimulation that a muscle receives based on these two nonlinear mechanisms? Likewise, in response to neural stimulation, the sarcoplasmic reticulum releases calcium ions that result in activation (force potential) at the level of the crossbridges. Activation should therefore be measured as the number of attached crossbridges that can produce force. But how does one measure this quantity? To get around these quantification difficulties, for modeling purposes both stimulation and activation are placed on relative scales that range from 0 to 100%. The exact shape of the relationship between stimulation and activation is, of course, difficult to ascertain, but for the present we can consider that it is a direct linear relation, although there is evidence from the neural literature that it is in fact nonlinear.

When a motor unit is initially activated, there is a time delay between the onset of the neural AP and the activation at the crossbridge level. This time delay has two components, the first of which is the transit time for the MUAP to travel from the myoneural junction to the sarcoplasmic reticulum. The second component is the length of time for the calcium ions to be released from the sarcoplasmic reticulum and become attached to the thin filaments, a process that, when completed, removes the inhibition for crossbridge attachment imposed by the troponin-tropomyosin complex. When the force response from the motor unit is no longer necessary, the α motor neuron stops sending impulses. However, for a brief period, there is still a supply of calcium ions within the sarcomeres, allowing the crossbridges to remain activated even in the absence of stimulation. The duration of this *deactivation* process is longer than the activation process, and it is dictated by the time it takes for the sarcoplasmic reticulum to reabsorb the free calcium ions within the sarcomeres. The time periods for both activation and deactivation are shown schematically in figure 9.2.

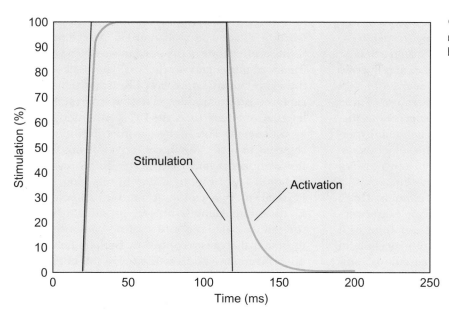

The CC SA temporal relationship. Stimulation shown as black line, activation as blue line.

Force-Activation

In the previous discussion, the term *force potential* was used to emphasize the fact that activation is a state in which force can be produced, rather than an actual force level. The importance of this distinction will become clear later, when we consider that the actual force level in newtons depends not only on activation, but also on the kinematic state of the CC. However, we need to convert the state of activation (in percentage of force potential) to an actual force level, expressed either in newtons or as a percentage of a particular muscle's maximal force. To keep this force independent from any specific kinematic state, the force-activation relation must be conceptual only;

it is impossible to measure the force produced in the absence of a specific CC kinematic state. Therefore, the force-activation relation is direct and linear (e.g., 10, 20, or 50% activation represents 10, 20, or 50% force, respectively).

Force-Velocity

Perhaps the most important CC mechanical property is the influence of CC velocity on force production, a fact well established in the first half of the 20th century. This relationship, shown in figure 9.3, is expressed mathematically by the famous Hill (1938) equation for a rectangular hyperbola:

$$(P + a) = P_o(v + b) \qquad (9.1)$$

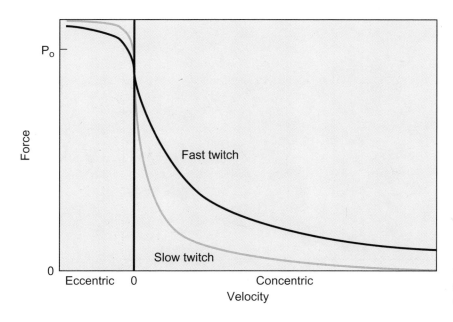

The CC FV relationship, at maximal activation for both slow- and fast-twitch muscles.

In this equation, P and v represent the CC force and velocity at a given instant in time, respectively. P_o represents the force level the CC would attain at that instant if it were isometric. The muscular *dynamic constants* a and b were originally conceived by Hill to represent constants of energy liberation. These dynamic constants are muscle- and species-specific, and dictate the exact shape of the rectangular hyperbola and its intercepts on the force and velocity axes. For example, predominantly slow-twitch muscles have different dynamic constants than those with a high proportion of fast-twitch fibers. Edman (1988) showed that a double hyperbolic shape better represents the concentric force-velocity relation, with a clear deviation from the single Hill hyperbola in low velocity/high force concentric situations. This presents no major conceptual problem for the Hill model, but it does require the use of a different equation to represent the CC's force-velocity characteristics.

Note that the Hill equation refers only to isometric or concentric CC velocities. The relationship can be extended to include CC eccentric (lengthening) conditions (figure 9.3), but the Hill equation must be modified. Finally, the relationship shown in figure 9.3 represents full activation of the muscle, and therefore the maximal force output possible across the range of CC velocities.

Force-Length

Another important CC property is the dependence of isometric force production on CC length, described in 1940 by Ramsey and Street. The best-known descrip-

tion of the force-length relation is in the paper by Gordon, Huxley, and Julian (1966), which proved to be the cornerstone of the well-known sliding filament theory of muscular contraction. The basic shape of the FL relation (figure 9.4) illustrates that isometric force production is greatest at intermediate CC lengths and declines as the CC is either lengthened or shortened. The highest isometric force level is referred to as P_o, which can be confusing because the same acronym found in the Hill force-velocity equation has a slightly different meaning. For the FL relation, P_o is said to occur at the muscle length L_o, the optimal length for force production, which is sometimes called the "rest length"—an unfortunate term that also causes confusion. Both of these points of confusion are addressed later in this chapter.

Contractile Component Properties Viewed Together

The four CC properties (SA, FA, FV, and FL) must be viewed together to understand CC function. When force output is required, the central nervous system (CNS) initiates this process by sending a stimulation signal to the muscle. This stimulation signal causes the activation of the CC, according to the stimulation-activation relationship and the temporal characteristics illustrated in figure 9.2. The CNS modifies the stimulation signal as needed to modify the CC's activation level and therefore its force capability. Thus the CNS exerts control over the CC FL through its regulation of stimulation, but this control is somewhat indirect because of the intermediate influence of the SA

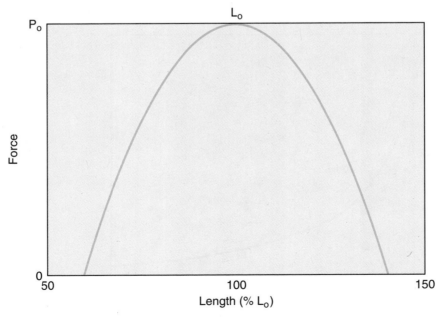

○ **Figure 9.4**   The CC FL relationship, maximal activation.

and FA relations. This control of force is more complex than it at first may seem, considering that the actual force is dictated by the kinematic state of the CC that results from the FV (figure 9.3) and FL (figure 9.4) relations. The combined effect of these two mechanical characteristics can be seen in figure 9.5, which depicts a force-length-velocity surface. Individual points on this surface represent the force output by the CC for any combination of CC length and velocity if the CC activation level is maximal (100%). In many cases, the CNS is operating at submaximal levels, with the actual force for a given CC length and velocity being less than the force depicted on this surface. One way to visualize these submaximal conditions is to consider an activation line segment joining the surface (100%) and the floor (0%) of this three-dimensional (3-D) plot. All submaximal activation levels lie somewhere along this line segment (figure 9.5). The exact CC force at a given time can be determined by first finding the CC length and velocity on the surface and then following the line segment to the current level of activation. Recent evidence suggests that figure 9.5 is too simplistic in that submaximal activation changes the shape of the FL relation, with L_o shifting toward greater muscle lengths (Huijing 1998).

SERIES ELASTIC COMPONENT

Muscle includes materials that display a degree of elasticity and are related to passive connective tissue rather than the active contractile proteins that produce the CC force in response to CNS stimulation. The existence of the series elastic component—elastic elements *in series* with the active force-producing structures in the muscle—has been known since the days of Hill's experiments on frog muscle over 50 years ago. Any force the CC produces is expressed across the SEC. One obvious contributor to the SEC elasticity is the tendon that joins the muscle fibers to the skeleton. Other structures also contribute to the elasticity, including the aponeurosis, or "inner tendon," which connects the tendon to the muscle fibers, and connective elements within the muscle fibers (e.g., Z-lines). Thus, although some authors characterize the CC as the equivalent of muscle fibers and the SEC as the equivalent of the tendon, those assertions are not quite true. Remember also that the SEC is a behavioral model component, and an exact correspondence with anatomical structures is unnecessary; SEC elasticity results from all elastic elements in series with the active force-generating elements in the muscle. Recent work indicates that

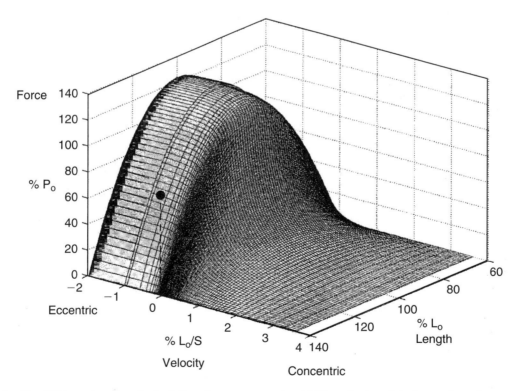

○ **Figure 9.5** The CC force-length-velocity 3-D surface, representing 100% activation. ● indicates force for a particular CC length and velocity. Submaximal activation results in forces beneath the surface; the broken vertical line indicates forces with a continuously decreasing activation level.

the aponeuroses contribute to a large extent to the series elasticity (Kawakami and Lieber 2000).

The SEC elastic behavior is described by its force-extension (FΔL) relationship (Bahler 1967), shown in figure 9.6. In physics, the elasticity of a material is usually quantified by its stiffness, k, calculated as the change in applied force divided by the resulting change in the length of the material $(k = \Delta F/\Delta L)$. If the material displays linear FΔL characteristics, the stiffness is the slope of the FΔL line. However, figure 9.6 illustrates that the SEC FΔL relation is highly nonlinear, which is common for biological materials. The slope of the FΔL relation increases as the SEC extends, meaning that at low force levels the SEC is quite compliant (low stiffness), whereas at higher force levels the SEC becomes stiffer and unit increases in force produce less extension. This nonlinear SEC elasticity is an extremely important feature of the Hill model, one that imparts a powerful influence on the muscular force response. Some experiments indicate that the SEC is lightly damped, meaning that the rate at which the force is applied changes its elastic nature. This has a relatively small influence on the overall model behavior and will be ignored in the present discussion.

PARALLEL ELASTIC COMPONENT

Muscles display elastic behavior even if the CC is inactive and producing no force. If an external force is applied across an inactive, passive muscle, it resists, but stretches to a longer length. This is not a response of the SEC, because no force is being produced by the inactive CC. Instead, this inactive elastic response is produced by structures that are *in parallel* to the CC. The parallel elastic component (PEC) is usually correlated with the fascia that surrounds the outside of the muscle and separates muscle fibers into distinct compartments. Like the SEC, the FΔL relationship of the PEC is highly nonlinear in nature, with increasing stiffness as the muscle lengthens.

The PEC elasticity is considered a passive response, yet it can play a role during active force production, too. In an active isometric force situation, the measured force response is a combination of the active CC force and the passive PEC force associated with the isometric length of the muscle (figure 9.7). Figure 9.7 depicts the FL relations of the active CC, the passive PEC, and the summed CC and PEC responses. At shorter lengths, the PEC is not stretched and thus the muscle force response will be entirely caused by the active CC. As the muscle is placed at longer lengths, the PEC is stretched and its force response is added to the active CC response. The exact shape of the summated response across the range

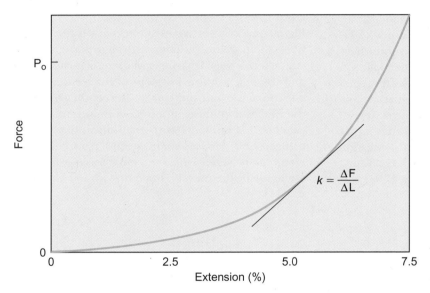

Figure 9.6 The SEC FΔL relationship.

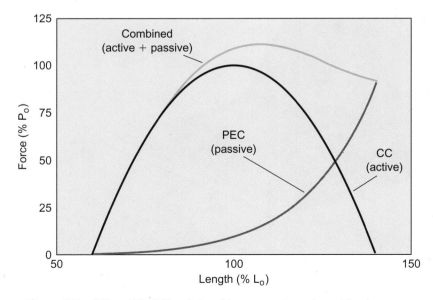

Figure 9.7 CC and PEC FL relationships, separate and combined.

From the Scientific Literature

Alexander, R.M. 1990. Optimum take-off techniques for high and long jumps. *Philosophical Transactions of the Royal Society of London B* **329:3-10.**

Alexander, R.M. 1992. Simple models of walking and jumping. *Human Movement Science* **11: 3-9.**

Selbie, W.S., and G.E. Caldwell. 1996. A simulation study of vertical jumping from different starting postures. *Journal of Biomechanics* **29:1137-46.**

Originally, the Hill model was developed to help elucidate the basic mechanical characteristics of striated muscle function. More recently, it has been used to replicate the mechanical properties of individual muscles or muscle groups during studies of specific human movements. Muscle models can play different roles in these investigations, but in all cases they are used to provide physiological information concerning the muscles involved. The exact manner in which the models are used depends on the specific aims of and questions addressed by the study. In some cases, the muscle models are rotational in nature, representing the total effect of all muscles contributing to a resultant joint moment. In other instances, the models represent individual muscles or a group of synergists.

A good example of a relatively simple muscle model is found in the work of Alexander (1990, 1992). In it, a rigid trunk mass supported by a massless, two-segment (thigh and leg) lower extremity represented the human body (figure 9.8). The trunk center of mass was located at the hip, and a single knee extensor "muscle" controlled the motion of the model. This rotational muscle model generated torque based on the angular velocity of the knee joint and was fully activated at all times. Therefore, the muscle model lacked both elasticity and SA dynamics. Alexander used this model to simulate the push-off phase of both long jumping and high jumping. Two initial conditions concerning the start of the contact phase before takeoff were specified: the center of mass horizontal velocity and angle of the almost fully extended lower extremity. The simulation calculated the torques and motion of the knee joint and center of mass during the contact phase and calculated the height and distance of the subsequent flight phase. After confirming that the model would predict realistic ground reaction forces and jump performances, Alexander used the model to study the effects of systematic alteration of the approach speed and touchdown angle. The model predicted different optimal characteristics for the two jumps and was useful in understanding why high jumpers might use a lower approach velocity than a long jumper. The physiological constraints imposed by the FV characteristics of the knee extensors limited the approach speed to provide the appropriate combination of vertical and horizontal impulses in the two jumping tasks.

Selbie and Caldwell (1996) also used rotational muscle models as torque generators in simulations of vertical jumping. They modeled the body as four linked rigid segments with torque

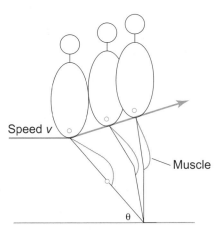

° **Figure 9.8** Alexander's jumping model (1990, 1992).

(continued)

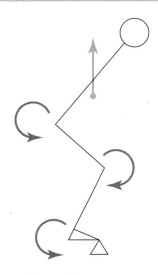

° **Figure 9.9** Selbie and Caldwell's vertical jumping model with three joint torque actuators (1996).

generators acting at the hip, knee, and ankle joints (figure 9.9). At each joint, the muscle model obeyed torque-velocity, torque-angle, and SA relationships, but did not include elastic characteristics. The jumping motion was initiated from a given static posture, and the simulation produced segmental and center-of-mass kinematics throughout the jumping motion until the moment of takeoff. The model kinematics at takeoff were used to calculate vertical jump height using projectile motion equations. This forward dynamics model was optimized to find the combination of hip, knee, and ankle stimulation-onset times that produced the highest jump from a given initial posture. The model was able to produce realistic jump heights and segmental kinematics. The researchers' question was whether the model could generate jumps to similar heights from widely varying initial postures, and the results illustrated that it could from most of the 125 initial positions simulated. However, some discrepancies from actual jumping performances were noted, perhaps because of the simplicity of the muscular representation.

of muscle lengths depends on the relative overlap of the CC and PEC responses.

COMPONENT INTERACTIONS DURING ACTIVE FORCE PRODUCTION

A real understanding of the Hill model can only be gained by considering the dynamic interactions among the components during an active force-producing situation. For this purpose, we will examine a simplified version of the Hill model that includes the CC and SEC but omits the PEC. This is akin to performing muscle experiments at shorter lengths in which the passive PEC plays no role. We begin by considering two often-studied muscular responses, the *isometric tetanus* and *isometric twitch*. These two muscular responses are easily produced in isolated muscle preparations and have been used to characterize muscles as either slow-twitch or fast-twitch, depending on the time course of the force response. The experimental setup consists of an isolated muscle kept viable in a bath of Ringer's solution, with the muscle fixed at one end and attached to a force transducer at the other (figure 9.10). Stimulation in the form of pulses of electric current is applied to the muscle through an electrode, with the pulse width, magnitude, and frequency determined by the muscle and protocol

under examination. For the isometric twitch, a single brief pulse is applied while the muscle length is kept constant. For the isometric tetanus, a continuous train of pulses is applied. The characteristic force responses elicited by these stimulation patterns are shown in figure 9.11. Note that for the tetanus condition, the frequency of stimulation pulses determines whether the force response is an incomplete summation of individual twitches (low frequency) or a smooth, complete tetanus (high frequency).

The shapes of the tetanic and twitch force responses raise a number of questions. Why do the force responses have these characteristic shapes? Why are the force responses for slow-twitch and fast-twitch muscles different? Why does the peak force in the isometric twitch reach only a fraction of the force level produced in the isometric tetanus? Why does the isometric twitch force peak occur so long after the end of the single stimulation pulse? The Hill muscle model can greatly help our understanding of these (and other) force responses.

CC–SEC Interactions During an Isometric Twitch

In response to the brief maximal impulse of stimulation, the muscle force exhibits a relatively slow rise to a submaximal peak, followed by a slow decline back to zero force (figure 9.11). Whereas the stimulation pulse may last only a few milliseconds, the time to

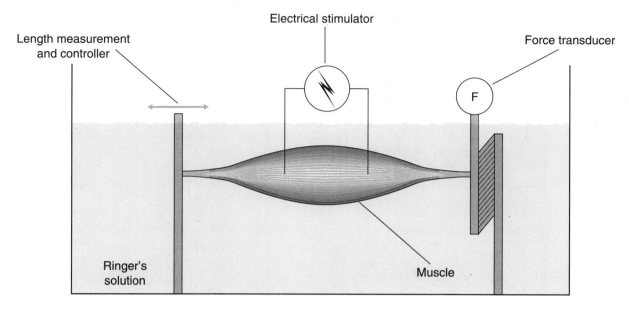

○ **Figure 9.10**   Schematic of isolated muscle preparation.

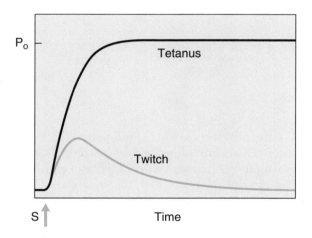

○ **Figure 9.11**   Isometric twitch and tetanus force responses.

peak twitch force is 25 to 50 ms. This discrepancy in time course results from CC and SEC dynamic interaction and the interplay of their mechanical properties. At the beginning, before the arrival of the stimulation pulse, the CC is inactive and producing no force and the SEC is at its unloaded length (figure 9.12a). The stimulation pulse causes the CC to become active and begin to produce force, according to the CC SA, FA, FV, and FL relations. This force is expressed across the SEC, which responds by extending according to its FΔL relation. This basic CC–SEC interaction continues throughout the isometric twitch condition, following a sequence of events associated with the magnitude and time course of the stimulation pulse. The only times that

the relative stimulation and force values coincide are at the beginning (figure 9.12a) and end (figure 9.12d), when both equal zero.

When the stimulation increases from zero to maximal, the CC activation rises according to the time course associated with the transport of calcium ions from the sarcoplasmic reticulum to the crossbridge binding sites on the thin filaments (figure 9.2). As activation rises, the CC produces force according to its instantaneous values of activation, velocity, and length (figures 9.3, 9.4, and 9.5). The force rises throughout the early part of the twitch, meaning that the SEC is continually lengthening until the peak twitch force is achieved (figure 9.12b). Consequently, the CC must be shortening by a concomitant amount, because the total muscle length (CC length plus SEC length) is being held constant (isometric). After the peak twitch force, the force continually falls (figure 9.12c), meaning that the SEC is recoiling (shortening) and the CC is therefore lengthening. Only at the peak, where $dF/dt = 0$, are the SEC and CC instantaneously in an isometric state, neither lengthening nor shortening.

The discrepancy between the rise time in stimulation (almost instantaneous) and in force (25 to 50 ms) could be related to either the time course of calcium dynamics that dictates the SA temporal relation or the ability of the CC to produce force according to its kinematic conditions (length and velocity). During the rising phase of force, the CC shortens (concentric) and during the falling force phase, the CC lengthens (eccentric). Only at the instant of peak twitch force is the CC isometric. Note that the

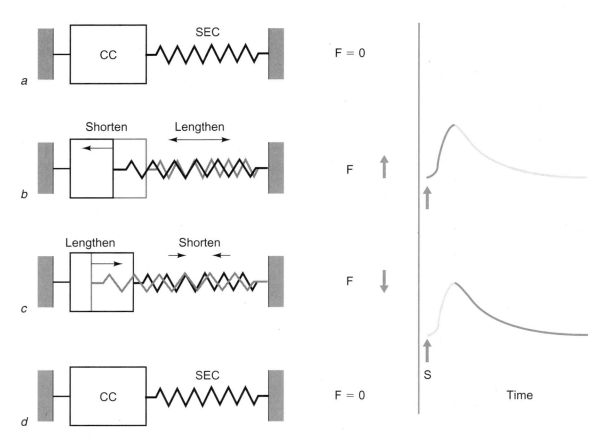

Figure 9.12   CC–SEC dynamic interaction during isometric twitch.

SEC does not directly influence the amount of force produced, but merely responds to the CC force by changing its length according to its FΔL relation.

Active State

Hill found that it was possible to delineate the effects of calcium dynamics vs. those from CC kinematics by describing the time course of the *active state* throughout the isometric twitch. Hill defined active state as the instantaneous force capability of the CC if it is neither shortening nor lengthening. While in this isometric

CC state, only the level of activation and the instantaneous CC length influence the force level attained. Keeping the CC length on the FL plateau eliminates FL effects, meaning that FL is dictated by the activation alone. By using experimental protocols known as *quick release* and *quick stretch*, researchers were able to follow the time course of the active state (figure 9.13a) and thus account for the effects of SA dynamics throughout the isometric twitch. By comparing the time course of the active state with that of the isometric twitch force, it is clear that if the CC is isometric,

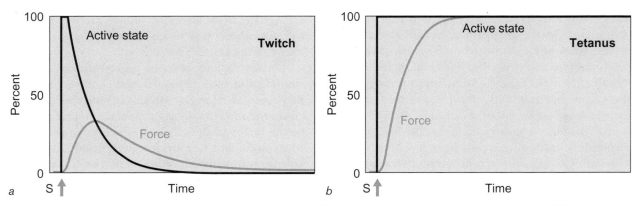

Figure 9.13   CC active state (black line) and force (blue line) response during *(a)* an isometric twitch and *(b)* an isometric tetanus.

it is indeed capable of producing large forces shortly after the onset of the stimulation pulse. The rapid rise in the active state with the onset of stimulation means that the CC is fully capable of producing maximal force almost immediately. Until the time of peak twitch force, the actual force produced by the CC is less than its capability if it were isometric. By deduction, we are left to conclude that the shortening of the CC, associated with the rising force and lengthening of the SEC, imposes a strong depressive effect on CC force production. To fully understand the shape of the force twitch, we must examine more closely the dynamic interaction of the CC and SEC.

Importance of CC–SEC Dynamics

The keys to understanding the shape of the twitch force profile are the sequence of CC–SEC dynamics shown in figure 9.12, the CC FV relation (figure 9.3), and the SEC FΔL relation (figure 9.6). As the stimulation pulse causes the CC activation to rise, the CC begins to produce force across the SEC. In response to this force, the SEC elongates according to its FΔL relation. Because of its highly compliant nature at low force levels, the SEC stretches very rapidly. The CC in turn shortens at the same high velocity, as the CC and SEC velocities must offset each other to keep the total muscle in an isometric state. As a result of the nature of the concentric FV relation, this high shortening velocity keeps the actual force production far below the CC isometric capability. As the force rises slowly, the SEC undergoes more extension, and because of its nonlinear stiffness characteristics, it becomes less compliant. This slows its stretching velocity, which in turn slows the CC shortening velocity and allows the force to rise.

In the isometric tetanus, the activation of the CC rises rapidly to its maximal level and remains there (figure 9.13b) in response to the continual train of stimulation pulses. Therefore, the force continues to rise, bringing the SEC to the stiffest portion of its FΔL curve. This dramatically slows the SEC lengthening and brings the CC shortening velocity toward zero. Eventually, the CC and SEC equilibrate and stop changing length, bringing the CC to its isometric value and the SEC to the constant length associated with that force level. As long as the stimulation pulses keep the CC fully activated, the force will remain at this isometric plateau level.

In the isometric twitch, the story differs in that the activation of the CC begins to decay after the end of the single stimulation pulse, as seen in the falling phase of the active state profile (figure 9.13a). Despite the falling activation, the twitch force continues to rise in response to the slowing of CC shortening. However, the falling activation results in a slower force rise, and

at the peak of the twitch force the active state and twitch force intersect. At this peak, the force is neither rising nor falling, the SEC ceases to lengthen, and both CC and SEC become instantaneously isometric. After the peak in the force profile, the SEC starts to shorten as the force begins to fall. Because of the reciprocal CC–SEC interaction, the CC begins to lengthen and produce force according to its eccentric FV relation. This keeps the force higher than the active state that reflects the CC isometric capability. However, because the activation continues to fall, the force level also drops. This falling, submaximal, eccentric CC state continues until the activation and force return to zero at the end of the twitch contraction.

General CC–SEC Interactions

Whenever a muscle is stimulated, either externally or by the CNS, the CC–SEC interaction has a direct bearing on how much force is generated. In the description of the isometric twitch and tetanus, we noted that the CC–SEC interaction can be understood using the relationship

$$muscle\ length = CC\ length + SEC\ length \qquad (9.2)$$

In these isometric conditions, the muscle length is constant, emphasizing the inverse relation between the CC and SEC lengths during isometric muscle contractions. Another kinematic relation that expresses the CC–SEC interaction is

$$muscle\ velocity = CC\ velocity + SEC\ velocity \qquad (9.3)$$

In the isometric case, muscle velocity equals zero, meaning that the CC and SEC have equal but opposite velocities during an isometric situation, and emphasizing the importance of the nonlinear SEC FΔL and CC FV relations in dictating muscle force-time profiles.

These two equations hold true during nonisometric muscle contractions as well, although now the constraints of constant muscle length and zero muscle velocity are removed and the CC and SEC kinematics lose their inverse, equal-but-opposite relationship. The SEC always serves to mediate the CC velocity so that, in general, the muscle velocity is not equal to the CC velocity. However, when the muscle changes length, it is much more difficult to envision the exact nature of the CC velocity than in the isometric case. For example, it has been shown experimentally that during eccentric muscle contractions, the muscle fibers (a crude representative for the CC) may either shorten or lengthen depending on the exact state of the elastic elements in series (Biewener, Konieczynski, and Baudinette 1998; Griffiths 1991). In general, the only way to predict the CC length and velocity during active contractions is by using a two-component Hill muscle model.

EXAMPLE 9.1

MUSCLE MODEL ALGORITHM

A Hill-type muscle model can be implemented in several ways. The most flexible arrangement is to write a software subroutine that contains the code for the muscle model in a general form. The model for any specific muscle is then implemented by writing a calling subroutine that contains the model parameters (i.e., P_o, FL information, a and b dynamic constants, SEC elasticity, and so on) for that particular muscle. The calling subroutine passes the specific parameters to the general model subroutine so that the model output represents that individual muscle. This permits the general model subroutine to be used for more than one specific muscle, which is advantageous when using a musculoskeletal model in which several different muscles are represented and are required to produce force simultaneously. This scheme can use the muscle model algorithm and software language of the researcher's choice.

The iterative algorithm for the Hill model that is presented here (Baildon and Chapman 1983; Caldwell and Chapman 1989, 1991) has explicit equations representing the CC and SEC properties, giving the user a sense of how the CC and SEC interact dynamically (figure 9.14). It is similar in concept (but not necessarily implementation) to others found in the literature (van den Bogert, Gerritsen, and Cole 1998; Bobbert, Huijing, and van Ingen Schenau 1986; Pandy et al. 1990; Winters and Stark 1985; Zajac 1989). The description of the algorithm indicates exactly where functions representing the CC and SEC properties

- For each time step Δt

    Need predicted muscle forces for N different muscles
    ($i = 1, 2, \ldots, i + 1, \ldots N$)

    - For muscle i

        Get muscle parameters for CC and SEC
        Get muscle length and stimulation inputs
        Get CC activation and length from previous time step
        Get SEC length and force from previous time step

        Call muscle model subroutine

        1. SEC force = f (SEC length)
           CC force = SEC force
        2. CC activation = f (stimulation)
        3. CC velocity = f (CC force, activation, length)
        4. CC length = ∫ CC velocity dt
        5. SEC length = muscle length — CC length

        Repeat for next muscle i + 1

    Repeat for each time step Δt

○ **Figure 9.14**   Flowchart of muscle model algorithm.

are needed, and table 9.1 gives examples of specific equations for these functions (Caldwell and Chapman 1991). Researchers building their own models may want to use different equations for these component properties.

The model subroutine needs two types of inputs: (1) model parameters to specify the particular muscle of interest, and (2) time histories of the muscle kinematics and stimulation for a given movement situation. For example, we can attempt to simulate an isometric twitch experiment for a rat gastrocnemius or human

Table 9.1   Model Equations and Parameters

Model component	Relation	Equation	Parameters
CC	FV concentric	$(P + a) = P_o(v + b)$	P_o = maximal isometric force a = Hill dynamic constant b = Hill dynamic constant
	FV eccentric	$S = b(P_s - P_o)/(P_o + a)$ $P = P_s - S(P_s - P_o)/(S - v)$	P_s = eccentric force saturation level (% of P_o)
	FL (parabola)	$RF = [c0(RL - 100)^2] + 100$	c0 = parabola width coefficient
SEC	FΔL	$RF = 0.0258 [\exp (stiff) (RLS)] - 0.0258$	stiff = nonlinear stiffness coefficient

P = force; v = CC velocity; RF = relative force; RL = relative CC length; RLS = relative SEC length.

(continued)

tibialis anterior. In either case, we need specific muscle parameters (more easily obtained directly for the rat gastrocnemius), such as its fiber-type composition, maximal isometric force output (P_o), and so on. For the isometric twitch conditions, the overall muscle length is held constant at a specified length for the entire time of simulation (perhaps 300 ms). The stimulation history is represented by a series of zeros interrupted for only a few milliseconds during the supramaximal pulse (see figure 9.11), when the stimulation value equals 100%. The model subroutine uses the muscle parameters and the kinematic and stimulation inputs to calculate histories of the force output and the CC and SEC dynamics that produced this force profile.

CALLING SUBROUTINE

1. Muscle parameters:
   CC: P_o, L_o, width of FL parabola, FV dynamic constants a and b, activation constants for temporal SA relation
   SEC: FΔL relation
2. Experimental conditions: Muscle kinematics and stimulation inputs

For each time point of the simulated contraction, specify the muscle length and input stimulation value. For this example, we will assume that we have 3,000 time steps so that time t will progress from $t = 0$ to $t = 3,000$ in steps of 1, each time step indicating a change of 0.1 ms for a total simulation time of 300 ms.

MUSCLE MODEL SUBROUTINE

For each time point of the simulated contraction, starting from $t = 0$:

1. Use the SEC FΔL relation to predict the instantaneous CC force. Because the CC and SEC are in series, the CC and SEC forces are equal. The SEC FΔL equation must therefore be written in the form

$$\text{SEC force is equal to} \atop \text{a function of its length, or} \atop \text{SEC force = f(SEC length)} \quad (9.4)$$

Note that for $t = 0$, the stimulation and activation equal zero, the CC is producing no force, and the SEC is at its unloaded length. Therefore, the SEC force will be predicted to be zero, as is consistent with CC "reality."

2. Use the CC FV relation to predict the CC velocity from the force predicted in step 1. Recognize that the CC force depends on both the CC FL and FA relations (see figure 9.4). Therefore, first calculate

and account for the (possibly) submaximal activation level of the CC, using the input stimulation value. Remember that there is both a magnitude and a temporal relation (figure 9.2) between stimulation and activation. Also account for the CC length and its effect on the P_o value used in the CC FV relation. Thus, the CC FV equation must be written in the form

$$\text{CC velocity = f(CC force,} \atop \text{activation, and length)} \quad (9.5)$$

Again, for $t = 0$, the stimulation and activation equal zero, so the CC is "passive" and producing no force. The equation will therefore work on the "zero activation" CC FV curve and predict a CC velocity of zero. Because the SEC has zero force exerted across it and the experimental conditions call for a constant muscle length, the SEC and CC are both isometric.

3. Use the predicted CC velocity from step 2 to predict the CC length at the next time increment. This numerical integration allows the simulation to progress from one time point to the next (in this case, from $t = 0$ to $t = 1$). For the simulation to be accurate, the time increments must be small (0.1 ms in this example). This is the same integration process described in chapter 10 for movement simulation.

$$\text{CC length} = \int \text{CC velocity } dt \quad (9.6)$$

With the CC velocity at zero at the beginning of the contraction, the CC length change is also equal to zero.

4. Use the predicted CC length from step 3 to estimate a new SEC length for the next time point ($t = 1$). Here, use the muscle length input from the calling subroutine along with the predicted CC length. Because the CC and SEC are in series and make up the total muscle length,

$$\text{SEC length = Muscle length − CC length} \quad (9.7)$$

At $t = 1$, given the CC and SEC dynamics described in steps 1 through 3, this step will predict the same SEC length as before (SEC unloaded, force = zero).

5. With the new SEC length for the next time point from step 4, return to step 1 with the time increment shifted to $t = 1$. The steps in the model algorithm are repeated continuously until $t = 3,000$ to give values for CC force, CC length, CC velocity, and SEC length at each time step throughout the entire contraction sequence. These values are consistent with the equations that dictate the behavior of each component of the Hill model.

(continued)

How does this algorithm produce force profiles consistent with experimental results such as those in an isometric twitch or tetanus? As the time moves forward from $t = 0$, soon the time increment representing the stimulation pulse is reached (stimulation becomes 1 instead of zero). The stimulation pulse produces a nonzero activation level that puts the CC on a submaximal FV relationship, and the equation in step 2 now predicts a nonzero CC shortening velocity. When the CC velocity is integrated (step 3), the CC length becomes shorter. In step 4, this shorter CC length results in SEC lengthening. The longer SEC length is fed back into step 1 and therefore predicts a nonzero SEC force. The CC force (equal to the SEC force) goes forward into the CC FV equation in step 2, along with the new activation level. This predicts a new CC velocity that in turn produces more shortening of the CC (step 3) and a longer SEC length (step 4). The cycle repeats as stimulation and activation change, and CC force climbs as the activation level increases. In the tetanus, the activation becomes maximal and stays there, resulting in a continuous rise in force along with a continually lengthening SEC and shortening CC (as force rises). When the force plateaus at the P_o level for the current CC length, the CC and SEC both become static, with the SEC extended and the CC at a shortened length. For the twitch condition, the stimulation becomes zero after the pulse, which causes the activation to fall back to zero. This activation drop causes another set of changes through the submaximal FV relationships seen in step 2, resulting in CC and SEC dynamics that produce the characteristic twitch force response.

Implement this technique by writing the computer code for the general muscle model and then testing the model using a calling subroutine set up to mimic a specific muscle under isometric twitch and tetanus conditions. The model subroutine should replicate these well-known force responses. Once the model has passed these validation tests, other well-documented conditions should be implemented (e.g., trains of stimulation pulses at different frequencies to produce unfused and fused tetanus, isovelocity contractions at different shortening and lengthening speeds, inertial contractions with different inertial loads). The model should also react correctly when the model parameters are altered (i.e., slow-twitch vs. fast-twitch dynamic constants, alteration of P_o, location of L_o, and so on). After demonstrating the validity of the model, move on to the more challenging task of implementing the model subroutine for musculoskeletal models in which each muscle of interest is represented by a specific version of the general Hill muscle model.

MUSCULOSKELETAL MODELS

In chapter 5, the concept of resultant joint moments was presented in the description of inverse dynamics movement analysis. During a movement sequence, the joint moments represent the calculated rotational kinetics that must have been present to generate the observed motion of body segments. This calculation implies that a separate entity is responsible for the moment at each joint involved in the motion. In reality, the human body consists of a multitude of individual muscles that span one or more joints of the rigid, bony skeleton. For example, the human lower extremity can be represented as a series of articulating bones (pelvis, femur, tibia, fibula, calcaneus, talus, and others). The bones form the basis for modeling the body as a set of linked rigid segments. Overlying these bones are individual muscles (vastus lateralis, tibialis anterior, soleus, and so on) that produce the forces that cause the resultant joint moments and, therefore, the movement of the skeleton and body segments. As discussed in chapter 8, muscle forces are produced in response to control signals arising from and delivered by the nervous system. Inverse dynamics modeling allows one to estimate the resultant moment at a joint that arises in response to the sum total of all muscle forces, but it cannot resolve the joint moment into individual muscular forces.

If the body is viewed as a mechanical movement system, it is clear that the system is redundant in that it has more muscles than are necessary to produce any given lower-extremity segmental motion. For example, only one muscle is needed to produce the movement of knee flexion, but the human body has multiple muscles (biceps femoris, semitendinosus, semimembranosus, gastrocnemius, popliteus) that directly produce this movement. Exactly why the system is overdetermined is a matter of some debate among movement scientists in the fields of biomechanics and motor control. Other questions of interest include why some muscles are biarticular (span two adjacent joints) whereas others are monoarticular (van Ingen Schenau 1989); why some muscles have long, thin tendons whereas others have short, thick ones (Biewener and Roberts 2000; Caldwell 1995); and what contributions individual muscles make during specific movements (Anderson and Pandy 2001; Zajac, Neptune, and Kautz 2002, 2003).

To fully understand skeletal movement during a movement sequence, it would be useful to measure the performance of each of these constituent parts (the control signals to each muscle, the resulting muscle forces and the moments they provide at each joint, and the movement of each body segment). Although EMG data can be used to estimate the muscle control signals in some cases and body segment motion is routinely measured, with today's technology one cannot easily measure muscle forces directly in humans. In this context, muscle models can help our understanding of musculoskeletal function in several ways. Rotational muscle models representing the properties of all muscles crossing a joint can be useful either in understanding calculated joint moment patterns or as simple torque generators in forward dynamics simulation models (chapter 10). Alternatively, muscle models can be used to represent the function of individual muscles and also can include biarticular muscles, which may be important for coupling the actions of adjacent joints. Direct muscle force measurement is possible in animals, and this approach has been used to investigate the force predictions from muscle models (e.g., Herzog and Leonard 1991; Perreault, Heckman, and Sandercock 2003). There are isolated human examples of direct muscle force measurement (e.g., Gregor et al. 1991; Komi et al. 1996), but such invasive techniques are unlikely to be widely used in human subjects.

To estimate the muscle forces during a motion sequence, first construct a representation of the entire movement system; this is known as a *musculoskeletal model*. Such a model should accurately represent the anatomy of the skeleton and the inertial characteristics of each body segment, along with neural and muscular systems controlling the skeletal motion. As with the Hill muscle model, the model's constituent parts are represented in software through mathematical expressions. For example, the motion of the skeleton is described by the equations of motion applied to each body segment, with angular segmental motion being produced by torques acting at proximal and distal joints. If the model is used in an inverse dynamics approach, the motion of the segments is controlled by estimated resultant joint moments, and additional modeling is necessary to distribute the calculated joint moment into its constituent individual forces from muscles, ligaments, and joint surfaces. Forward dynamics models, which simulate body segment motion from given kinetic inputs, are driven either by joint torque actuators or individual muscle actuators. In either case, muscle models of some description can represent the physiological attributes of these kinetic actuators.

It is clear that resultant joint moments are mathematically associated with the (unknown) forces produced by the individual muscles. These unknown muscle forces can be estimated with a musculoskeletal model that uses a Hill model to represent each specific muscle. The mathematical relation between a given muscle force and a given joint moment is dictated by the geometrical relation between the muscle's position and the center of rotation of the joint in question. The individual muscle's force output is determined in part by the stimulation it receives from the nervous system, its so-called *control signal*. This control signal can be generated from an experimentally derived EMG signal or from a theoretically derived control algorithm. In the following sections, we outline the basics of a musculoskeletal model.

JOINT/SKELETAL GEOMETRY

From an anatomical viewpoint, the skeleton is the framework that physically supports the muscles. An accurate skeletal model permits the correct assignment of muscle origin and insertion locations, as well as the path that the muscle follows between its origin and insertion site. This information is often drawn from cadaver studies in the literature and then scaled to the size of a given individual. Inaccuracies in model predictions could arise from (1) unknown variations in skeletal shape between individuals, (2) unknown variations in muscle origin and insertions between individuals, and (3) inaccurate scaling. These problems can be overcome by using imaging techniques such as MRI to directly obtain subject-specific data.

From a mechanical viewpoint, the skeletal geometry is important because the forces transmitted by the muscles to the skeleton have direct bearing on the linear and angular motion of the skeleton and individual body segments. As the muscles run their course from origin to insertion, they pass over the joints formed by adjacent bones abutting. The mechanical nature of each joint is determined by the local skeletal geometry of the contact surfaces between the bones. For example, at the hip, the head of the proximal femur and the pelvic acetabulum form a stable union that allows rotation of the femur with very little translation of its head. Mechanically, this can be modeled as a ball-and-socket joint with three rotational degrees of freedom (DOF). This bony union dictates the mechanical nature of soft tissues such as ligaments and muscles crossing the joint, for these tissues must create forces that provide for both skeletal motion and joint integrity. Clearly, the motion and stability afforded by the bony surfaces play a role in the degree to which the muscles and ligaments must provide additional support. For example, the relatively flat surfaces of the distal femur and proximal tibia grant much less bony stability at the knee, and it is no surprise that the ligaments and muscles there must provide a much higher degree of joint stability than their counterparts do at the hip. In 3-D, the knee should be viewed as having six DOF (three rotations and three translations). However, because motion is restricted in some directions, it is sometimes modeled simply as a planar hinge with some anterior/posterior translation, or even as a basic one DOF hinge.

MUSCLE KINEMATICS AND MOMENT ARMS

Muscles change length as a function of angular changes in the joint or joints they cross (figure 9.15). The exact degree of the length change depends on the architecture and geometry of each bone and joint. The ability to model these length changes is important for ascertaining the muscle kinematics needed to determine how much force a given muscle can produce. One technique is to express the length for each muscle as a function of the angular displacement for all joints spanned by that muscle. For example, Grieve, Pheasant, and Cavanagh (1978) used measurements from cadavers to estimate the muscle length changes for the soleus and gastrocnemius at a range of given angular positions for both the ankle and knee joints. Their results were used to form equations that predict muscle length, normalized to segment length, as a function of joint angles. Their data are easily applied

to other studies by scaling to specific subjects based on measured segment lengths. Another method is to measure the muscle length at a series of joint angles and form subject-specific equations directly using an imaging technique such as MRI. This method has the advantage of accounting for variability in individual subject anatomy, rather than assuming similar muscle length-joint angle relations for all subjects.

A different approach is to calculate the desired muscle lengths directly from the instantaneous positions of the skeletal origin and insertion points during the movement under study. The first step is to measure the locations of these points on the bones of a skeleton. During a motion, the positions of the points can be computed by mathematically imbedding the bones into the body segments that define the biomechanical model. Because the segment positions in space are defined throughout a movement by digitized marker locations, the bone and muscle attachment positions can be calculated using the transformation techniques described in chapters 1 and 2. A simple approach is to model the muscle kinematics based on a straight line from the origin on one bone to the insertion on another. This approach fails, however, when a muscle is constrained in its motion because it overlies other muscles or wraps around a bony prominence in the course of its path. For some muscles, this wrapping effect can be severe at a joint, particularly in highly flexed positions. A more complex model views the muscle path as a set of line segments that run between constraining *via points* that represent anatomical locations where the muscle motion is restricted (e.g., the location where the path of the vasti muscles are constrained by the patella). This

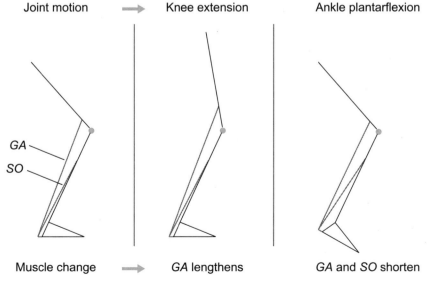

° **Figure 9.15**   Muscle lengths change as a function of joint angles. GA = gastrocnemius; SO = soleus.

method can also give better estimates of the bone-on-bone forces acting at each joint, because the locations of muscle attachment points and the orientation of the tendon can be computed, permitting calculated muscle forces to be added to segmental free-body diagrams.

Another consideration is the muscle and bone geometry in the local vicinity of the joint, a factor that determines the mechanical consequence of a muscle's force production. If the joint allows some translation, the direction of the muscle force may accelerate one of the adjacent bones with respect to the other. If the joint allows rotation, the location of the force line of action dictates how much muscle torque is produced about the joint axis of rotation according to the equation $\tau = Fd$. This rotational torque (τ) or moment of force varies with the muscle moment arm, the perpendicular distance (d) between the force's (F) line of action and the axis of rotation. Because the origin and insertion of a muscle are fixed on the adjacent bones that make up a joint, as the joint angle changes, the moment arm also changes. This partially accounts for the variation in joint moment potential as a function of joint angle that is commonly seen in human muscle strength studies (chapter 4).

There are various ways to determine moment arms for use in musculoskeletal models. For simplicity, the moment arm for a muscle can be quantified with a static measurement from a cadaver or skeleton and assumed constant at every joint angle. Although this may be reasonable when the joint range of motion is limited, in general the moment arm changes as the joint angle is modified. As with muscle length changes, moment arm data can be drawn from a series of measurements from a subject or cadaver while the joint is manipulated through a range of angles. Another method called *tendon excursion* calculates the moment arm based on the amount of tendon length change observed during joint alteration. The measured or calculated moment arms are expressed as a mathematical function (polynomial, quintic spline, or others) of joint angle that can be incorporated into a musculoskeletal model. Alternatively, if the muscle paths in the model are represented by their origins, insertions, and constraining via points, the moment arms can be computed during a motion by using the locations of joint centers and muscle lines of action.

MUSCLE ARCHITECTURE

Each muscle is unique in its morphology and its ability to produce force under different situations. A variety of studies have linked the morphological and physiological characteristics of muscles with their mechanical properties, so estimates of appropriate values for muscle models can be made from measurable physiological characteristics. Together, the characteristics that influence the mechanical force-producing properties of muscle are known as *muscle architecture*. Certain architectural parameters can be used to characterize individual muscles for modeling purposes, including pennation angle, physiological cross-sectional area, fiber length, fiber-type composition, tendon length, and tendon elasticity. These architectural elements are important because they affect the mechanical properties of the Hill model CC and SEC.

Pennation Angle

Muscles are sometimes characterized according to the direction in which their fibers are oriented. The fibers of fusiform muscles run in the same direction as their tendon, parallel to a line joining the muscle origin and insertion. In contrast, pennate muscles have fibers oriented at an angle to the tendon, and therefore they produce force with vector components both parallel and perpendicular to the tendon. The obvious consequence of this pennation angle, θ_p, is that when the fibers produce X newtons of force, only $X \cos \theta_p$ newtons are transmitted via the tendon to the connected bones. If the angle θ_p is small, this effect is not dramatic (e.g., $\cos 10° = 0.985$). However, in some situations (short fiber lengths, very compliant aponeurosis), the fibers rotate and the pennation angle increases dramatically, so the ability of the fibers to provide force in the intended direction may be severely compromised. Also important in pennate muscles is the phenomenon of fiber packing. A pennate muscle has shorter fibers than a fusiform muscle of equal volume, but more of the fibers are arranged in parallel. Thus, the pennate muscle can produce more force than a similar fusiform muscle, which suggests that pennate muscles are "designed" for high force production.

Cross-Sectional Area

The concept of fiber packing can be understood by considering the arrangement of sarcomeres within a muscle. Sarcomeres arranged in series, such as in a single myofibril, transmit force only to the extent that each individual sarcomere can support. If each sarcomere produces 1 N of force, 100 sarcomeres in series would produce a total of 1 N. In contrast, sarcomeres arranged in parallel each transmit their own force so that the more sarcomeres there are in parallel, the greater the force transmitted. The same 100 sarcomeres arranged in parallel could produce a total force of 100 N. A muscle's cross-sectional area (CSA) indicates the number of fibers in parallel, and thus can be used as an indicator of the muscle's maximal force capacity. Several factors must be considered when predicting a muscle's force capability in this way.

One of these factors is muscle shape, because the muscle belly's girth is often larger than the girth at either end. At what point along the length of the muscle do you measure the CSA? One possibility is to take several girth measures in an attempt to get an "average" CSA. However, this assumes that all of the fibers run in parallel and are oriented with respect to the muscle origin and insertion, which is not true for pennate muscles. To account for pennation angle and the fact that not all fibers run the entire length of a muscle belly, the term *physiological cross-sectional area* (PCSA) was introduced. PCSA is calculated by dividing muscle volume by fiber length. Originally, PCSA was applicable only to animal or cadaver studies in which these measurements were possible. However, the advent of ultrasound and MRI made it possible to obtain this information from living human subjects as well.

Another factor to consider when using the PCSA is the muscle FL relation. When a muscle fiber changes length, its isometric force capabilities are modified by the alteration in the myofilament overlap. At what muscle length should measurements be taken for the calculation of PCSA? The obvious answer is the length at which the fibers are at L_o, the optimal force-producing length. However, it is difficult to know the optimal length for any given muscle, and PCSA data taken from cadavers were almost certainly obtained with the fibers at a nonoptimal length. Considering all of the factors outlined in the preceding discussion, it is obvious that the maximal isometric force (P_o) capabilities of individual human muscles are not well established, despite the straightforward observation that force production is dependent on muscle size.

Fiber-Type Composition

It is well known that muscle fibers and motor units vary in histochemical and mechanical characteristics. For many musculoskeletal models, it is sufficient to recognize two types: slow-twitch (ST) and fast-twitch (FT) fibers. A further distinction between fatigable and fatigue-resistant FT fibers may be warranted in models that consider the metabolic consequences of a motion. The importance of the ST/FT dichotomy arises from their separate FV characteristics and the modifications in force-time profiles that result from these differing characteristics. The time for force to rise during an isometric tetanus is longer for an ST muscle than for a comparable FT muscle. In isometric twitch conditions, an ST muscle also exhibits a reduced peak twitch force as a consequence of the delayed time at which peak force occurs.

Fiber-Tendon Morphology

Each muscle–tendon unit in the human body is unique in the relative length of the muscle fibers compared to its tendon. This architectural detail dictates the relative excursion of each portion during a movement sequence. The number of sarcomeres in series determines the length of the fibers. Because each sarcomere can shorten across its FL range by about 2.2 μm (from 3.65 to 1.45 μm), the total number of sarcomeres in series dictates the maximal excursion the fiber can undergo. If a fiber contains more sarcomeres, it can undergo a greater total excursion. The fiber length also determines the absolute fiber velocity during a movement and therefore has a significant impact because of the FV relation.

The length of the tendon is important for similar reasons, because of the amount it will extend as force is applied across its length. Some tendons are short and thick, while others are long and slender. These morphological features affect overall tendon elasticity and seem to be correlated with muscle–tendon function. For example, more-distal extensor muscles of the lower extremity often have long, compliant tendons, which allow the fibers to undergo less excursion. This design may reduce the metabolic cost during locomotion, because the fibers have lower contractile velocities. More-proximal muscles have shorter, thicker tendons attached to relatively long fibers that have large excursion capabilities. It has been suggested that these muscles are ideally designed for producing large amounts of mechanical work.

MODEL PARAMETER ESTIMATES

Differences in morphology and fiber-type composition suggest that individual muscles have unique mechanical characteristics. The values used within muscle models to represent specific properties are known as *muscle parameters*. At the level of individual joints, the properties of interest are the combined torque-angle and torque-angular velocity relations for the muscle groups crossing each joint. Measuring the torque capabilities under different kinematic conditions establishes these isometric and dynamic relationships for each joint in the musculoskeletal model. This is routinely done with an isovelocity (isokinetic) dynamometer that can control the isometric position or joint velocity while a subject applies maximal effort torque (chapter 4). A shortcoming of this approach is that the action of biarticular muscles that couple the torque capabilities at adjacent joints cannot be fully appreciated.

FL and FV properties of specific human muscles that are needed for Hill-type muscle models are difficult to ascertain because of the difficulty in measuring individual muscle forces. Nevertheless, modeling studies in the literature report these properties for individual muscle actuators in data usually estimated from muscle morphological and architectural considerations. FL characteristics can be estimated by considering the number of sarcomeres in series in a fiber and extrapolating the sarcomere FL relation to the fiber as a whole (Bobbert, Huijing, and van Ingen Schenau 1986). This method has the advantage of predicting the optimal force length, L_o. However, there are relatively few estimates of sarcomere number for any given human muscle, and this value may widely vary between individuals. Another method for estimating FL parameters is to extend animal results (Woittiez, Huijing, and Rozendal 1983) to humans based on the *index of architecture* (Ia) of a muscle, which is defined by the ratio of fiber length to muscle belly length. In general, the pennation angle dictates the length of muscle fibers in relation to the muscle belly length. Woittiez and colleagues found a strong association between the width of a muscle's FL relation and its Ia, with higher Ia (i.e., less pennate) values indicating a wider FL relationship. In this case, the exact L_o length and the physiological range of the FL relationship in which the muscle operates are both unknown.

FV properties are often estimated from morphological and fiber-type considerations. Animal studies indicate that the dynamic constants a and b that shape the Hill rectangular hyperbola are related to the fiber-type profile of a muscle. This can be observed by contrasting the force rise-time profile in ST and FT muscles. Researchers have therefore extrapolated from animal studies and used fiber-typing to estimate a and b values for specific human muscles (Bobbert, Huijing, and van Ingen Schenau 1986). Another feature that dictates the muscle model rise-time response is the elasticity of the SEC. Because the SEC is based on structures within the fibers, aponeurosis, and tendon, there is no clear morphological measurement that can be used to estimate total SEC elasticity. In the literature, the tendon CSA and length are often used, but use of caution is necessary because the tendon is stiffer than the other series elastic structures. Ultrasound studies that measure aponeurosis excursion may better predict SEC elasticity.

Overall, some model parameters can be estimated with confidence, whereas others are in general unknown for specific human muscles. One method used to overcome this limitation is to use an optimization approach to predict unknown model parameters. For example, assume that five muscles are responsible for a given joint moment. Using Hill models to represent each muscle, force and moment estimates for a given isometric joint position can be obtained. When summated, a model joint torque estimate for that position can be found. By repeating this process across a full range of joint motion, a model torque-angle relation can be formed. Using an optimization algorithm and the actual measured joint torque-angle data, the best values of L_o for each muscle model can be predicted. The optimization procedure would be formulated to find the set of five unknown L_o values that gave the best fit between measured and model torque-angle relationships. This method also can be used to help find other unknown muscle parameters, such as estimates of a and b dynamic constants that produce model torque-angular velocity relations that closely approximate experimental data. Examples of selected muscle parameters drawn from the literature are shown in table 9.2.

Table 9.2   Human Muscle Parameters

Muscle	P_o (N)	L_o (mm)	Pennation θ (°)	Tendon, unloaded length (mm)
Gluteus maximus	1050–4490	144–250	0	90–150
Hamstrings	2000–3900	104–107	9	350–390
Vasti	4500–6375	84–93	3–10	160–225
Rectus femoris	925–1700	75–81	5–14	323–410
Soleus	3550–4235	24–55	20–25	238–360
Gastrocnemius	1375–3000	45–62	12–17	48–425

CONTROL MODELS

As observed earlier, the human body is redundant in that more muscles than are necessary cross each joint. What rules concerning muscle selection does the CNS use? How do researchers formulate muscle model control signals in their musculoskeletal models? The answer to the first question is a matter of great interest to motor control scientists, and it is beyond the scope of this text to answer in the detail that the topic deserves. The answer to the second question is central to the use of musculoskeletal models, and in this section we discuss possible approaches. The type of control invoked depends on the nature of the research question under investigation and on the structure of the specific musculoskeletal model. In many cases, the goal is to supply the muscle models with patterns of stimulation that mirror the control signals that the CNS sends to individual muscles during an actual movement sequence. Alternatively, the goal might be to supply the muscle models with hypothetical signals representing an optimal or even supramaximal performance. The approaches that are used to generate control signals can be grouped into three general categories: (1) modeling from measured EMG signals, (2) theoretical neural models,

and (3) optimization models. Hybrid models that combine these approaches in various ways have also been described.

Electromyographic Models

The most straightforward approach, it seems, would be to measure the "real" CNS control signals with EMG during a movement performance and use these EMG signals as the model control signals. Several issues make this approach difficult, however, including the fact that many deep muscles are impossible to monitor with surface electrodes. Indwelling electrodes can be used, but they are selective and may not provide an accurate assessment of the activity level of the muscle as a whole. Furthermore, indwelling electrodes are invasive and may interfere with a subject's ability to perform the movement. Even with surface electrodes, technical considerations make it questionable whether the measured signal truly represents the CNS control signal (see chapter 8).

Another problem is that musculoskeletal models often simplify reality and only explicitly represent some muscles. For example, models of human jumping commonly use 6 or 8 muscle models (figure 9.16) to represent the muscular capabilities of the human lower extremity when in fact there

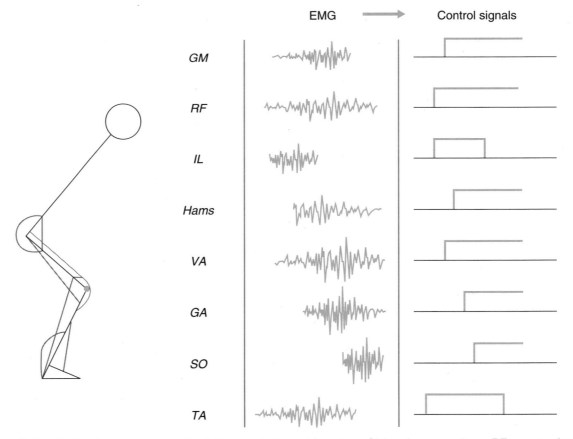

Figure 9.16 Vertical jumping model with eight musculoskeletal actuators. GM = gluteus maximus; RF = rectus femoris; IL = iliopsoas; Hams = hamstrings; VA = vasti; GA = gastrocnemius; SO = soleus; TA = tibialis anterior.

are more than 40. This means that model control signals must represent the activity of several actual muscles, and it is unclear how measured EMG from several individual muscles should be combined to form a control signal for one muscle model. Finally, the raw EMG signals have high-frequency content and must be preprocessed to form control signals for Hill-type muscle models (figure 9.16). This process must transform the EMG signal—a manifestation of CNS motor unit recruitment and rate coding—into a simplified, low-frequency control signal that varies from 0.0 (inactivity) to 1.0 (full activity), and the nature of this transformation is unclear. Because the musculoskeletal model is not a perfect representation of reality, measured EMG signals will not match model control signals exactly.

Theoretical Control Models

Another approach to finding appropriate control signals is to build a neural control model that is based on some theoretical understanding of how the CNS controls and coordinates motion. These models might be based on first principles (i.e., they might attempt to predict the signals the CNS would send to each muscle by modeling brain and spinal neuronal action) or on some motor control principle (e.g., a general principle of proximal-to-distal sequencing in lower-extremity extensions). Another possibility is to predict stimulation to muscle models based on some hypothetical control basis. For example, Pierrynowski and Morrison (1985) generated lower-extremity muscle stimulation patterns during walking by considering which muscles were in the best anatomical position to create the required joint torques in 3-D at each joint. In a similar fashion, Caldwell and Chapman (1991) used a neural model with control "rules" based on EMG measurements to predict forces in the individual muscles controlling the elbow joint.

One problem with the theoretical approach is that, in general, the manner in which the brain formulates control signals is unknown. How can we check that the output signals from a neural model are reasonable? One method is to compare the timing (onset and offset) and patterns of predicted control signals with measured EMG signals, but as described above, the use of EMG can be problematic and requires caution in interpreting these signals. Despite the problem of confirming whether the control model predicts the "correct" signals, this theoretical approach is potentially powerful because it allows testing for the consequences of different control strategies. For example, a model of arm motion that is driven by an open-loop control mechanism could be built and compared to a model that relies on proprioceptive feedback to guide the muscle stimulation patterns. Comparisons between the two models might provide insight into the role of peripheral feedback in a specific motor task.

Optimization Models

The most popular method for generating control signals in musculoskeletal models is to use some form of optimization model. In some cases, this is tantamount to forming a theoretical control model, because the optimization criterion is derived from motor control principles. The optimization approach tries to answer the question, Which set of model control signals will produce a result that optimizes (minimizes or maximizes) some given criterion measure? The criterion measure is known as a *cost function, objective function,* or *performance criterion.* The cost function can be relatively simple (e.g., find the solution that gives maximal total muscle torque or minimal total muscle stress [force per unit PCSA]) or complex (e.g., to determine a nonlinear combination of maximal muscle force at minimal metabolic cost). The objective function may be directly related to muscle function (e.g., minimizing muscular work) or to some aspect of the motion under study (e.g., maximizing vertical jump height). The cost function serves as a guiding constraint that determines the selection of one particular set of "optimal" muscular controls from many different possible solutions.

Initial attempts to predict individual muscle forces used *static optimization* models in conjunction with inverse dynamics analysis of an actual motor performance. The inverse dynamics analysis permits the calculation of observed resultant joint moments at incremental times throughout the movement (see chapter 5). Optimization is used to find the set of muscle forces that balances these joint moments and satisfies a selected cost function. Static optimization models treat each time interval independently, and early models were plagued by *nonphysiological* results of three kinds. The first was the prediction of model forces too high for actual muscles to produce, a difficulty that can be solved by adding maximal force constraints to each muscle under study. The second problem was that the model would often select only one muscle to balance the resultant moment, rather than choosing a more realistic muscular synergy. Mathematically, synergism can be produced by nonlinear cost functions (e.g., minimizing the sum of muscle forces squared or cubed), even though there is no neural or physiological reason to suggest such a nonlinear cost. The third problem was sudden, nonphysiological switching on and off of muscle forces caused by the independence of solutions for sequential time increments (i.e., the optimization model separately balanced the joint moments for each time interval). This problem has been addressed by using muscle models that invoke physiological realism through FL and FV relations and time-dependent SA dynamics.

Ongoing work with static optimization models led to the development of *dynamic optimization* or *optimal control* models, which are applied in conjunction with forward dynamics models of human motion. In contrast to inverse dynamics approaches that use experimental data from an actual performance to calculate resultant joint moments, forward dynamics analysis simulates the motion of body segments based on a given set of joint torques or muscle forces (see chapter 10). Dynamic optimization models are therefore set up to find the set of muscle stimulation patterns that will result in an optimal motion pattern (e.g., to find the set of muscle controls that will produce the highest vertical jump). The stimulation patterns are used as the input to individual (usually Hill-type) muscle models that predict both muscle forces and resultant joint moments. In addition, the model computes the body segment kinematics associated with optimal performance. Because the model includes muscle models and seeks a solution based on the whole movement performance and not separate time increments, many of the problems associated with static optimization models are overcome.

Dynamic optimization has been most successful when applied to movements in which the performance criterion to be optimized is clear, such as vertical jumping. In jumping, the cost function can be stated as the maximization of vertical displacement of the body's center of mass during the flight phase after the model leaves the ground. If the model is constructed with appropriate constraints (realistic muscle properties and limits on joint motion), jump heights and body segment motions approximating those of human jumpers may be attained (Pandy and

Zajac 1991; van Soest et al. 1993). However, many human movements (e.g., walking) do not have clear performance criteria. In these cases, the optimal control model may be used to explore various proposed theoretical criteria, or a solution may be found by setting up a *tracking cost function* based on how well the model's results match the kinematic, kinetic, or control signal (EMG) data from actual performances. In this latter case, it is unclear which is the most important experimental measure for the model to track, because it is unlikely that the simplified model structure will be able to produce a complete set of kinematic, kinetic, and control signal results that match those from real performances.

Although optimization has evolved as a popular method for predicting individual muscle forces, several issues concerning its use must be addressed. Do humans actually produce movements based on one given performance criterion, or does the performance objective change during the movement? If the optimal model's results differ from those of a human performer, is it because the model is too simple or not constrained properly, or is the human not performing optimally? Finally, EMG data have illustrated various degrees of simultaneous antagonistic cocontraction during some movement sequences. Optimization models tend to lead away from solutions predicting muscle antagonism around a joint, although some antagonism is predicted when models contain muscles that contribute to more than one joint moment. However, optimization models that seek to minimize or maximize specific cost functions do not predict antagonistic cocontraction associated with joint stability or stiffness. Despite these drawbacks, optimization models have increased the understanding of human biomechanics and will continue to do so in the future.

From the Scientific Literature

Bobbert, M.F., P.A. Huijing, and G.J. van Ingen Schenau. 1986. A model of the human triceps surae muscle-tendon complex applied to jumping. *Journal of Biomechanics* 19:887-98.

Pandy, M.G., and F.E. Zajac. 1991. Optimal muscular coordination strategies for jumping. *Journal of Biomechanics* 24:1-10.

Soest, A.J. van, A.L. Schwab, M.F. Bobbert, and G.J. van Ingen Schenau. 1993. The influence of the biarticularity of the gastrocnemius muscle on vertical-jumping achievement. *Journal of Biomechanics* 26:1-8.

Studies of vertical jumping have used muscle models to examine the roles played by individual muscles during the jumping movement. Bobbert, Huijing, and van Ingen

(continued)

Schenau (1986) used Hill-type models to represent the soleus (SO) and gastrocnemius (GA) muscles during vertical jumping (figure 9.17). Their study examined the reasons that the maximal ankle plantarflexion velocity from isovelocity dynamometer studies is much lower than the velocities seen during natural movements such as jumping. The SO and GA muscle models included FV, FL, series elasticity, and SA dynamics. Kinematics from actual jumping performances were imposed on the SO and GA models, with muscle lengths and moment arms at the ankle calculated using equations based on cadaver studies. The SO and GA muscle force predictions were then used to estimate the plantarflexor torque at the ankle joint throughout the push-off phase prior to takeoff. The model ankle torque demonstrated good agreement with the torque calculated using inverse dynamics from the actual jumping performances. Analysis of the SO and GA muscle forces and contractile kinematics uncovered two reasons the ankle muscles could produce high plantarflexor torques at velocities much higher than in dynamometer studies. The first was that the series elasticity of the muscles kept the CC shortening velocity much lower than the measured ankle velocity. Second, the GA is a biarticular muscle that acts as both a knee flexor and ankle plantarflexor. In jumping, simultaneous ankle plantar flexion and knee extension result in a much lower GA shortening velocity than had occurred in dynamometer studies of isolated plantar flexion. This study demonstrates clearly that muscle modeling can lead to a greater appreciation of muscular usage than that provided by inverse dynamics analysis.

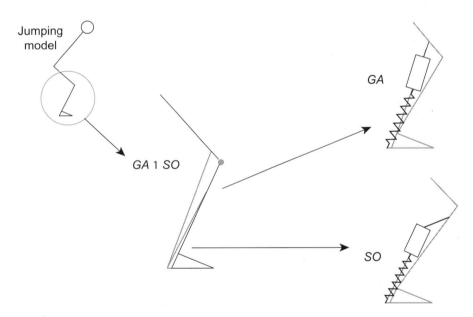

○ **Figure 9.17**   Bobbert, Huijing, and van Ingen Schenau (1986) GA and SO model of plantarflexor torque in vertical jumping. Each muscle is represented by a two-component Hill model (CC + SEC).

Subsequent modeling studies of vertical jumping investigated the overall segmental motion and coordination of muscles, which required a more complete representation of the jumper. Both Pandy and Zajac (1991) and van Soest et al. (1993) presented dynamic optimization studies of simulated jumping in which the segmental motion was driven by Hill-type models representing specific individual muscles (figure 9.18). In contrast to models that use independent joint torque generators, the use of individual muscle models can uncover the effect of biarticular muscles that contribute to torque

(continued)

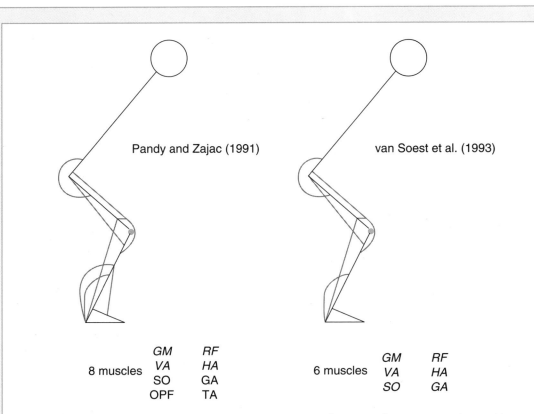

Pandy and Zajac (1991)   van Soest et al. (1993)

	GM RF		GM RF
8 muscles	VA HA	6 muscles	VA HA
	SO GA		SO GA
	OPF TA		

○ **Figure 9.18** Vertical jumping models of Pandy and Zajac (1991, *left*) and van Soest et al. (1993, *right*). GM = gluteus maximus; RF = rectus femoris; HA = hamstrings; VA = vasti; GA = gastrocnemius; SO = soleus; OPF = other plantarflexors; TA = tibialis anterior.

at adjacent joints. The optimization problem was to find the set of muscle stimulation onset times that produced the maximal jump height.

Although both models produced realistic jump heights and segmental kinematics, their interpretations of results were quite different. Van Soest and colleagues (1993) used six muscle models to provide motion, representing both mono- and biarticular muscles of the lower extremities. They emphasized the contributions to jumping performance made by individual muscles, with particular emphasis given to the ability of biarticular muscles to transport energy between the adjacent joints that they span. They considered the energy transport mechanism to be an important feature that helps to dictate the overall muscular coordination during the jump and contributes substantially to a proximal-to-distal energy flow. In contrast, Pandy and Zajac (1991) characterized the energy flow as distal-to-proximal and identified the contributions of eight individual muscle models to the increased energy of the body's center of mass. They discounted the importance of biarticular energy transport, and to emphasize this point they performed "virtual surgery" by moving the origin of the normally biarticular GA from the distal femur to the proximal tibia, limiting the GA model action to ankle plantar flexion alone. Simulations with the now monoarticular GA demonstrated only slight reductions in the model's jump-height ability, casting doubt on the importance of its biarticular nature. It was subsequently suggested that the discrepancies between the van Soest and Pandy models might be the result of differences in the GA knee flexion moment arm relations used by each group, with the Pandy model underestimating the moment arm, particularly at highly extended knee angles. This underestimation would limit the ability

(continued)

of their GA model to participate in the energy transport mechanism, and therefore the switch to a monoarticular GA would have relatively little effect.

Examination of the similar yet conflicting natures of these two jumping models illustrates several important modeling issues. First, both models demonstrated results for individual muscle use during jumping that could not be obtained with experimental techniques—one of the main strengths of this musculoskeletal modeling approach. Second, both groups used their models to investigate questions about movement coordination, rather than to merely describe muscle-specific results. Therefore, they both increased our understanding of the biomechanics of human jumping. However, the discrepancies between the two models highlight the importance of musculoskeletal model development. Should the lower extremity be represented with six muscles or eight? What are the consequences of the model parameters chosen, such as muscle moment arm relations at a specific joint? There is a degree of uncertainty in model parameters used to represent any given subject, and it behooves researchers to investigate the sensitivity of their model results to errors in these parameters.

SUMMARY

This chapter introduced the basic concepts of muscle mechanical properties and modeling. The major mechanical characteristics of muscle were presented within the framework of the Hill muscle model. The latter part of the chapter described the use of such muscle models in the construction of musculoskeletal models used to study segmental motion during human movement. Students interested in muscle mechanics and modeling should realize that many issues and topics concerning muscle were omitted from this chapter. Students should read and study other materials once they are comfortable with the material presented here; see Suggested Readings at the end of the chapter. Some important topics were alluded to early in the chapter, such as the Huxley muscle model. Other issues worthy of pursuit include detailed analysis of force transmission within and between muscle fibers (e.g., Huijing 1999; Monti et al. 1999), force enhancement and depression related to contractile history (Edman 1975; Edman, Elzinga, and Noble 1978; Herzog, Leonard, and Wu 2000), measurement of force production at the level of individual myosin heads (Finer, Simmons, and Spudich 1994), and detailed descriptions of crossbridge dynamics (Lutz and Lieber 1999).

SUGGESTED READINGS

Muscle Mechanics

Chapman, A.E. (1985). The mechanical properties of human muscle. *Exercise and Sports Science Reviews* 13: 443-501.

Chapman, A.E., G.E. Caldwell, and W.S. Selbie. 1985. Mechanical output following muscle stretch in forearm supination against inertial loads. *Journal of Applied Physiology* 59:78-86.

Ettema, G.J.C. 1996. Contractile behaviour in skeletal muscle-tendon unit during small amplitude sine wave perturbations. *Journal of Biomechanics* 29:1147-55.

Ettema, G.J.C., and P.A. Huijing. 1990. Architecture and elastic properties of the series element of muscle tendon complex. In *Multiple Muscle Systems*, ed. J.M. Winters and S.L.-Y. Woo, 57-68. New York: Springer-Verlag.

Heckman, C.J., and T.G. Sandercock. 1996. From motor unit to whole muscle properties during locomotor movements. *Exercise and Sports Science Reviews* 24:109-33.

Herzog, W., and H.E.D.J. ter Keurs. 1988. Force-length relation of in-vivo human rectus femoris muscles. *Pflugers Archive: European Journal of Physiology* 411:642-7.

Hill, A.V. 1970. *First and Last Experiments in Muscle Mechanics.* Cambridge: Cambridge University Press.

Joyce, G.C., P.M.H. Rack, and D.R. Westbury. 1969. The mechanical properties of cat soleus muscle during controlled lengthening and shortening movements. *Journal of Physiology* 204:461-74.

Lieber, R.L., G.J. Loren, and J. Friden. 1994. In vivo measurement of human wrist extensor muscle sarcomere length changes. *Journal of Neurophysiology* 71:874-81.

Pollack, G.H. 1990. *Muscles and Molecules: Uncovering the Principles of Biological Motion.* Seattle: Ebner and Sons.

Rack, P.M.H., and D.R. Westbury. 1969. The effects of length and stimulus rate on tension in the isometric cat soleus muscle. *Journal of Physiology* 204:443-60.

Rack, P.M.H., and D.R. Westbury. 1984. Elastic properties of the cat soleus tendon and their functional importance. *Journal of Physiology* 347:479.

Wilkie, D.R. 1950. The relation between force and velocity in human muscle. *Journal of Physiology* 110:249-80.

Winters, J.M., and L. Stark. 1988. Estimated mechanical properties of synergistic muscles involved in movements of a variety of human joints. *Journal of Biomechanics* 21:1027-42.

Hill Muscle Model

Audu, M.L., and D.T. Davy. 1985. The influence of muscle model complexity in musculoskeletal motion modeling. *Journal of Biomedical Engineering* 107:147-57.

Bobbert, M.F., and G.J. van Ingen Schenau. 1990. Isokinetic plantar flexion: Experimental results and model calculations. *Journal of Biomechanics* 23:105-19.

Garner, B.A., and M.G. Pandy. 2003. Estimation of musculotendon properties in the human upper limb. *Annals of Biomedical Engineering* 31:207-20.

Hof, A.L., and J. van den Berg. 1981a. EMG to force processing I: An electrical analogue of the Hill muscle model. *Journal of Biomechanics* 14:747-58.

Hof, A.L., and J. van den Berg. 1981b. EMG to force processing II: Estimation of parameters of the Hill muscle model for the human triceps surae by means of a calf ergometer. *Journal of Biomechanics* 14:759-70.

Hof, A.L., and J. van den Berg. 1981c. EMG to force processing III: Estimation of parameters of the Hill muscle model for the human triceps surae muscle and assessment of the accuracy by means of a torque plate. *Journal of Biomechanics* 14:771-85.

Hof, A.L., and J. van den Berg. 1981d. EMG to force processing IV: Eccentric-concentric contractions on a spring-flywheel set up. *Journal of Biomechanics* 14:787-92.

Lieber, R.L., C.G. Brown, and C.L. Trestik. 1992. Model of muscle-tendon interaction during frog semitendinosus fixed-end contractions. *Journal of Biomechanics* 25:421-8.

Winters, J.M., and L. Stark. 1987. Muscle models: What is gained and what is lost by varying model complexity. *Biological Cybernetics* 55:403-20.

Muscle Architecture

Alexander, R.M., and A. Vernon. 1975. The dimensions of knee and ankle muscles and the forces they exert. *Journal of Human Movement Studies* 1:115-23.

Biewener, A.A. 1991. Musculoskeletal design in relation to body size. *Journal of Biomechanics* 24:19-29.

Gans, C., and A.S. Gaunt. 1991. Muscle architecture in relation to function. *Journal of Biomechanics* 24:53-65.

Huijing, P.A. 1981. Bundle length, fibre length and sarcomere number in human gastrocnemius. *Journal of Anatomy* 133:132.

Huijing, P.A. 1985. Architecture of human gastrocnemius muscle and some functional consequences. *Acta Anatomica* 123:101-7.

Huijing, P.A., and R.D. Woittiez. 1984. The effect of architecture on skeletal muscle performance: A simple planimetric model. *Netherlands Journal of Zoology* 34:21-32.

Kaufman, K.R., K.N. An, and E.Y.S. Chao. 1989. Incorporation of muscle architecture into the muscle length-tension relationship. *Journal of Biomechanics* 22:943-8.

Otten, E. 1988. Concepts and models of functional architecture in skeletal muscle. *Exercise and Sports Science Reviews* 16:89-139.

Spector, S.A., P.F. Gardiner, R.F. Zernicke, R.R. Roy, and V.R. Edgerton. 1980. Muscle architecture and force-velocity characteristics of cat soleus and medial gastrocnemius: Implications for motor control. *Journal of Neurophysiology* 44:951-60.

Musculoskeletal Geometry

Fukashiro, S., M. Itoh, Y. Ichinose, Y. Kawakami, and T. Fukunaga. 1995. Ultrasonography gives directly but noninvasively elastic characteristic of human tendon in vivo. *European Journal of Applied Physiology* 71:555-7.

Fukunaga, T., M. Ito, Y. Ichinose, S. Kuno, Y. Kawakami, and S. Fukashiro. 1996. Tendinous movement of a human muscle during voluntary contractions determined by real-time ultrasonography. *Journal of Applied Physiology* 81:1430-3.

Fukunaga, T., Y. Ichinose, M. Ito, Y. Kawakami, and S. Fukashiro. 1997. Determination of fascicle length and pennation in a contracting human muscle in vivo. *Journal of Applied Physiology* 82:354-8.

Herzog, W., and L. Read. 1993. Lines of action and moment arms of the major force-carrying structures crossing the human knee joint. *Journal of Anatomy* 182:213-30.

Kawakami, Y., T. Muraoka, S. Ito, H. Kanehisa, and T. Fukunaga. 2002. In vivo muscle fibre behaviour during counter-movement exercise in humans reveals a significant role for tendon elasticity. *Journal of Physiology* 540(Pt. 2): 635-46.

Kellis, E., and V. Baltzopoulos. 1999. In vivo determination of the patella tendon and hamstrings moment arms in adult males using videofluoroscopy during submaximal knee extension and flexion. *Clinical Biomechanics* 14:118-24.

Maganaris, C.N., V. Baltzopoulos, and A.J. Sargeant. 1999. Changes in the tibialis anterior tendon moment arm from rest to maximum isometric dorsiflexion: In vivo observations in man. *Clinical Biomechanics* 14:661-6.

Murray, W.M., S.L. Delp, and T.S. Buchanan. 1995. Variation of muscle moment arms with elbow and forearm position. *Journal of Biomechanics* 28:513-25.

Rugg, S.G., R.J. Gregor, B.R. Mandelbaum, and L. Chiu. 1990. In vivo moment arm calculations at the ankle using magnetic resonance imaging (MRI). *Journal of Biomechanics* 23:495-501.

Spoor, C.W., and J.L. van Leeuwen. 1992. Knee muscle moment arms from MRI and from tendon travel. *Journal of Biomechanics* 25:201-6.

Visser, J.J., J.E. Hoogkamer, M.F. Bobbert, and P.A. Huijing. 1990. Length and moment arm of human leg muscles as a function of knee and hip-joint angles. *European Journal of Applied Physiology* 61:453-60.

Yamaguchi, G.T., A.G.U. Sawa, D.W. Moran, M.J. Fessler, and J.M. Winters. 1990. A survey of human musculo-tendon actuator parameters. In *Multiple Muscle Systems*, ed. J.M. Winters and S.L.-Y. Woo, 717-73. New York: Springer-Verlag.

Musculoskeletal/Optimization Models

Anderson, F.C., and M.G. Pandy. 1993. Storage and utilization of elastic strain energy during jumping. *Journal of Biomechanics* 26:1413-27.

Anderson, F.C., and M.G. Pandy. 2001. Static and dynamic optimization solutions for gait are practically equivalent. *Journal of Biomechanics* 34:153-61.

Crowninshield, R.D., and R.A. Brand. 1981. The prediction of forces in joint structures: Distribution of inter-segmental resultants. *Exercise and Sports Science Reviews* 9:159-81.

Davy, D.T., and M.L. Audu. 1987. A dynamic optimization technique for predicting muscle forces in the swing phase of gait. *Journal of Biomechanics* 20:187-201.

Delp, S.L., J.P. Loan, M.G. Hoy, F.E. Zajac, E.L. Topp, and J.M. Rosen. 1990. An interactive graphics-based model of the lower extremity to study orthopaedic surgical procedures. *IEEE Transactions on Biomedical Engineering* 37:757-67.

Dul, J., M.A. Townsend, R. Shiavi, and G.E. Johnson. 1984a. Muscular synergism: I. On criteria for load sharing between synergistic muscles. *Journal of Biomechanics* 17:663-73.

Dul, J., M.A. Townsend, R. Shiavi, and G.E. Johnson. 1984b. Muscular synergism: II. A minimum-fatigue criterion for load sharing between synergistic muscles. *Journal of Biomechanics* 17:675-84.

Gerritsen, K.G.M., A.J. van den Bogert, M. Hulliger, and R.F. Zernicke. 1998. Intrinsic muscle properties facilitate locomotor control: A computer simulation study. *Motor Control* 2:206-20.

Gerritsen, K.G.M., A.J. van den Bogert, and B.M. Nigg. 1998. Direct dynamics simulation of the impact phase in heel-toe running. *Journal of Biomechanics* 28:661-8.

Hoy, M.G., F.E. Zajac, and M.E. Gordon. 1990. A musculoskeletal model of the human lower extremity: The effect of muscle, tendon, and moment arm on the moment-angle relationship of musculotendon actuators at the hip, knee and ankle. *Journal of Biomechanics* 23:157-69.

Neptune, R.R., and M.L. Hull. 1998. Evaluation of performance criteria for simulation of sub-maximal steady-state cycling using a forward dynamic model. *Journal of Biomechanical Engineering* 120:334-41.

Soest, A.J. van, and M.F. Bobbert. 1993. The contribution of muscle properties in the control of explosive movements. *Biological Cybernetics* 69:195-204.

Muscular Force Production

Gregoire, L., H.E. Veeger, P.A. Huijing, and G.J. van Ingen Schenau. 1984. Role of mono- and biarticular muscles in explosive movements. *International Journal of Sports Medicine* 5:301-5.

Gregor, R.J., R.R. Roy, W.C. Whiting, R.G. Lovely, J.A. Hodgson, and V.R. Edgerton. 1988. Mechanical output of cat soleus during treadmill locomotion: *In vivo* vs *in situ* characteristics. *Journal of Biomechanics* 21:721-32.

Herzog, W. 1996. Force-sharing among synergistic muscles: Theoretical considerations and experimental approaches. *Exercise and Sports Science Reviews* 24: 173-202.

Ingen Schenau, G.J. van. 1994. Proposed actions of bi-articular muscles and the design of hindlimbs of bi- and quadrupeds. *Human Movement Science* 13:665-81.

Jacobs, R., M.F. Bobbert, and G.J. van Ingen Schenau. 1993. Function of mono- and biarticular muscles in running. *Medicine and Science in Sports and Exercise* 25:1163-73.

Jacobs, R., and G.J. van Ingen Schenau. 1992. Control of an external force in leg extensions in humans. *Journal of Physiology* 457:611-26.

Prilutsky, B.I. 2000. Coordination of two- and one-joint muscles: Functional consequences and implications for motor control. *Motor Control* 4:1-44.

Prilutsky, B.I., and R.J. Gregor. 1997. Strategy of coordination of two- and one-joint leg muscles in controlling an external force. *Motor Control* 1:92-116.

CHAPTER 10

Computer Simulation of Human Movement

S.N. Whittlesey and J. Hamill

Modeling of physical systems can be divided into two categories. *Physical modeling* is a process in which we construct tangible scale models that look very much like the real system. The Greeks and Romans long ago recognized the advantages of building small models of ships, buildings, and bridges before constructing the large, real-life structures. In the past century, consider the model wings flown by the Wright brothers, the model hydrofoil boat of Alexander Graham Bell, the laboratory crash-testing of automobiles, and the animal models that have significantly influenced the development of disease treatments and artificial joints alike. However, scale models and crash-test dummies require a great deal of time and resources to develop, and there are limits to what can be learned from them. *Behavioral* or *mathematical modeling* is a more abstract system used for studying a research question that does not necessarily lend itself to physical modeling. In the modern era, this means that researchers construct a set of equations and run them on a computer. Behavioral models are used for researching weather systems (including models pioneered by Konrad Lorenz), animal populations and the spread of disease, and economic conditions to help governments set interest rates. In these models, the system is simplified by limiting the number of components so that they represent the net effect of many parts. For example, a model of a country's economy does not have millions of individuals; rather, it has a relatively small number of components representing the behavior of different socioeconomic and geographical groups. Biomechanists take the same approach in their research: Instead of modeling every bone and motor unit in the body, they simplify the body to have only as many parts as needed to answer their research question.

The pioneers in human movement studies, such as E. Muybridge, E.-J. Marey, and A.V. Hill, were largely experimentalists. They formulated theories by collecting large amounts of subject data, because they were not able to construct physical or computer models of the human body. Behavioral models such as Hill's equation came into common usage with the advent of digital computing. Computers, of course, changed human movement study forever, because models of human movement could be constructed, tested, and studied within a reasonable amount of time.

In this chapter, we explore a variety of topics related to movement simulation:

- how to use simulation for research,
- why simulation is a powerful tool,
- the general procedure for creating and using a simulation,
- free-body diagrams,

- differential equations of motion,
- numerical integration,
- control theory,
- examples of human movement simulation, and
- limitations of computer simulations.

OVERVIEW: MODELING AS A PROCESS

Readers of this chapter will learn about the tools they will need to construct their own computer models of a human movement. However, an equally important—and perhaps more fundamental—purpose of this chapter is to convey the fact that modeling is a process used to enhance our understanding of human movement. We demonstrate that modeling is both enormously powerful and greatly limited as a tool. As with any tool, the most important thing is to use it correctly.

How do we use mechanical models in human movement science? The answer, of course, is that researchers have their own styles, but nonetheless, a basic process should occur. This process, diagrammed in figure 10.1, is a variant of the scientific method. It starts with a research question, proceeds to constructing the model, and then progresses through repeatedly improving the model. Eventually, the model is capable of making predictions that in turn lead to new understanding and new questions. We now discuss these steps in order.

All studies in human movement should start with a research question. Modeling in particular does not

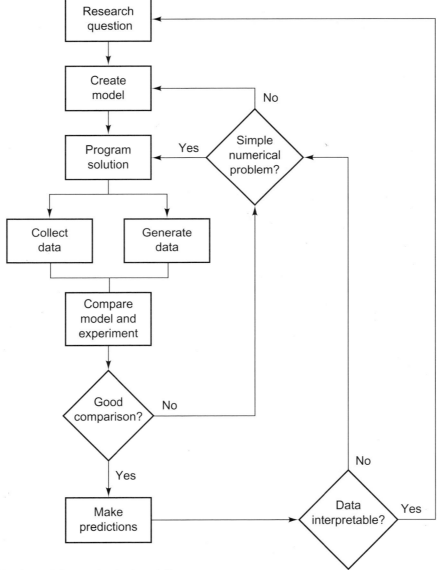

○ Figure 10.1   Flowchart of the mechanical modeling process.

stand up well to testing if there is not a specific question to test for. Not only does the basic purpose of the model become unclear, the proper design for the model will also be unclear, because it depends on the question asked. For example, there is no all-encompassing model of a bicyclist. If the question is, What is the effect of body mass when climbing hills on a bicycle?, the body can be modeled as a point mass. In contrast, if the question is, How does the activation of two-joint muscles change between cycling on level ground and up a steep hill?, a complex model that includes multiple lower-extremity segments and muscles is needed. Every model must start with a particular research question so that the relevant parameters are included in the model. This point cannot be overemphasized.

Once the research question has been formulated, the next steps are to create a mechanical model of the movement and program it into a computer. These steps are detailed later in this chapter. Then, the model is evaluated to see if it seems to work. This is actually a three-step process that involves using the model to generate data, collecting corresponding experimental data, and then comparing the two data sets to establish whether the simulation data are reasonable. There are no set criteria for this, and the evaluative process is more of an art than a science. In addition, the data needed to verify a model vary widely: Sometimes a full, three-dimensional (3-D) kinematic data set is necessary, but in other cases, the results sheet from a recent athletic competition will suffice. If the model data are not realistic, it must be determined if there is a problem with the model itself or with the way the model program runs. In either case, development must begin again.

Once a model can replicate experimental data, it is ready to make predictions about the human movement to answer the research question. Researchers again have to ask themselves if the data the model generates are reasonable, and perhaps compare them with experimental data. An entire series of studies may have to be performed to establish the model's validity, and in many cases, models have been debated for years in the literature. For instance, the model of the running human as a "bouncing ball" (Cavagna, Heglund, and Taylor 1977) has been revisited for over two decades. When model data are not reasonable, development must be undertaken again until the predictions appear to be reasonable and lead to new questions.

There are different categories of human movement simulations. *Rigid-body* models represent the human system in whole or part as a set of rigid segments controlled in their movements by joint moments. Rigid-body models are the most common in human movement studies. *Mass-spring* models comprise one or more masses linked to one or more springs; they are commonly used to model running, hopping, and other repetitive movements. *Wobbling-mass* and *mass-spring-damper* models consist of linked masses, springs, and dampers (also called *dashpots*). Mass-spring-damper models are most often used to model human movements in which impacts occur. *Musculo-skeletal models* also have rigid segments, but instead of having simple moments at each joint, they have sub-models of individual muscles. The details of muscle models are discussed in chapter 9 of this text.

Simulating human movement requires drawing from a variety of technical areas, especially from such mathematical techniques as differential equations and numerical analysis, but also from control theory, advanced dynamics, and computer programming. In this chapter, we explore each of these areas separately and then integrate them. Readers should be able to construct their own simulations by using the information in this chapter. However, effective modeling does not depend exclusively on the mastery of these disciplines. As with any tool, proper implementation is essential. Modeling is riddled with approximations and interpretations, and thus the proper use of models depends on recognizing its limitations. Before we delve into such details, however, we will first discuss the applications of computer simulation.

WHY SIMULATE HUMAN MOVEMENT?

The best way to obtain human movement data, it seems, would be to collect it from human subjects. In other words, no data can be more accurate, more real than those collected from human subjects, so why would we want to generate data artificially? The most general answer is that human subjects are extremely complex and have intrinsic limitations. They are randomly variable, subject to fatigue, have limits of strength and coordination, and must be treated with due consideration for safety and ethics. A computer model eliminates these constraints, and so there are many studies for which computer simulation is a good tool. Consider these examples:

- Suppose we wanted to determine the best position in which to perform an athletic task such as jumping or cycling. We would have many different positions to test, and we would have to collect several trials of each position because of subject variability. In this study, the best subject performance probably would occur early in the data collection, before the subjects became fatigued. Subjects also tend to perform best under the conditions in which they normally train (Selbie and Caldwell 1996; Yoshihuku and Herzog 1990).

• One exciting use for simulations is in virtual surgery, in which multiple surgical outcomes can be reviewed. Suppose a surgeon has a patient with certain spastic muscles (as do people with cerebral palsy, for example). A common procedure is to lengthen or relocate the insertion points of these muscles. A sophisticated computer simulation could predict the proper amount to lengthen specific muscles, eliminating the trial and error from the surgical process (Delp et al. 1990; Delp, Arnold, and Piazza 1998; Lieber and Friden 1997).

• Risks to workers in many occupations could be alleviated by using simulations. Military paratroopers, for example, have trained for landings by jumping from platforms of up to 4.3 m high. Suppose that a military unit wanted to establish safe limits for jumping from various heights with backpacks of various weights onto surfaces such as concrete, soil, and sand, and into shallow water. If human subjects were used, the study could establish limits only by injuring the subjects. Clearly, that is unacceptable, so this study is feasible and ethical only as a computer simulation.

Computer models also allow researchers to experiment with conditions that cannot be tested on human subjects because they exceed practical limits. For example, locomotion patterns under an increased and decreased gravitational constant (9.81 m/s^2) could be tested to study the effects of gravity on human movement. The mass of model limbs could be changed to study the effect of limb mass on a movement (Bach 1995; Tsai and Mansour 1986). A single parameter in a model can also be changed and immediately tested. For example, a researcher who wants to study how increasing elbow extensor strength might improve throwing performance can change the model thrower and test it immediately, whereas a human subject would have to undergo a training regimen in which strength might improve in other muscles as well. Another example is to test the effects of torso mass on ground reaction forces. This, too, is a matter of changing a single model parameter. However, if we attempted this study with human subjects, the results might be confounded by concomitant changes in walking velocity, stride length, and perhaps even footwear function.

Another important and commonly used feature of computer simulations is that they can be adapted to search for optimal solutions. This is very useful for athletic performance, clinical applications, and other large issues in human movement. For example, we could determine the optimal movement pattern for a football kick, the most economical walking cadence

for individuals with different body types, and the optimum flexibility of a prosthetic foot for an amputee's body weight, age, and stride length.

Models are often used simply to test our understanding of movement. This is a key reason for the popularity of rigid-body models in biomechanics despite the fact that none of them are perfect. If we understood a movement perfectly, we could replicate it on a computer. However, models eventually fall short in their predictive capabilities, and we are left to assess why. The search for answers often develops into a new movement theory. This topic is discussed in more detail later in this chapter. Now, we will discuss the how-to of computer simulation.

GENERAL PROCEDURE FOR SIMULATIONS

In general, a simulation is a computer program that generates a movement pattern (kinematics). In some cases, a simulation might also generate the kinetics of the movement. The procedure for developing a simulation is to

• create and diagram the mechanical model,

• derive the equations of motion for the mechanical model,

• program a numerical solution of the equations of motion,

• determine the starting and ending conditions for the simulations,

• run the program, generating model kinematics, and

• interpret the model data and compare them with experimental data.

This oversimplifies the modeling process by reflecting only the most important steps. The first step is to diagram the model so that it approximates the human system. These free-body diagrams (FBDs) are then used for the second step, the derivation of model equations, which are used in the third step, and so on.

FREE-BODY DIAGRAMS

FBDs are discussed at length in chapters 4 and 5 of this book. For the purposes of this chapter, they are used as visual aids for deriving the equations of motion and to establish the *type of movement* that the model can perform. Specifically, researchers must answer these questions:

• How many components does the system have?

- Does each system component move linearly, rotationally, or both?
- Can the system's movements be described by a one- or two-dimensional (2-D) model, or must a 3-D representation be used?
- How many coordinates (degrees of freedom, or DOF) are needed to locate each component's position?
- Are there any means to reduce the number of coordinates?

One-dimensional (1-D) models have been used to study locomotion cadence (for example, Holt, Hamill, and Andres 1990), impacts between a runner and the ground (Derrick, Caldwell, and Hamill 2000; McMahon, Valiant, and Frederick 1987), deformation of the leg (Mizrahi and Susak 1982), and even the influence of the viscera on breathing patterns (Minetti and Belli 1994). However, most models

are either 2-D or 3-D, because most human movements occur in multiple directions with multiple body segments. Three-dimensional models remain less common because of the complexity of controlling the model. Models of aerial movements such as diving and gymnastics (for example, Yeadon 1993) have been constructed in 3-D because of the multiaxial nature of the movements. Certain walking models (such as Pandy and Berme 1989) have also been constructed in 3-D to include the effects of 3-D pelvic movements on gait.

The complexity of the computer model depends on the answer to each question. If a system has n components and each component requires k DOF, then there will be n times k equations of motion unless there are factors (called *constraints*) that reduce their number. In this regard, the answers to the last two questions impact the size of the computer simulation code.

EXAMPLE 10.1

Consider the following double pendulum representing the lower extremity.

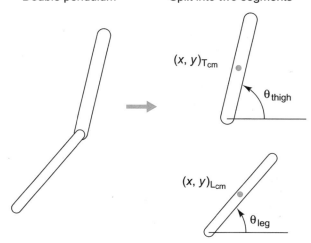

Double pendulum   Split into two segments

In response to the five questions:

1. *Components:* The system has two components, a leg and thigh.
2. *Motion:* Each segment moves both linearly and rotationally.
3. *Dimensions:* This can be a planar model. It clearly is not a 1-D model, and the planar movement is analogous to our use of 2-D lower-extremity kinematics.
4. *Coordinates:* Like any 2-D segment, each component requires an (x,y) coordinate and an angular position to completely describe its position in 2-D

space. In this case the (x,y) coordinates are the coordinates of the centers of mass of the thigh (T_{CM}) and the leg (L_{CM}), respectively.
5. *Constraints:* This is tricky, but the answer is that we can simplify the coordinates of the system. The constraint is that the leg must join the thigh at the knee; wherever the thigh is located, we know where the knee is in relation to it and thus where the leg is located. In other words, the coordinates of the leg mass center are functions of the thigh mass center position, thigh angle, and leg angle. Therefore, we can completely describe the position of the system with four coordinates—the (x,y) position of the thigh, the angle of the thigh, and the angle of the leg. There is no set procedure for identifying constraints; instead, it is a process learned by example. Our double pendulum can now be drawn like this:

Double pendulum   Split into two segments

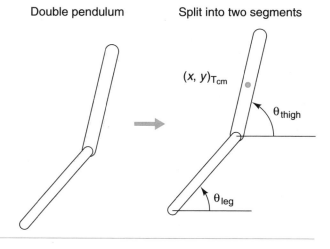

DIFFERENTIAL EQUATIONS

Most physics textbooks deal with the motions of very simple systems, such as projectiles, pendulums, and springs. The positions of these simple systems may be described as polynomial or trigonometric functions. For example, the trajectory of a stone thrown from a cliff is a parabola. The position of a simple pendulum as a function of time is a sinusoid. In human movement simulations, simple systems are rare. We cannot describe most movements with polynomials, trigonometric functions, or any of the common functions. Instead, we have to describe how the positions in the system change. Specifically, the motions of body segments are described with *differential equations* that quantify the segments' changes in position over time. In other words, every kinetic factor that alters the movement of a body segment must be expressed in the differential equation of motion for that segment. For example, a typical body segment is acted upon by muscle forces, joint forces, gravity, and frictional forces. Each of these factors must be expressed mathematically.

Students of calculus may wish for a more precise reason that we must express human movements as differential equations rather than as the more understandable functions of time. The answer is that the equations of motion for body segments are too complex to be solved into positions as functions of time. Although the movement of one segment alone is often solvable, even that requires simplification (i.e., $\sin \theta \approx \theta$). A system of two segments (a double pendulum) becomes hopelessly complex, because the segments interact. When we simulate an entire human body, the equations become very complex, indeed. Consider the following examples.

EXAMPLE 10.2

The height of a projectile launched from the ground can be written as

$$y = y_o + v_o t + 0.5g\, t^2$$

where y is the height above ground, y_o is the starting height over the ground, v_o is the velocity at which the projectile was launched, g is the gravitational constant, and t is the time from launching.

Alternatively, the differential equation for the height of this projectile is

$$\frac{d^2 y}{dt^2} = g$$

subject to initial conditions y_o and v_o. Both of these equations are short, and the latter is readily integrated twice to yield the former.

EXAMPLE 10.3

The differential equation of motion for the lower segment of a double pendulum is

$$\frac{d^2 \theta_L}{dt^2} = \frac{1}{I_L} M_K + m_L d_L \left(L_T \alpha_T \cos\left(\theta_L - \theta_T \right) \right.$$

$$+ L_T \omega_T^2 \sin\left(\left(\theta_L - \theta_T \right) + a_{Hx} \right) \cos \theta_L + \left(a_{Hy} + g \right) \sin \theta_L$$

The notation is listed in appendix G, table G.1. Because this equation cannot be solved by any known analytical technique, we cannot write an equation for the segment's position. Interested readers can refer to Putnam (1991) and Cappozzo, Leo, and Pedotti (1975).

MODEL DERIVATION: LAGRANGE'S EQUATION OF MOTION

The equations of motion of simple systems can often be derived via visual examination. However, multiseg-mented human systems typically defy such analysis because of the complex interactions between segments. There are several advanced techniques used to perform such derivations, of which *Lagrange's equation of motion* is the most popular. Unlike Newton's second law, Lagrange's equation uses the mechanical energies of the system; its formulation is

$$\frac{d}{dt} \frac{\partial KE}{\partial V_i} - \frac{\partial KE}{\partial Y_i} + \frac{\partial PE}{\partial Y_i} = F_i \qquad (10.1)$$

where KE and PE are, respectively, the translational and rotational kinetic energies of the entire system being simulated, V_i and Y_i are, respectively, the velocity and position of the ith component of the system in the y direction, and F_i is the external force acting on the ith component. This equation can be written for the x and z directions as well. This equation may at first appear to be a messy set of partial derivatives, but it is, in fact, a more general form of Newton's second law, $F = ma$. Consider a simple projectile moving in the vertical direction that is acted on by the force of gravity and wind resistance, as shown in figure 10.2.

Applying Newton's second law to this FBD yields

$$F_{wind} - mg = ma \qquad (10.2)$$

Let us now apply Lagrange's equation to this same system. The energies of the complete system are given by

$$KE = 0.5mv^2 \text{ and } PE = mgy \qquad (10.3)$$

These are simple, of course, because the system has only one component. Applying the terms of Lagrange's equation,

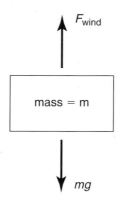

F_{wind}

mass = m

mg

○ **Figure 10.2**   FBD of an object in free fall.

$$\frac{\partial PE}{\partial Y_i} = mg \qquad (10.4)$$

$$\frac{\partial KE}{\partial Y_i} = 0 \qquad (10.5)$$

$$\frac{\partial KE}{\partial V_i} = mv \Rightarrow \frac{d}{dt}\frac{\partial KE}{\partial V_i} = ma \qquad (10.6)$$

Collecting these terms into Lagrange's equation yields

$$ma - 0 + mg = F_i$$
$$\Box \; F_i - mg = ma \qquad (10.7)$$

The only F_i in our FBD is F_{wind}. Substituting this for F_i in these last equations yields the same equation of motion that we obtained by applying Newton's second law to the FBD.

The middle term of Lagrange's equation is usually zero for single-component systems. However, it is usually nonzero in multicomponent systems in which components interact with each other, so it is referred to as the *interaction term*.

The angular form of Lagrange's equation is

$$\frac{d}{dt}\frac{\partial KE}{\partial \omega_i} - \frac{\partial KE}{\partial \theta_i} + \frac{\partial PE}{\partial \theta_i} = M_i \qquad (10.8)$$

where ω is an angular velocity, θ is an angular position, and M_i is a moment. Because most simulations of human movement involve body segments that rotate, it is this angular form of Lagrange's equation that is most commonly used. The moment M is generally the sum of the moments acting on a segment. In the case of the thigh, for example, M would equal the hip moment minus the knee moment.

To implement Lagrange's equation, one must first establish the DOF needed to completely describe the position of the system. This is something of an art, but in general, angular coordinates should be used to describe each segment's position and one linear coordinate should be identified to locate either the most proximal or most distal joint of the entire system.

EXAMPLE 10.4

A suspended bar (physical pendulum).

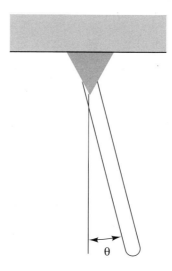

θ

The position of a pendulum can be completely described by one angle, which for convenience can be referenced to the vertical. The pendulum has mass m, moment of inertia I about its pivot, and mass center location d. The energies of the system are

$$KE = 0.5I\,\omega^2 \text{ and } PE = mgd(1 - \cos\theta)$$

Applying Lagrange's equation yields

$$\frac{\partial PE}{\partial \theta_i} = -mgd\sin\theta \qquad \frac{\partial KE}{\partial \theta_i} = 0$$

$$\frac{\partial KE}{\partial \omega_i} = I\omega \qquad \frac{d}{dt}\frac{\partial KE}{\partial \omega_i} = I\alpha$$

Compare this short derivation to the two-segment double pendulum derivation in appendix G. When referring to it, note the dramatic increase in complexity. Also note that the term $\frac{\partial KE}{\partial \theta}$ is now nonzero.

Lagrange's equation of motion is an extensive area of inquiry by itself. Interested readers should refer to books such as *Schaum's Outline* (Hill 1967) and texts by Kilmister (1967) or Marion and Thornton (1995).

NUMERICAL SOLUTION TECHNIQUES

Generating model kinematics is a complex process generally referred to as *solving* the model equations. It is simply a glorified extrapolation process in which a current position, velocity, and acceleration are used to calculate a new position and velocity. For example, if you were in your car and driving at 100 km/h, you could *estimate* that you would reach a city 200 km away in 2 h. Note that your estimate would not be exact, but that the closer you got to the destination, in general, the better you would be able to estimate your arrival time. The same principles operate in human movement simulation. We use the current kinematics of the system to estimate future kinematics, and in general, we are more accurate when we estimate into the very near future. As shown in figure 10.3, the velocity at P_1 (given by the slope of the line V_1 tangent to the position-time profile) is a good estimator of the position at P_2. It is slightly less accurate in determining the position at P_3, so we use the velocity V_3 to find P_4. Neither velocity V_1 nor V_3 will generate a reasonable estimate of position P_5.

An *analytical solution* to a differential equation can be determined in the form of equations, as demonstrated earlier in the simple projectile motion example. A *numerical solution*, as the name suggests, is developed using numerical values. It is an *iterative* process in which the differential equations are solved over and over again to estimate how much the

position of the body will change over small periods of time. A flowchart of this process is presented in figure 10.4. It starts with the *initial conditions,* the starting values for the position and velocity of each segment. These specify the total starting energy in the system. The values are inserted into the differential equations, and the result is a new set of positions and velocities. These new positions and velocities are inserted into the differential equations, and the process is repeated for the next instant in time. This continues until one or more of the values being calculated reaches the desired ending value. This process is often referred to as a *forward simulation, forward solution,* or *numerical integration.*

An important part of the numerical solution is determining the starting and ending conditions. Typically, choosing initial conditions is fairly simple. For example, when simulating walking, the conditions of the body segments at the start of a gait cycle could be used. When simulating a jump, a logical starting posture would be used. However, end conditions are not as simple, because the exact position of the simulated body becomes less and less predictable as the simulation proceeds. For example, heel contact may seem like a logical ending condi-

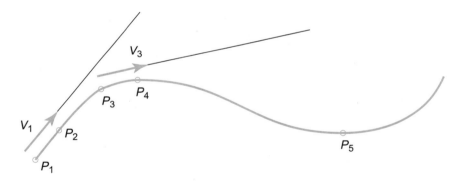

○ **Figure 10.3**   An arbitrary path of an object, with velocity vectors noted along tangent lines.

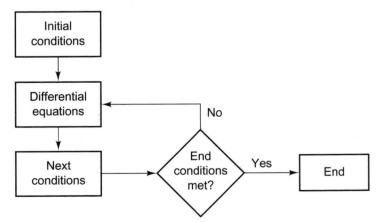

○ **Figure 10.4**   Flowchart of an iterative solution process.

tion for a walking simulation. However, there can be times when a simulated heel drags the ground early in the swing phase or a toe contacts the ground before the heel. Thus, emphasis must be placed on determining a robust algorithm for properly ending the simulations.

EXAMPLE 10.5

The differential equation of motion is

$$F(v) - mg = ma$$

Note that the external force *F*—the wind resistance—has been written as a function of the object's velocity. This equation must be solved for the acceleration, *a:*

$$a = \frac{F(v)}{m} - g$$

The forward solution code then takes the following form:

- Set the initial value for the position of the object.
- Set the initial value for the velocity of the object.
- Set the time step of the iteration.
- Set the ending condition of the simulation.
- Repeat the following:
- Calculate the acceleration *(a)*.
- Estimate a new velocity *(v)* for some small time step later.
- Estimate a new position at this small time step later.
- Record the new position and velocity.
- If the ending condition is not met, continue iterating.

In this example, the initial position of the object is its height above the ground. The ending condition could be any number or combination of variables. For example, we might stop the simulation when the object hits the ground, when it has fallen a certain distance, when it reaches a given velocity, or some other condition. The ending condition is chosen according to the goal of the simulation.

Complete expositions on a variety of numerical solution techniques appear in most textbooks on differential equations or numerical analysis. The simplest technique, called the *Euler method,* often is not precise enough for human movement simulation. A more precise and popular technique is the *Runge-Kutta algorithm.* Because it is a four-step process, it runs slower than the Euler method in computer code, but it returns more precise data.

Numerical solutions require that the *time step* be defined. Often denoted as Δt or *h,* it is the time between data points in the simulation. The size of the time step depends on the complexity of the model equations, but normally, it is between 0.01 and 0.001 s. Too large a time step can lead to an inaccurate result after a number of iterations. Increasingly smaller time steps generally do not cause problems because most computers use extended-precision numbers, but the time step should remain large enough so that the simulation does not take an inordinate amount of time to run.

A feature common to human movement simulations is that they start with a differential equation for the acceleration. If we know the acceleration of the object at some point in time, then we can estimate how much the velocity will change over some small time step. Similarly, we can then estimate how much the object's position will change over that period of time. The remainder of the solution is then just a matter of repetition. This is demonstrated in example 10.6.

EXAMPLE 10.6

Again, the differential equation is

$$a = \frac{F(v)}{m} - g$$

Let $F(v) = -0.2 \, v$. Note that the right-hand side of this equation is negative because this force, like all frictional forces, opposes the direction of movement. In mathematical/computer code, the Euler solution would be written as

$g = 9.81$ (acceleration due to gravity)

$m = 10$ (mass of object in kg)

$h_o = 100$ (initial height, 100 m over ground)

$v_o = 0$ (initial velocity, zero)

$\Delta t = 0.01$ (time step)

while $h > 0$, do (stop when object hits ground)

$a = F(v)/m - 9.81,$

(calculate acceleration; note value of *g*)

$v = v_o + a \, \Delta t$ (calculate new velocity)

$h = h_o + v \, \Delta t + 0.5 \, a \, \Delta t^2$

(calculate new position)

$h_o = h$ (set starting position for next iteration)

$v_o = v$ (set starting velocity for next iteration)

end while loop.

(continued)

Carrying this example out numerically, the first iteration is:

$$F(v) = -0.2(0 \text{ m/s}) = 0 \text{ N}$$

$$a = 0 \text{ N}/10 \text{ kg} - 9.81 \text{ m/s}^2 = -9.81 \text{ m/s}^2$$

$$v = 0 \text{ m/s}^2 + (-9.81 \text{ m/s}^2) \, 0.01 \text{ s} = -0.0981 \text{ m/s}$$

$$h = 100 \text{ m} + (-0.0981 \text{ m/s}) \, 0.01 \text{ s}$$
$$+ \, 0.5(-9.81 \text{ m/s}^2)(0.01 \text{ s})^2 = 99.99853 \text{ m}$$

In the first hundredth of a second, the object has fallen only half a millimeter. The second iteration uses the velocity and position values calculated in the first iteration:

$$F(v) = -0.2(-0.0981 \text{ m/s}) = 0.01962 \text{ N}$$

$$a = 0.01962 \text{ N}/10 \text{ kg} - 9.81 \text{ m/s}^2 = -9.80804 \text{ m/s}^2$$

$$v = -0.0981 \text{ m/s}^2 + (-9.80804 \text{ m/s}^2) \, 0.01 \text{ s}$$
$$= -0.19618 \text{ m/s}$$

$$h = 99.99853 \text{ m} + (-0.19618 \text{ m/s}) \, 0.01 \text{ s}$$
$$+ \, 0.5(-9.80804 \text{ m/s}^2)(0.01 \text{ s})^2 = 99.99608 \text{ m}$$

Now the object has fallen 3.9 mm. Because this process uses small time steps, the distance traversed between each step is very small. Iterations can be carried out in a spreadsheet or with computer code. The first 10 iterations return the following values for wind resistance, acceleration, velocity, and height.

F(v)	a	v	h
0	−9.81	0	100
0	−9.81	−0.0981	99.99853
−0.01962	−9.80804	−0.19618	99.99608
0.039236	−9.80608	−0.29424	99.99264
0.058848	−9.80412	−0.39228	99.98823
0.078456	−9.80215	−0.4903	99.98284
0.098061	−9.80019	−0.58831	99.97646
0.117661	−9.79823	−0.68629	99.96911
0.137258	−9.79627	−0.78425	99.96078
0.15685	−9.79431	−0.88219	99.95147
0.176439	−9.79236	−0.98012	99.94118

The iteration process is more complicated for multi-segmented systems, because each iteration must be performed for each segment with the more sophisticated Runge-Kutta algorithm. However, it is still analogous to the example presented above. Readers may refer to appendix F for a complete set of equations and steps for a two-segment pendulum.

CONTROL THEORY

Note that there are many different ways to model the forces and moments acting on the human body. This becomes a concern as models become more sophisticated, because in effect we begin to model the manner in which the human body controls itself. *Control theory* is a branch of engineering that studies the control of systems, both natural and man-made. Its applications are broad: autopilot controls for airplanes, temperature regulation in a nuclear power plant, antilock brakes for cars, and robotic controls on a factory production line. Human movement researchers explore the means by which a particular task is controlled and thus draw from control theory. We mention it in this chapter only to alert readers to its complexities and the elements that it shares with human movement simulation.

One aspect of control theory that is of particular interest is *feedback*. Feedback is information collected by the organism about either itself or the environment. When we walk, for example, we receive many different forms of feedback, including sensory information from our vision, proprioception, and vestibular system, to name but a few of the channels. We use this information to determine how to control our gait, altering our control as it becomes necessary. This type of control is called *closed-loop control*. In contrast, a control strategy in which a preprogrammed routine is enacted without using feedback is referred to as *open-loop* or *feedforward control*. In computer simulations, it is difficult to implement closed-loop control. Typically, researchers program the model controls in advance, run the simulation, and then change the controls afterward. The simplicity of this control may seem unrealistic, but every tool is limited, and we just have to use them while respecting their limitations. The following examples of human movement simulations show that there is still much to learn even without perfect controls.

LIMITATIONS OF COMPUTER MODELS

No matter how sophisticated a computer model becomes, many limitations remain, and they arise from a number of factors:

* the numerical imperfections that exist in the solution process,
* the difficulty of accurately modeling impacts,
* the approximation of complex human structures,

From the Scientific Literature

Mochon, S., and T.A. McMahon. 1980. Ballistic walking. *Journal of Biomechanics* **13:49-57.**

A classic computer simulation was the ballistic walking model of Mochon and McMahon (see also McMahon 1984). This model had three segments representing the human body: a straight, rigid stance limb that supported a swinging thigh and leg (figure 10.5). The mass of the head, trunk, and arms was lumped together as a point mass at the hip. The model had no muscles. It started "walking" at the instant of right toe-off, and the simulation ended at right heel contact. This model was, of course, too simple to predict many features of human walking, but the authors noted that their goal was only to make simple statements. In addition to the fact that their model could locomote without falling down, they found that their model reflected the change in walking speed that occurs with increasing body height. Also, as is the case with humans, the model's speed was limited by the need for the swing leg to clear the ground.

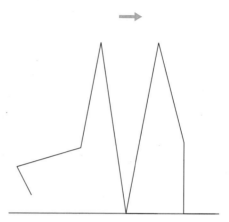

○ **Figure 10.5**   Stick figure of the walking model of Mochon and McMahon (1980).

From the Scientific Literature

Onyshko, S., and D.A. Winter. 1980. A mathematical model for the dynamics of human locomotion. *Journal of Biomechanics* **13:361-8.**

Onyshko and Winter developed a more advanced model than that of Mochon and McMahon in this 1980 paper. As shown in figure 10.6, it had seven segments corresponding to a torso with two thighs, legs, and feet. Six joint moments controlled this system, and it could perform a full gait cycle. Although this model appears to be similar to Mochon and McMahon's, it was in fact much more complicated. The equations of motion, derived via Lagrange's method, required 22 anthropometric constants and had literally dozens of terms. The equation of motion for each segment also required joint moments determined from experimental data. Moreover, the model had to operate differently

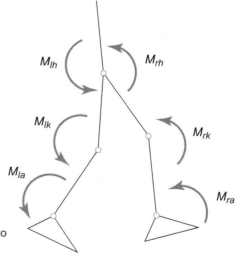

○ **Figure 10.6**   Stick figure of the walking model of Onyshko and Winter (1980), showing its joint moment controls.

(continued)

depending on the phase of the gait cycle. In other words, single support on the right foot required the equations of motion opposite to those for single support on the left foot, and these were different from the double-support equations. In effect, the simulation included four separate programs run by a master program that constantly determined what phase of the gait cycle the model was in.

From the Scientific Literature

Derrick, T.R., G.E. Caldwell, and J. Hamill. 2000. Modeling the stiffness characteristics of the human body while running with various stride lengths. *Journal of Applied Biomechanics* **16**:36-51.

Sometimes the simplest models to implement are the hardest to interpret. Earlier, we mentioned mass-spring models of human running, in which a mass corresponding to body mass bounces on a large spring. Interested readers should refer to McMahon, Valiant, and Frederick (1987) for further reading. Alexander, Bennett, and Ker (1986) noted that mass-spring models generate ground reaction force (GRF) curves shaped like inverted parabolas, whereas human GRFs are more complex, typically showing an initial sharp peak followed by a larger, more rounded peak. Therefore, Alexander proposed a more sophisticated mass-spring-damper model of running. In this model, a large mass rested atop a spring, which in turn rested on a second, smaller mass. The smaller mass then rested on a spring and a damper (or dashpot), which is an energy-dissipation element much like a shock absorber in a car. Alexander demonstrated that this model produced the more lifelike two-peak GRF. Diagrams of the two model types and their GRFs are shown in figure 10.7.

Derrick, Caldwell, and Hamill (2000) used a mass-spring-damper model to study the influence of different stride characteristics on GRF characteristics. In the model, the spring between the two masses tended to influence the time course of the GRF (i.e., a stiffer spring had less contact time with the ground), and the lower spring tended to influence only the first impact peak. In the study, 10 experimental subjects ran at their preferred stride length

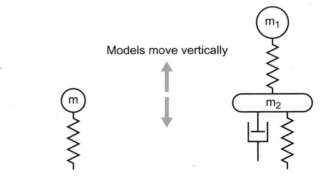

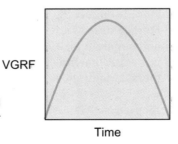

○ **Figure 10.7** Mass-spring and mass-spring-damper models with their respective vertical GRF curves.

(continued)

and at ±10 and ±20% of their preferred stride length. Then, for each experimental condition, the two model springs' stiffness was adjusted using an optimization algorithm to match the model GRF to the experimental. In these experimental curves, as stride length decreased, the GRF impact peak decreased by 33% and stance time decreased by 7%. In the model parameters, the upper spring was twice as stiff with the short strides as it was with the long strides. The lower spring had the opposite effect, being about 80% as stiff with the short strides.

Interpretation of mass-spring-damper data is not straightforward, because the model is very abstract. The lower mass was 20% of body mass, which might mean that it corresponded to the mass of the stance limb. Therefore, we can think of the upper spring as a measure of whole-body compliance. As much as we might like to, we cannot say that the upper spring represents the stiffness of a specific joint, such as the knee. The fact is that multiple joints are involved in absorbing the impacts of running, so we cannot draw such a direct analogy. The correct interpretation is that when human subjects run with longer strides, their bodies become more compliant. Both Derrick, Caldwell, and Hamill (2000) and McMahon, Valiant, and Frederick (1987) noted that human subjects achieved greater compliance with greater degrees of knee flexion. However, neither is this something we can extract from the mass-spring-damper model, simply because it does not have knees.

The mass-spring-damper model demonstrates clearly that models must be interpreted carefully. The lower extremity is not a spring, but we can glean some useful information by modeling it as a simple spring.

- the enormous adaptability of the human system, and
- the inevitability of having to make assumptions, because we can never account for all of the human and environmental factors.

Numerical errors are often slight and can often be resolved (see, for example, Risher, Schutte, and Runge 1997). Nonetheless, models become increasingly sensitive as the number of components increases. In addition, small errors that arise early in a simulated movement can propagate and grow to measurable levels by the end of the simulation. Also, models can have parameters that differ by many orders of magnitude, a situation referred to as a *stiff system* of equations. This can lead to a sensitive model that generates inconsistent results.

Impacts are very difficult to model. One reason for this is that they incur sharp changes in the accelerations of the body segments, which leads to numerical errors. Also, impacts generally violate the assumption that the human body consists of rigid segments. During walking, for example, the feet deform, storing and dissipating mechanical energy in a manner that is extremely difficult to model. When the human body runs or performs a drop landing, the impacts make computer simulation very difficult, indeed. Often, these issues can be resolved by reducing the time step of the solution, but sometimes it is necessary to increase the complexity of the model. For example, Derrick, Caldwell, and Hamill's (2000) model used a time step of 0.0001 s. However, modeling heel contact in running and walking is still a very difficult problem, especially for different locomotion speeds and subject anthropometries. Refer to Gilchrist and Winter (1997) for more details.

Human structures themselves are much more complicated than our models of them. For example, most simulations use simple representations of joints and assume that segments are rigid; for most purposes, these work well. However, joints are not simple pivots, muscles are not simple torque actuators, and segments are not rigid. Thus, researchers must be alert to the effects of such simplifications. For example, if a discrepancy were observed in the ankle moment data of a running model, a researcher might first question whether the problem stemmed from the assumption that the foot segment was rigid. A second question might be whether the model ankle reasonably approximated the orientation of the ankle (subtalar) joint.

When modeling, we must remember that we are simulating a human system in which the sensory and control systems and their organization are extraordinarily complex. A complete model of the moving human—even if we understood the entire human system to begin with—would be too much for the most sophisticated computers to handle. Therefore, every model of the human system is an oversimplification. This is an accepted part of the science. For example, as was mentioned earlier in this chapter, virtually all walking simulations have been performed on level ground, yet researchers have difficulty completely modeling the movement in even this simple situation. In addition, humans can adapt to myriad surfaces, turns, steps, bumps, and other conditions while walking, and each of these would require a different—and likely complex—adjustment to the model. Our models also neglect the vast array of senses that influence our movements with their messages. Even if we put aside these concerns, the issue of adaptability under simple test conditions is also pertinent. In short, replicating a human movement is generally straightforward, but predicting how that movement will change under different test conditions is difficult. Suppose, for example, that we have a highly accurate simulation of normal walking. What will happen when the torso mass increases, as it would if we had to account for a backpack? Will the stride lengthen or shorten? Will the torso angle change? Will the load be too heavy altogether?

In light of these limitations, it is clear that models should be interpreted with caution. It is more instructive to look at the larger results of the study than at small differences in joint moments or segment kinematics. Interpretations must also be limited to the specific circumstances of a model, even when the risks of extrapolating results may seem inconsequential. For example, a model of normal walking should not be used to infer the gait of an amputee, nor should a simulation performed at one speed of movement be assumed to apply to all speeds. Environmental conditions and constraints strongly affect subject behavior, as do such subject characteristics as age, gender, physical training, and body type; all of these can introduce limitations. Obviously, one cannot be too careful in deciding how much a model says about the real world. The beauty of simulation lies in its great flexibility and controllability, but it is because of those factors that we cannot draw direct inferences to the real world. All conclusions are implied and subject to further experimentation. In short, although model interpretation may be more art than science, its importance cannot be overemphasized. When interpreting model data, use the following guidelines:

- The exact magnitudes of quantitative data should be treated conservatively, because of their sensitivity to model assumptions.

- Analogies between model and human structures should remain loose.

- Never state that the way in which the model works is the way in which the human system works; discuss the model and humans separately.

Regarding the first point, it is shown throughout this text that many assumptions go into our calculations of such data as joint moments and segment kinematics. These same problems exist in computer models. In fact, models have additional assumptions, and so we need to be even more reserved in making our quantitative assessments. In chapters 2, 5, and 11, we relate that anthropometric estimates and various treatments of data such as filtering and differentiation limit us to about a 10% threshold of interpretation of joint moment data. So, for example, for a jumping model that generates knee joint moment data, we must consider that 10% threshold of interpretation. However, the simulation model also makes other assumptions about how the subject generated that joint moment. Perhaps our model would produce slightly different joint moments if we imposed different constraints on the subject or if we employed different optimization techniques or numerical integration methods.

In reference to the second guideline, we generally should not state that a particular model component *is* a human structure. Almost any physical system in the world can be modeled as a second-order differential equation, and a system of masses, springs, and dampers can be constructed to model almost any human movement. This should not lead us to state that some part of the human body *is* a mass, spring, or damper; instead, state that the body part *acts* like a mass, spring, or damper in our model. For example, do not state that the spring of a mass-spring running model *is* the lower extremity, but rather that the spring *behaves* like the lower extremity. This may seem to be only a matter of semantics, but it is not. Biomechanists construct *behavioral models* of human movement, and therefore, we believe that it is best to discuss a model separately from the real world.

SUMMARY

This chapter began with a discussion of why one would want to simulate human movement on a computer. It concluded by discussing how limited our current models are, which might seem to suggest that simulation is too grossly limited to be of much use. This is not the case, however. The fact is that *we do not understand* how we perform even a "simple" task like walking. Simulation is a fertile ground for testing and developing what we understand, a powerful tool that has great promise for developing many practical applications in the medical and industrial fields. The key, as with all tools, is to use it properly. Constructing a model is relatively straightforward, but heeding the limits of interpretation is the most important part.

SUGGESTED READINGS

Bobbert, M.F., K.G. Gerritsen, M.C. Litjens, and A.J. van Soest. 1996. Why is countermovement jump height greater than squat jump height? *Medicine and Science in Sports and Exercise* 28:1402-12.

Gerritsen, K.G.M., A.J. van den Bogert, and B.M. Nigg. 1995. Direct dynamics simulation of the impact phase in heel-toe running. *Journal of Biomechanics* 28:661-8.

Hill, D.A. 1967. *Schaum's Outline of Theory and Problems of Lagrangian Dynamics.* New York: McGraw-Hill.

Hull, M.L., H.K. Gonzalez, and R. Redfield. 1988. Optimization of pedaling rate in cycling using a muscle stress-based objective function. *International Journal of Sports Biomechanics* 4:1-20.

Kilmister, C.W. 1967. *Lagrangian Dynamics: An Introduction for Students.* New York: Plenum Press.

Marion, J.B., and S.T. Thornton. 1995. *Classical Dynamics of Particles and Systems.* 4th ed. New York: Harcourt Brace College.

Soest, A.J. van, M.F. Bobbert, and G.J. van Ingen Schenau. 1994. A control strategy for the execution of explosive movements from varying starting positions. *Journal of Neurophysiology* 71:1390-1402.

CHAPTER 11

Signal Processing

Timothy R. Derrick

A signal is a time- or space-varying quantity that conveys information. It may take the form of a sound wave, voltage, current, magnetic field, displacement, or a host of other physical quantities. These are examples of continuous signals (meaning that the signal is present at all instances of time or space). For convenience and to enable manipulation by computer software, we often convert continuous signals into a series of discrete values by sampling the phenomena at specific time intervals (figure 11.1). In this chapter, we

- define how to characterize a signal,
- examine the Fourier analysis of signals,
- outline wavelet analysis,
- explain the sampling theorem,
- discuss how to ensure cyclic continuity, and
- review various data-smoothing techniques.

CHARACTERISTICS OF A SIGNAL

A sinusoidal time-varying signal has four characteristics: frequency (f), amplitude (a), offset (a_0), and phase angle (θ). These characteristics are depicted in the schematics in figure 11.2. The *frequency* represents how rapidly the signal oscillates and usually is measured in cycles per second (s) or hertz (Hz).

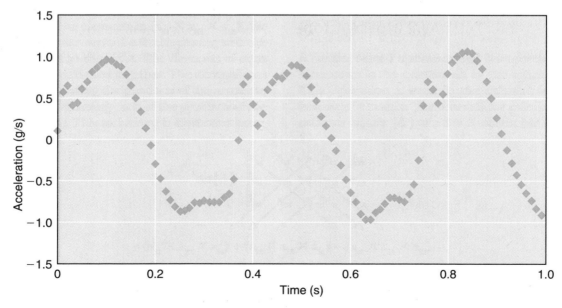

° **Figure 11.1** The digitized discrete representation of the acceleration of the head while running. The signal was sampled at 100 hertz (100 samples per second).

Frequency: *f*

a

Amplitude: *a*

b

Offset: a_0

0

c

Phase angle (shift): θ

d

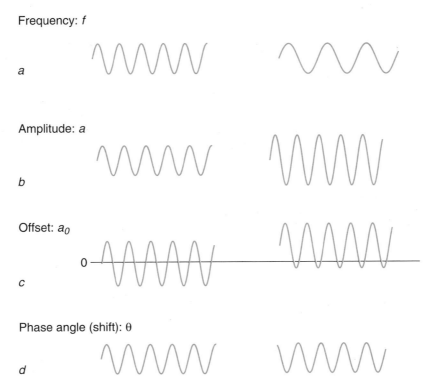

○ **Figure 11.2**   The four essential components of a time-varying signal.

One hertz (Hz) is equal to 1 cycle per second. For example, the second hand of a clock completes one cycle every 60 seconds. Its frequency is one cycle per 60 seconds, or 1/60 Hz. The frequency of a signal is easy to determine in a single sine wave (figure 11.2a), but more difficult to visualize in noncyclic signals with multiple frequencies. The *amplitude* of a signal quantifies the magnitude of the oscillations (figure 11.2b). The *offset* (or direct current [DC] offset or DC bias) represents the average value of the signal (figure 11.2c). The phase angle (or phase shift) in

the signal is the amount of time the signal may be delayed or time shifted (figure 11.2d).

Any time-varying signal, *h(t)*, is made up of these four characteristics. The following equation incorporates each of the four variables:

$$h(t) = a_0 + a\sin(2\pi ft + \theta) \qquad (11.1)$$

but $2\pi f = \omega$ (because *f* is in cycles/s or Hz, ω is in radians per second, and there are 2π radians in a cycle), so another way to write this is

$$h(t) = a_0 + a\sin(\omega t + \theta) \qquad (11.2)$$

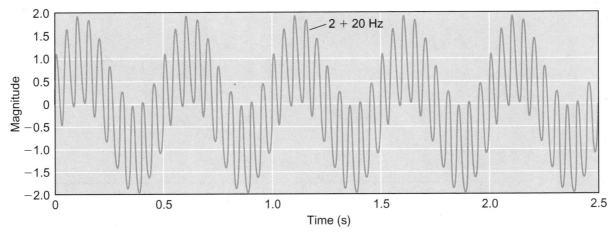

○ **Figure 11.3**   A 2 Hz and a 20 Hz sine wave summed over a 2.5 s period. The offset (a_0) and angle (θ) are zero for both waves, and the amplitude is 1.

The time (t) is a discrete time value that depends on how frequently the signal is to be sampled. If the sampling frequency is 100 Hz (100 samples/s), then the sampling interval is the inverse (1/100th or 0.01). This means that there will be a sample or datum registered every 0.01 s. So, t is one of the discrete values in the set (0, 0.01, 0.02, 0.03, . . . , T). The variable T represents the duration of the digitized signal. For example, by adding the equations for a 2 Hz sine wave to a 20 Hz sine wave, the following waveform is created (as illustrated in figure 11.3):

$$h(t) = \sin(2\pi 2t) + \sin(2\pi 20t) \tag{11.3}$$

FOURIER TRANSFORM

Any time-varying signal can be represented by successively adding the individual frequencies present in the signal (Winter 1990). The a_n and θ_n values may be different for each frequency (f_n) and may be zero for any given frequency.

$$h(t) = a_0 + \Sigma \, a_n \sin(2\pi f_n t + \theta_n) \tag{11.4}$$

By using the cosine and sine functions, this series can be rewritten without the phase variable as

$$h(t) = a_0 + \Sigma[b_n \sin(2\pi f_n t) + c_n \cos(2\pi f_n t)] \tag{11.5}$$

This series is referred to as the *Fourier series*. The b_n and c_n coefficients are called the *Fourier coefficients*. They can be calculated using the following formulae:

$$a_0 = \frac{1}{T}\int_0^T h(t)\,dt \tag{11.6}$$

$$b_n = \frac{2}{T}\int_0^T h(t)\sin(2\pi f_n t)\,dt \tag{11.7}$$

$$c_n = \frac{2}{T}\int_0^T h(t)\cos(2\pi f_n t)\,dt \tag{11.8}$$

Here is another way of looking at it: If you want to know how much of a certain frequency (f_n) is present in a signal h(t), you can multiply your signal by the sine wave $[\sin(2\pi f_n t)]$, take the mean value, and multiply it by 2. If you repeat this process for a cosine wave and then add the squares of the sine and cosine values together, you will get an indication of how much of the signal is composed of the frequency f_n. This is called the *power* at frequency f_n.

The Fourier coefficients can be calculated from the equally spaced time-varying points with the use of a *discrete Fourier transformation* (DFT) algorithm (appendix H). Given the Fourier coefficients, the original signal can be reconstructed using an inverse DFT algorithm. The DFT is a calculation-intensive algorithm. Faster and more commonly used are the *fast Fourier transformations* (FFTs). An FFT requires that the number of original data points be a power of 2 (. . . 16, 32, 64, 128, 256, 512, 1,024, . . .). If this is not the case, the usual method of obtaining a "power of 2" number of samples is to pad the data with zeros (add zeros until the number of points is a power of two). This creates two problems:

- Padding reduces the power of the signal. Parseval's theorem implies that the power in the time domain must equal the power in the frequency domain (Proakis and Manolakis 1988). When you pad with zeros, you reduce the power (a straight line at zero has no power). You can restore the original power by multiplying the power at each frequency by $(N+L)/N$, where N is the number of nonzero values and L is the number of padded zeros.

- Padding can introduce discontinuities between the data and the padded zero values if the signal does not end at zero. This discontinuity shows up in the resulting spectrum as increased power in the higher frequencies. To ensure that your data start and end at zero, you can apply a windowing function or subtract the trend line before performing the transformation. Windowing functions begin at zero, rise to 1 and then return to zero again. By multiplying your signal by a windowing function, you reduce the endpoints to zero in a gradual manner. Windowing should not be performed on data unless there are multiple cycles. Subtracting a trend line that connects the first point to the last point can be used as an alternative.

Most software packages give the result of an FFT in terms of a real portion and an imaginary portion. For a real discrete signal, the real portion corresponds to the cosine coefficient and the imaginary portion corresponds to the sine coefficient of the Fourier series equation. An FFT results in as many coefficients as there are data points (N), but half of these coefficients are a reflection of the other half. Therefore, the N/2 points represent frequencies from zero

$$a_0 = \text{mean}\big[h(t)\big] \tag{11.9}$$

$$b_n = 2 \times \text{mean}\big[h(t) \times \sin(2\pi f_n t)\big] \tag{11.10}$$

$$c_n = 2 \times \text{mean}\big[h(t) \times \cos(2\pi f_n t)\big] \tag{11.11}$$

$$\text{power}(f_n) = b_n^2 + c_n^2 \tag{11.12}$$

to one-half of the sampling frequency $(f_s/2)$. Each frequency bin has a width of f_s/N Hz. By increasing the number of data points (by padding with zeros or collecting for a longer period of time), you can decrease the bin width. This does not increase the resolution of the FFT; rather, it is analogous to interpolating more points from a curve.

Researchers often adjust the bin width so that each bin is 1 Hz wide. This is referred to as *normalizing the spectrum*. Adjusting the bin width changes the magnitude because the sum of the power frequency bins must equal the power in the time domain. Normalizing the spectrum allows data of different durations or sampling rates to be compared. The magnitude of a normalized spectrum is in units of (original units)^2/Hz.

A plot of the power at each frequency is referred to as the *power spectral density* (PSD) plot or simply the *power spectrum*. A PSD curve contains the same information as its time-domain counterpart, but it is rearranged to emphasize the frequencies that contain the greatest power rather than the point in time in the cycle at which the most power occurs. Figure 11.4 shows a leg acceleration curve along with its PSD.

TIME-DEPENDENT FOURIER TRANSFORM

The DFT has the advantage that frequencies can be separated no matter when they occur in the signal. Even frequencies that occur at the same time can be separated and quantified. A major disadvantage is that we do not know *when* those frequencies are present. We could overcome this difficulty by separating the signal into sections and applying the DFT to each section. We would then have a better idea of when a particular frequency occurred in the signal. This process is called a time-dependent Fourier transform.

Because we are already able to separate frequencies, we will use that technique to build an intuitive feeling for how this transform works. If we take a signal that contains frequencies from 0 to 100 Hz, the first step is to separate the frequencies into two portions, 50 Hz and below and 50 Hz and above. Next, we take these two sections and separate them into two portions each. We now have sections of 0 to 50, 50 to 100, 0 to 25, 25 to 50, 50 to 75, and 75 to 100 Hz. This procedure—called *decomposition*—continues to a predefined level. At this point, we have several time-series representations of the

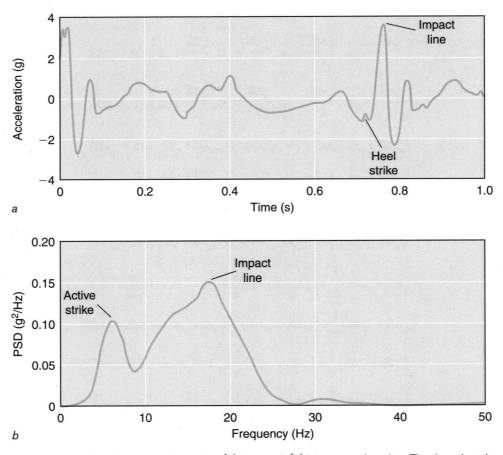

● Figure 11.4 Leg acceleration during running in the *(a)* time and *(b)* frequency domains. The time domain graph shows two ground impacts, whereas the frequency domain graph is for a single stance phase.

From the Scientific Literature

Shorten, M.R., and D.S. Winslow. 1992. Spectral analysis of impact shock during running. *International Journal of Sport Biomechanics* **8:288-304.**

The purpose of this study was to determine the effects of increasing impact shock levels on the spectral characteristics of impact shock and impact shock wave attenuation in the body during treadmill running. Three frequency ranges were identified in leg acceleration curves collected during the stance phase of running. The lowest frequencies (4 to 8 Hz) were identified as the active region as a result of muscular activity. The midrange frequencies (12 to 20 Hz) resulted from the impact between the foot and ground. There was also a high-frequency component (60 to 90 Hz) resulting from the resonance of the accelerometer attachment. Because these frequencies all occurred at the same time, it was impossible to separately analyze them in the time domain. Head accelerations were also calculated so that impact attenuation could be calculated from the transfer functions (TFs). TFs were calculated from the power spectral densities at the head (PSD_{head}) and the leg (PSD_{leg}) using the following formula:

$$TF = 10 \log_{10} (PSD_{head} / PSD_{leg}) \qquad (11.13)$$

This formula resulted in positive values when there was a gain in the signal from the leg to the head and negative values when there was an attenuation of the signal from the leg to the head. The results indicated that the leg impact frequency increased as running speed increased. There was also an increase in the impact attenuation so that head impact frequencies remained relatively constant.

original signal, each containing different frequencies. We can plot these representations on a three-dimensional (3-D) graph with time on one axis, frequency on a second axis, and magnitude on the third (figure 11.5). There are problems with this process that involve Heisenberg's uncertainty principle (both the time and frequency cannot be known at a point on the time-frequency plane). A similar process (wavelet analysis) can be used to identfy the content of waves other than sines and cosines.

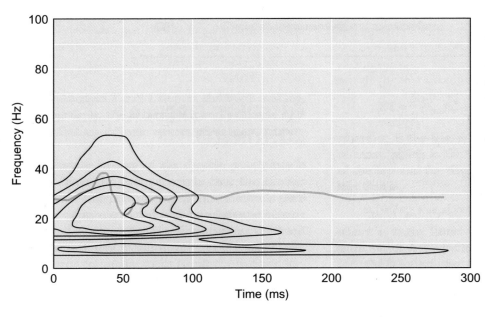

○ Figure 11.5 A 3-D contour map of the frequency-by-time values of a leg acceleration curve during running. The time domain curve is superimposed on the contour. There are two peaks in this curve. The high-frequency peak (approximately 20 Hz) occurs between 20 and 60 ms. The lower-frequency peak (approximately 8 Hz) occurs between 0 and 180 ms.

From the Scientific Literature

Wakeling, J.M., V. Von Tscharner, B.M. Nigg, and P. Stergiou. 2001. Muscle activity in the leg is tuned in response to ground reaction forces. *Journal of Applied Physiology* **91:1307-17.**

The purpose of this study was to investigate the response of muscle activity in the leg to different impact forces. A human pendulum apparatus was used to control leg geometry and initial conditions as the pendulum impacted a force platform. The loading rate was varied by changing the viscoelastic properties of the shoe midsole. Myoelectrical signals were recorded from the tibialis anterior, medial gastrocnemius, vastus medialis, and biceps femoris muscles. These signals were resolved by wavelet analysis into their magnitudes in time and frequency space. Traditional Fourier transformations are inadequate to describe a nonstationary signal like the one that would be anticipated during the impact. Differences occurred in the magnitude, time, and frequency content of the myoelectric signals during the period from 50 ms before impact until 50 ms after impact. These differences justified the use of the wavelet technique to accomplish the decomposition. The authors speculated that the change in myoelectric patterns that occurred with different loading rates resulted from differences in muscle fiber-type recruitment. Furthermore, they concluded that the levels of muscle activity adjusted in response to the loading rate of the impact forces.

SAMPLING THEOREM

The process signal must be sampled at a frequency greater than twice as high as the highest frequency present in the signal itself. This minimum sampling rate is called the *Nyquist sampling frequency* (f_N). In human locomotion, the highest voluntary frequency is less than 10 Hz, so a 20 Hz sampling rate should be satisfactory; however, in reality, biomechanists usually sample at 5 to 10 times the highest frequency in the signal. This ensures that the signal is accurately portrayed in the time domain without missing peak values.

The sampling theorem holds that if the signal is sampled at greater than twice the highest frequency, then the signal is completely specified by the data. In fact, the original signal *(h)* is given explicitly by the following formula (also see appendix I):

$$h(t) = \Delta \sum h_n \left[\frac{\sin\left[2\pi f_c(t - n\Delta)\right]}{\pi(t - n\Delta)} \right] \quad (11.14)$$

where Δ is the sample period (1/sampling frequency), f_c is $1/(2\Delta h_n$ is the *n*th sampled datum, and *t* is the time. By using this formula (Shannon's reconstruction formula; Hamill, Caldwell and Derrick 1997), it is possible to collect data at slightly greater than twice the highest frequency and then apply the reconstruction formula to "resample" the data at a higher rate

(Marks 1993). Figure 11.6 illustrates the use of the resampling formula to reconstruct a running vertical GRF curve. The signal was originally sampled at 1000 Hz, and the impact peak was measured at 1345 N. Every 20th point was then extracted to simulate data sampled at 50 Hz. The peak value occurred between samples, with the nearest data point at 1316 N. This also changed the time of occurrence of the impact peak. After applying the reconstruction formula to the 50 Hz data, the peak was restored to 1344 N with the same time of occurrence as the originally sampled data. With modern computers, there is little reason to undersample a signal unless the hardware is somehow limited, as is often the case when collecting kinematic data from video cameras with a sampling rate limited to 60 or 120 Hz.

ENSURING CIRCULAR CONTINUITY

For the resampling formula to work correctly, the data must have circular continuity. To understand what circular continuity is, draw a curve from end to end on a piece of paper and then form a tube with the curve on the outside by rolling the paper across the curve. Circular continuity exists if there is no "discontinuity" where the start of the curve meets the end of the curve. This means that the first point on

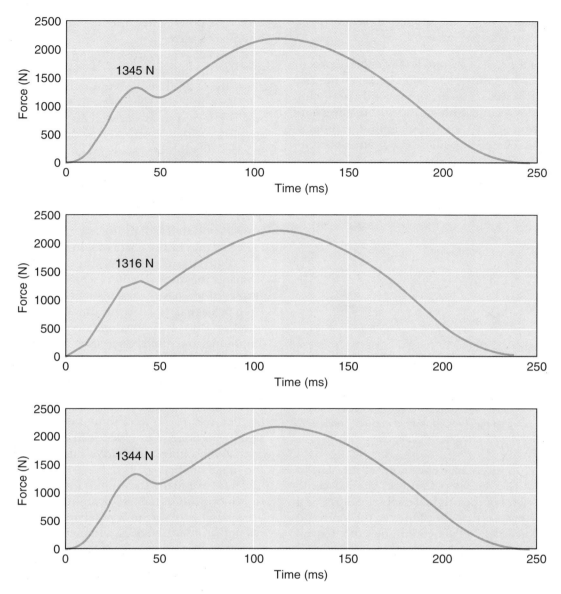

○ **Figure 11.6**   A running vertical GRF curve sampled at 1,000 Hz (top), sampled at 50 Hz (middle), and sampled at 50 Hz, then reconstructed at 1,000 Hz (bottom). The magnitude of the impact peak is identified in each graph. Reconstructing the signal results in a peak value very close to the original.

the curve must be equal to the last point. Nevertheless, the principle of circular continuity goes further: The slope of the curve at the start must equal the slope of the curve at the end. The slope of the slopes (the second derivative) must also be continuous. If you do not have circular continuity and you apply Shannon's reconstruction algorithm, you may be violating the assumption that only frequencies of less than half of the sampling frequency are present in the data. Discontinuities are by definition high-frequency changes in the data. If this occurs, you will see in the reconstructed data oscillations that have high amplitudes at the endpoints of the curve. These

oscillations become smaller (damped) the farther you get from the endpoints, and they become much more evident if derivatives are calculated.

Use the following steps (Derrick 1998) to approximate circular continuity (see figure 11.7):

- Split the data into two halves.
- Copy the first half of the data in reverse order and invert them. Attach this segment to the front of the original data.
- Copy the second half of the data in reverse order and invert them. Attach this segment to the back of the original data.

- Subtract the trend line from the first data point to the last data point.

Reversal of the first or second half of data is a procedure by which the first data point becomes the last data point of the segment, the second data point becomes the second to last, and so on. Inversion is a procedure that flips the magnitudes about a pivot point. The pivot point is the point closest in proximity to the original data. Figure 11.7 shows a schematic diagram of the data after summing the front and back segments and before subtracting the trend line.

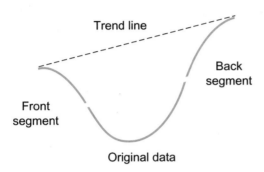

° Figure 11.7 Schematic diagram of the procedure used to ensure that a signal has the property of circular continuity.

Step 2 ensures circular continuity at the start of the original data set. Step 3 ensures circular continuity at the end of the original data set. Steps 2 and 3 together ensure that the slopes at the start and end of the new data set are continuous, but it is still possible to have a gap between the magnitude of the first point and the magnitude of the last point of the new data set. Step 4 removes this gap by calculating the difference between the trend line and

each data point. Thus, the first and last points will be equal to zero.

If you fail to heed the sampling theorem, you not only lose the higher frequencies, but the frequencies above the $1/2f_N$ (half the Nyquist frequency) actually fold back into the spectrum. In the time domain, this is referred to as *aliasing*. An anti-aliasing low-pass filter with a cutoff greater than $1/2f_N$ applied to a signal before it is analyzed ensures that there is no aliasing.

SMOOTHING DATA

Errors associated with the measurement of a biological signal may be the result of skin movement, incorrect digitization, electrical interference, artifacts from moving wires, or other factors. These errors, or "noise," often have characteristics that are different from the signal. Noise is any unwanted portion of a waveform. It is typically nondeterministic, lower in amplitude, and often in a frequency range different from that of the signal. For instance, errors associated with the digitizing process are generally higher in frequency than human movement. Noise that has a frequency different from those in the signal can be removed. If you were to plot the signal and the signal plus noise, it would look like figure 11.8. The goal of smoothing is to eliminate the noise, but leave the signal unaffected.

There are many techniques for smoothing data to remove the influence of noise. Outlined below are a number of the most popular. Each has its own strengths and weaknesses, and none is best for every situation. Researchers must be aware of how each method affects both the signal and the noise components of a waveform. Ideally, the signal would be unaffected by the smoothing process used to remove

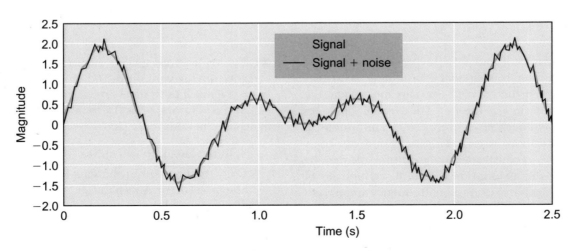

° Figure 11.8 A biological signal with and without noise.

the noise, but most smoothing techniques affect the signal component to some extent.

POLYNOMIAL SMOOTHING

Any n data points can be fitted with a polynomial of degree $n - 1$ of the following form:

$$x(t) = a_0 + a_1 t + a_2 t^2 + a_3 t^3 + \ldots + a_{n-1} t^{n-1} \tag{11.15}$$

This polynomial will go through each of the n data points, so no smoothing has been accomplished. Smoothing occurs by eliminating the higher-order terms. This restricts the polynomial to lower-frequency changes and thus it will not be able to go through all of the data points. Most human movements can be described by polynomials of the ninth order or less. Polynomials produce a single set of coefficients that represent the entire data set, resulting in large savings in computer storage space. The polynomial also has the advantages of allowing you to interpolate points at different time intervals and of making the calculation of derivatives relatively easy. Unfortunately, they can distort a signal's true shape; see the article by Pezzack, Norman, and Winter (1977)—reviewed in chapter 1—for an example of this technique. In practice, avoid using polynomial fitting unless the signal is a known polynomial. For example, fitting a second-order (parabolic) polynomial to the vertical motion of the center of gravity of an airborne body is appropriate. The path of the center of gravity during walking follows no known polynomial function, however.

SPLINES

A spline function consists of a number of low-order polynomials that are pieced together in such a way that they form a smooth line. Cubic (third-order) and quintic (fifth-order) splines are the most popular for biomechanics applications (Wood 1982). Splines are particularly useful if there are missing data in the data stream that need interpolation. Many techniques, such as digital filtering (discussed below), require equally spaced data. Splines do not have this requirement.

FOURIER SMOOTHING

Fourier smoothing consists of transforming the data into the frequency domain, eliminating the unwanted frequency coefficients, and then performing an inverse transformation to reconstruct the original data without the noise. Hatze (1981) outlined how to apply this method for smoothing displacement data.

MOVING AVERAGE

A three-point moving average is accomplished by replacing each data point (n) by the average of $n - 1$, n, and $n + 1$. A five-point moving average utilizes the data points $n - 2$, $n - 1$, n, $n + 1$, and $n + 2$ and results in more smoothing than a three-point moving average does. Note that there will be undefined values at the start and end of the series. This method is extremely easy to implement but is incapable of distinguishing signals from noise. It will attenuate valid signal components and may not affect invalid noise components. A better choice is the digital filter.

DIGITAL FILTERING

A digital filter is a type of weighted moving average. The points that are averaged are weighted by coefficients in such a manner that a cutoff frequency can be determined. In the case of a low-pass filter, frequencies below the cutoff are attenuated whereas frequencies above the cutoff are unaffected.

The type of digital filter is determined by the frequencies that are passed through without attenuation. The following digital filters are all implemented in the same manner, but the coefficients are adjusted for a particular cutoff frequency. Signals that are band passed or notch filtered are run through the filter with both low-pass and high-pass cutoff frequencies (figure 11.9).

Type of Filter

Filters can be constructed to attenuate different parts of the frequency spectrum. One or sometimes two cutoff frequencies are necessary to define which part of the frequency spectrum is attenuated and which part is left "as is," or passed unattenuated.

- Low-pass: The cutoff is selected so that low frequencies are unchanged, but higher frequencies are attenuated. This is the most common filter type. It is often used to remove high frequencies from digitized kinematic data and as a digital anti-aliasing filter.

- High-pass: The cutoff is selected so that high frequencies are unchanged, but lower frequencies are attenuated. It is used as a component in band-pass and band-reject filters or to remove low-frequency movement artifacts from low-voltage signals in wires that are attached to the body (e.g., electromyographic [EMG] signals).

- Band-pass: The frequencies between two cutoff frequencies are passed unattenuated. Frequencies below the lower cutoff and frequencies above the higher cutoff are attenuated. Such a filter is often

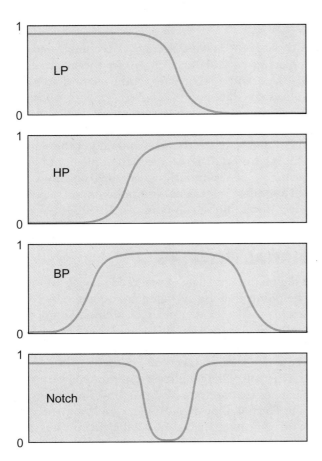

○ **Figure 11.9** Frequency responses of different types of digital filters. The digital filter is implemented in the time domain, but it can be visualized in the frequency domain. The frequency response function is multiplied by the signal in the frequency domain and then transformed back into the time domain. LP = low pass; HP = high pass; BP = band pass.

used in EMG when there is movement artifact in the low-frequency range and noise in the high-frequency range.

- Band-reject: The frequencies between the two cutoff frequencies are attenuated. The frequencies below the lower cutoff and above the higher cutoff are passed unattenuated. This filter has little use in biomechanics.

- Notch: A narrow band or single frequency is attenuated. It is used to remove power-line noise (60 or 50 Hz) or other specific frequencies from a signal. This filter generally is not recommended for EMG signals, because they have significant power at 50 to 60 Hz (for additional information, see chapter 8).

Filter Roll-Off

Roll-off is the rate of attenuation above the cutoff frequency. The higher the order (the more coefficients) or the greater the number of times the signal is passed through the filter, the sharper the roll-off.

Recursive and Nonrecursive Filters

Recursive filters use both raw data and data that were already filtered to calculate each new data point. They sometimes are called infinite impulse response (IIR) filters. Nonrecursive filters use only raw data points and are called finite impulse response (FIR) filters. It is theoretically possible that a recursive filter will show oscillations in some data sets, but they will have sharper roll-offs. Data that are smoothed using a recursive filter will have a phase lag, which can be removed by putting the data through the filter twice—once in the forward direction and once in reverse. The filter is considered a *zero lag* filter if the net phase shift is zero. Digital filters distort the data at the beginning and end of a signal. To minimize these distortions, extra data should be collected before and after the portion that will be analyzed.

Optimizing the Cutoff

The selection of a cutoff frequency is very important when filtering data. This is a somewhat subjective determination based on your knowledge of the signal and the noise. A number of algorithms try to find a more objective criterion for determining the cutoff frequency (Jackson 1979). These optimizing algorithms typically are based on an analysis of the residuals, which are what is left over when you subtract the filtered data from the raw data. As long as only noise is being filtered, some of these values should be greater than zero and some less than zero. The sum of all of the residuals should equal zero (or at least be close). When the filter starts affecting the signal, the sum of the residuals will no longer equal zero. Some optimization routines use this fact to determine which frequency best distinguishes the signal from the noise (figure 11.10). These algorithms are not completely objective, however, because you still must determine how close to zero the sum of the residuals is before selecting the optimal cutoff frequency.

Steps for Designing a Digital Filter

The following steps create a Butterworth low-pass, recursive digital filter. Modifications for a critically damped and a high-pass filter are also discussed. Butterworth filters are said to be optimally flat in the pass band, making them highly desirable for biomechanical variables. This means that the amplitudes of the frequencies that are to be passed are relatively unaffected by the filter. Some filters, such as the Chebyshev, have better roll-offs than the Butterworth filter, but they alter the amplitudes of the frequencies in the pass band.

1. Convert the cutoff frequency (f_c) from Hz to rad/s.

$$\omega_A = 2\pi f_c \qquad (11.16)$$

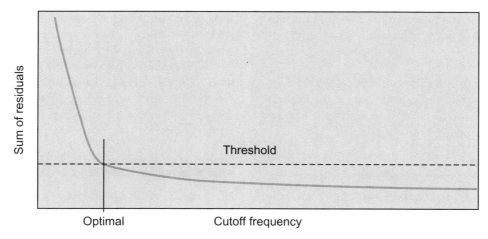

Selection of an optimal cutoff frequency using residual analysis.

2. Adjust the cutoff frequency to reduce "warping" that resulted from the bilinear transformation.

$$\Omega_A = \tan\left[\frac{\omega_A}{2 \times \text{sample rate}}\right] \tag{11.17}$$

3. Adjust the cutoff frequency for the number of passes *(P)*. A pass occurs each time the data are passed through the filter. For every pass through the data, a second pass must be made in the reverse direction to correct for the phase shift. Increasing the number of passes increases the sharpness of the roll-off.

$$\Omega_N = \frac{\Omega_A}{\sqrt[4]{2^{\left(\frac{1}{P}\right)} - 1}} \tag{11.18}$$

4. Calculate the coefficients.

$$C_1 = \frac{\Omega_N^2}{\left(1 + \sqrt{2}\Omega_N + \Omega_N^2\right)}$$

$$C_2 = \frac{2\Omega_N^2}{\left(1 + \sqrt{2}\Omega_N + \Omega_N^2\right)} = 2C_1$$

$$C_3 = \frac{\Omega_N^2}{\left(1 + \sqrt{2}\Omega_N + \Omega_N^2\right)} = C_1 \tag{11.19}$$

$$C_4 = \frac{2\left(1 - \Omega_N\right)^2}{\left(1 + \sqrt{2}\Omega_N + \Omega_N^2\right)}$$

$$C_5 = \frac{\left(\sqrt{2}\Omega_N - \Omega_N^2 - 1\right)}{\left(1 + \sqrt{2}\Omega_N + \Omega_N^2\right)} = 1 -$$

$$\left(C_1 + C_2 + C_3 + C_4\right)$$

5. Apply the coefficients to the data to implement the weighted moving average. Y_n values are filtered data and X_n values are unfiltered data.

The filter is recursive because previously filtered data—Y_{n-1} and Y_{n-2}—are used to calculate the filtered data point, Y_n.

$$Y_n = C_1 X_{n-2} + C_2 X_{n-1} + C_3 X_n + \\ C_4 Y_{n-1} + C_5 Y_{n-2} \tag{11.20}$$

This filter is underdamped (damping ratio = 0.707) and will therefore "overshoot" and "undershoot" the true signal whenever there is a rapid change in the data stream. For more information, consult the article by Robertson and Dowling (2003). A critically damped filter can be designed (damping ratio = 1) by changing $\sqrt{2}$ to 2 in each equation in step 4. The warping function must also be altered as follows:

$$\Omega_N = \frac{\Omega_A}{\sqrt{2^{\left(\frac{1}{2P}\right)} - 1}} \tag{11.21}$$

In practice, there is little difference between the underdamped and critically damped filter. The distinction can be seen in response to a step input (a function that transitions from 0 to 1 in a single step). The Butterworth filter will produce an artificial minimum before the step and an artificial maximum after the step (Robertson and Dowling 2003).

It is possible to calculate the coefficients so that the filter becomes a high-pass filter instead of a low-pass filter (Murphy and Robertson 1994). The first step is to adjust the cutoff frequency by the following:

$$f_c = \frac{f_s}{2} - f_{c-old} \tag{11.22}$$

where f_c is the new cutoff frequency, f_{c-old} is the old cutoff frequency, and f_s is the sampling frequency. The coefficients (C_1 through C_5) are then calculated in the same way that they were for the low-pass filter and then the following adjustments are made:

$$c_1 = C_1,\ c_2 = -C_2,\ c_3 = C_3,\ c_4 = -C_4,\ \text{and } c_5 = C_5 \qquad (11.23)$$

where c_1 through c_5 are the coefficients for the high-pass filter. The data can now be passed through the filter, forwards and backwards, just as in the low-pass filter case outlined previously.

SUMMARY

This chapter outlined the basic principles and rules for characterizing and processing signals acquired from any data-collection system. Special emphasis was given to the frequency or Fourier analysis of signals and data smoothing. Data-smoothing techniques are of particular interest to biomechanists because of the frequent need to perform double time differentiation of movement patterns to obtain accelerations. As illustrated in chapter 1, small errors in digitizing create high-frequency noise that, after double differentiating, dominates true data history. Removing high-frequency noise prior to differentiation prevents this problem. The biomechanist, however, should be aware of how these smoothing processes affect data so that an appropriate method can be applied without distorting the true signal.

SUGGESTED READINGS

Burrus, C.S., R.A. Gopinath, and H. Guo. 1998. *Introduction to Wavelets and Wavelet Transforms, A Primer.* Upper Saddle River, NJ: Prentice Hall.

Press, W.H., B.P. Flannery, S.A. Teukolsky, and W.T. Vetterling. 1989. *Numerical Recipes in Pascal,* 422-97. New York: Cambridge University Press.

Transnational College of LEX. 1995. *Who Is Fourier? A Mathematical Adventure.* Trans. Alan Gleason. Belmont, MA: Language Research Foundation.

International System of Units (System International, SI)

SOME RULES FOR REPORTING SI UNITS

- Do not use a decimal after the abbreviated versions of metric units unless the unit appears at the end of a sentence. For example, 35.6 N (short for newtons), 3.00 kg (short for kilograms), or 0.500 s (short for seconds) are correct forms; but 40.0 m., 20.5 kPa., or 20.5 sec. are incorrect forms.

- A centered dot (·) is used to separate abbreviated SI units involving combined quantities, such as N·s for newton seconds or kg·m² for kilogram meters squared. However, a decimal point (.) is also acceptable and is usually much easier to use.

- Do not capitalize a unit derived from a proper name when spelling out the unit even though the unit's abbreviation is a capital letter. Examples of such units include the watt (W), the newton (N), the hertz (Hz), the pascal (Pa), and the joule (J).

- Use a slash (/) to indicate an arithmetic division of units, such as, m/s for meters per second or N/m² for newtons per square meter.

- Do not mix abbreviations and unabbreviated forms in an expression. For instance, the following are incorrect forms: newtons per m, kg.meters, N.seconds, and watts/kg.

- The prefixes hecto, deca, deci, and centi should be avoided, except for the measurements of area, volume, and length, such as hectare, deciliter, and centimeter.

- When pronouncing metric units that have a prefix, the accent should always be placed on the complete prefix, that is, kilo'-meter versus ki-lo'-meter or kilo-meter'.

- Always type a space between the numeric part and the number except for °C, ° (angle) and %. Examples: 76.4 W, 20.4°C, 13.45%, 678 N·m, and 45.2°.

- When writing out numbers with greater than four digits on either side of the decimal point, use a space instead of a comma to separate digits into groups of three as, for example, 23 400 m or 0.002 63 m. This is because the comma is used in many countries as a decimal point. It is permissible to omit the blank in four digit numbers, such as 1002, 9980, and 0.1234.

Quantity*	Name	Symbol	Formula
Kinematic domain			
Length (*l, r, s, x, y, z*)	meter	m	
Area (*A*)	square meter	m^2	
	hectare	ha	$hm^2 = 10\ 000\ m^2$
Volume (*V*)	cubic meter	m^3	
	liter	L	$1\ dm^3$
Linear velocity, speed (*v*)	meter per second	m/s	
Linear acceleration (*a*)	meter per second squared	m/s^2	
Linear jerk (*j*)	meter per second cubed	m/s^3	
Plane angle ($\alpha, \beta, \gamma, \theta, \phi$)	radian	rad	m/m = 1
	degree	deg, °	i/180 rad
	minute	'	1/60°
	second	"	1/360°
	revolution	r	2 i rad, 360°
Angular velocity (ω)	radian per second	rad/s	
Angular acceleration (α)	radian per second squared	rad/s^2	
Solid angle (Ω)	steradian	sr	
Inertial property domain			
Mass (*m*)	kilogram	kg	
	metric ton or tonne	t	1 Mg = 1000 kg
Moment of inertia (*I, J*)	kilogram meter squared		$kg \cdot m^2$
Density (ρ)	kilogram per cubic meter		kg/m^3
Viscosity (η)	pascal second		$Pa \cdot s$
Time (temporal) domain			
Time (*t*)	second	s	
	minute	min	60 s
	hour	h	3600 s
	day	d	86 400 s
	year	a	31.536 Ms
Frequency ($\smallint$)	hertz	Hz	1/s
Kinetic domain			
Force (*F*)	newton	N	$kg \cdot m/s^2$
Moment of force (*M*), torque (τ)	newton meter		$N \cdot m$
Pressure (*p*)	pascal	Pa	N/m^2
	millibar	mbar	1 mbar = 100 Pa
Stress (σ or τ)	pascal	Pa	N/m^2
Energy (*E*), work (*W*)	joule	J	$kg \cdot m^2/s^2$
Power (*P*)	watt	W	J/s
Linear impulse	newton second		$N \cdot s$ or $kg \cdot m/s$
Linear momentum (*p*)	kilogram meter per second		$kg \cdot m/s$ or $N \cdot s$

Angular impulse	newton meter second		N·m·s or kg·m²/s
Angular momentum *(L)*	kilogram meter squared per second		kg·m²/s or N·m·s
Electrical domain			
Current *(I)*	ampere	A	
Voltage *(V)*	volt	V	W/A
Charge *(Q)*	coulomb	C	s·A
Power *(P)*	watt	W	J/s
Resistance *(R)*, impedance *(Z)*	ohm	Ω	V/A
Capacitance *(C)*	farad	F	C/V
Magnetic flux *(Φ)*	weber	Wb	V·s
Magnetic flux density *(B)*	tesla	T	Wb/m²
Inductance *(L)*	henry	H	Wb/A
Conductance *(G)*	siemens	S	A/V
Electric energy *(E)*	joule	J	W·s
Temperature domain			
Temperature (T)	kelvin	K	
	degree Celsius	°C	
Chemical domain			
Amount of substance *(n)*	mole	mol	
Concentration *(c)*	mole per cubic meter	mol/m³	
Light domain			
Luminous intensity *(I)*	candela	cd	
Luminous flux *(Φ)*	lumen	lm	cd·sr
Illuminance *(E)*	lux	lx	lm/m²
Radiant intensity *(I)*	watt per steradian		W/sr
Radiance *(L)*	watt per square meter per steradian		W·m²/sr

* Base units are in blue.

SI Prefixes

Multiplication factor	Prefix*	Symbol
1 000 000 000 000 000 000 000 000 = 10^{24}	yotta	Y
1 000 000 000 000 000 000 000 = 10^{21}	zetta	Z
1 000 000 000 000 000 000 = 10^{18}	exa	E
1 000 000 000 000 000 = 10^{15}	peta	P
1 000 000 000 000 = 10^{12}	tera	T
1 000 000 000 = 10^{9}	giga	G
1 000 000 = 10^{6}	mega	M
1000 = 10^{3}	kilo	k
100 = 10^{2}	hecto*	h
10 = 10^{1}	deca or deka*	da
0.1 = 10^{-1}	deci*	d
0.01 = 10^{-2}	centi*	c
0.001 = 10^{-3}	milli	m
0.000 001 = 10^{-6}	micro	μ
0.000 000 001 = 10^{-9}	nano	n
0.000 000 000 001 = 10^{-12}	pico	p
0.000 000 000 000 001 = 10^{-15}	femto	f
0.000 000 000 000 000 001 = 10^{-18}	atto	a
0.000 000 000 000 000 000 001 = 10^{-21}	zepto	z
0.000 000 000 000 000 000 000 001 = 10^{-24}	yocto	y

* These prefixes normally are not used except for special quantities, such as the hectare, the centimeter, and the decibel.

Selected Factors for Converting Between Units of Measure

Unit	Conversion factor
Area	
1 acre	= 0.405 ha
1 square inch	= 645.16 mm^2
Energy, work	
1 calorie	= 4.1868 J*
1 calorie (dietetic)	= 4.1855 kJ*
1 erg	= 0.1 μJ
1 foot pound-force	= 1.356 J
Force, weight	
1 dyne	= 10 μN
1 kilopond (kg force)	= 9.806 65 N
1 pound-force	= 4.448 N
Length	
1 foot	= 30.48 cm = 0.3048 m
1 inch	= 2.54 cm
1 yard	= 0.9144 m
1 mile	= 1.609 344 km

(continued)

Unit	Conversion factor
Mass	
1 ounce (avoirdupois)	= 28.35 g
1 pound (avoirdupois)	= 0.4536 kg
1 slug	= 14.59 kg
1 stone (14 lb, UK)	= 6.350 kg
1 ton (long, 2240 lb, UK)	= 1.016 Mg
1 ton (short, 2000 lb)	= 0.907 Mg
Power	
1 British thermal unit (BTU) per hour	= 0.293 W
1 horsepower (electric)	= 746 W
Pressure, stress	
1 atmosphere (standard)	= 101.325 Pa
1 mmHg (0°C)	= 133.3 Pa
1 pound per square inch (psi) (lb/in^2)	= 6.895 kPa
Temperature	
1 Fahrenheit degree	= 5/9 K**
Velocity	
1 mph	= 0.447 04 m/s = 1.609 344 km/h
Volume	
1 cubic foot	= 0.028 32 m^3
1 cubic inch	= 16.39 cm^3
1 gallon (imperial)	= 4.546 L
1 gallon (US)	= 3.785 L

All values in blue are exact conversions.
* The calorie commonly used by dietitians to refer to the amount of energy in foods is the same in international agreement.
** Add 32° after converting from Celsius to Fahrenheit or subtract 32° before converting from Fahrenheit to Celsius. Note that a Celsius degree is a kelvin (K).

Basic Electronics

S.N. Whittlesey and J. Hamill

This appendix gives a brief overview of the elementary electronic circuit concepts that are relevant to the collection of human movement data. The topics discussed include basic electronic components, Ohm's law, circuit diagrams, and the functions of several common lab instruments, such as amplifiers and electrogoniometers. Students interested in further detail or sample problems on any particular topic should refer to textbooks on electronics or linear circuits (such as *Schaum's Outline* [O'Malley 1992] or Winter and Patla 1997). The focus here is on simple, steady state circuit concepts and how they apply to common measurements of human movement.

Electronics notation and symbols are standardized across different fields. Here, we use the notations and symbols given in this table

Basic SI Electrical Units

Quantity	Symbol	SI unit	SI abbreviation
Current	I	ampere	A
Voltage	V	volt	V
Resistance	R	ohm	Ω
Capacitance	C	farad	F
Power	P	watt	W

CIRCUIT DIAGRAMS

A circuit diagram is a formal means of representing an electric circuit. We use these diagrams in this appendix to illustrate different examples. Circuit diagrams have many conventions, the most common of which are that

- components are represented by standard icons with their sizes noted,
- wires are represented by straight lines for zero resistance,
- wires are drawn only in north-south-east-west directions,
- a connection of two wires is indicated by a solid dot,
- one wire passing over another is indicated by a short loop, and
- interface points are indicated by open dots and labeled.

Circuit diagrams can, of course, become very complicated. The conventions just listed are displayed in figure C.1. This diagram shows the symbols for various components: a 9-volt (V) battery and its ground, a 100-ohm (Ω) resistor, a 10 Ω variable resistor, and a 1-microfarad (μF) capacitor. A discussion on these electrical components and several principles of electricity follows.

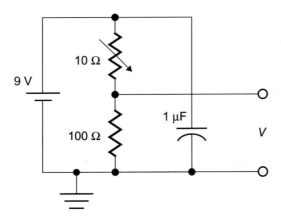

○ **Figure C.1**   Circuit diagram of a 9 V battery powering two resistors and a capacitor. The lower side of the battery is grounded. The voltage (V) is the quantity that we measure. The 10 Ω resistor has a variable resistance with a maximum of 10 Ω.

ELECTRIC CHARGE, CURRENT, AND VOLTAGE

Electric charge can be either positive or negative, depending on whether we are dealing with protons or electrons. Electricity is the flow of electrons through some medium, whether through a wire in a house or lightning through the air. The basic SI unit of electric charge is the coulomb (C). It represents about 6.25×10^{18} electrons. The rate of flow of electricity, or current, has units of amperes, or amps (A); 1 A is a flow rate of 1 C/s. As practical examples, consider that a handheld calculator requires a few microamperes (mA) to operate, a D-cell battery supplies about 100 mA, a car battery offers a maximum of about 2 A, and a typical house circuit provides 20 A. Current flow occurs when there is a difference between the electrical potential energy at two sites. This potential difference is called a voltage. One volt is defined as 1 joule (J) of energy per C of charge. A D-cell battery offers 1.25 V, a car battery offers 12 V, and house electricity averages 110 V. A human electromyograph (EMG), in contrast, is on the order of μV.

A point of zero voltage is called a ground. This is never an absolute quantity, but rather a defined reference point in a circuit. Thus, two circuits can have their own grounding references, but there may be a potential difference between the grounds of the two circuits. For example, in a small battery-powered circuit such as a clock or flashlight, ground is typically defined as the negative terminal of the battery powering the circuit. In house applications, ground is defined as the potential of the surrounding soil.

This is accomplished by connecting the circuit to a metal rod driven into the earth. This house ground is different from the ground in any battery-powered circuit unless a connection is made between them. As another example, jump-starting a car is dangerous because potential differences can exist between two cars; even though the battery in each car is 12 V, their tires insulate them from the road (which is the ground). In human movement, we often see these principles applied in EMG recording because different voltage potentials can exist over the skin surface of the body depending on what muscles are active. We often record EMG with a separate grounding plate on a bony landmark away from the musculature.

Voltage and current are related (as is discussed later in this appendix), and this is often a source of confusion. The basic principles, stated previously, must be remembered: Current is the flow of electrons and voltage is a potential energy difference that can cause electron flow. If current is flowing between two sites, then there must be a voltage difference between them. However, there can be a voltage difference without current flowing; in that case, there is no complete circuit for the current to flow through. For example, there is a voltage difference between the terminals of a wall outlet, regardless of whether an appliance is connected to it. Current only flows between the terminals when an appliance is connected to them and turned on. An extreme example is that birds can land on a high-voltage overhead power line without being harmed. The same principle applies to electrical line workers: As long as workers are highly insulated from the ground, it is possible for them to touch the wire with their bare hands. When contact is made, a person is thousands of volts higher than the ground, but because virtually no current can flow through the insulation, the worker is unharmed. However, when a power line is broken in a storm and one end falls to the ground, touching the wire can be fatal because making contact with the wire connects a circuit to the ground.

Circuits are often difficult to conceptualize because they cannot be visualized directly. A measurement instrument, such as a voltmeter, oscilloscope, or computer, must be used to establish the state of a circuit. This is an abstract task, and it can be helpful to use the flow of a fluid through a pipe system as an analogy. Electric current (amperage) is analogous to the rate of fluid flow through the pipe (i.e., liters per second). Voltage is analogous to the pressure in the pipe system. Thus, if water is flowing through a hose, there must be a pressure difference between the ends of the hose; however, we can have a closed, pressurized container with no water leaking out of it. Flow implies that a potential energy difference exists. The fact that a potential energy difference exists, however, does not imply that something is flowing. Other fluid examples will be offered throughout this appendix to illustrate key points.

We most often think of voltage as the strength of a power supply. However, it is also an important quantity that we measure. In biophysical systems, we almost always measure a voltage, not a current. This is primarily a matter of ease of use and the relative durability of voltmeters as compared to ammeters. When we speak of a biophysical signal, we are referring to a time-varying voltage produced by a human subject or some device attached to it.

RESISTORS

Electrical resistivity is a fundamental material property: As electrons pass through a material, energy is dissipated as heat. Resistance is a measure of this effect in a specific object. Resistance is measured in units of ohms (Ω), and thus resistivity has units of ohms per meter (Ω/m). In other words, the resistance of an object is a function of the resistivity of its material as well as the object's dimensions. In particular, resistance is directly proportional to the length of the material. Returning to fluid flow, resistivity is analogous to the friction that exists between a fluid and the pipe that it flows through; resistance is analogous to the total frictional force of the pipe system. The total resistance of a pipe depends on its frictional characteristics as well as its length. Electrical resistivities of materials vary over many orders of magnitude. For example, copper wire has a resistivity of about 10^{-4} Ω/m; human skin, 20 to 50 kΩ/m; semiconductors such as silicon are around 10^5 Ω/m; and wood, about 10^{13} Ω/m.

EXAMPLE C.1

Estimate the resistance of 1 cm of copper wire using the resistivity just given, 10^{-4} Ω/m.

See answer C.1 on page 280.

A resistor is a device that resists electricity. Typical resistor sizes vary from around 1 Ω to 1 MΩ. Knowledge of the resistances within a circuit is critical to understanding its behavior. Indeed, we typically use our knowledge of resistors to manipulate the flow of current and perform the desired function. Also, in human movement study, we often need to be aware of the resistances in both our instruments and the human body.

Some resistors have variable resistances. A common type of variable resistor is the potentiometer, often called a *pot*. Some potentiometers can be adjusted by turning them (a rotary pot), whereas others slide linearly. Volume controls on radios can take both forms, as can dimmer switches for indoor lighting.

Most circuits include multiple resistances. Thus, it is important to understand how resistors act when connected together. The two basic manners of connecting are in a series and parallel. In a series connection, there is one path. One resistor follows the other, and all current flowing through one resistor must also flow through the other (figure C.2). The total resistance of two resistors in series is equal to the sum of the resistances, that is,

$$R = R_1 + R_2 \qquad (C.1)$$

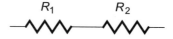

○ **Figure C.2**   Circuit diagram of two resistors in series.

If more resistors are added to the series, the total resistance is equal to the sum of each resistance:

$$R = R_1 + R_2 + R_3 + \ldots + R_n \qquad (C.2)$$

In a parallel connection, there is branching (figure C.3). The total current flowing through the system is divided between two or more resistors. The total resistance R of two or more resistors in parallel is given by

$$\frac{1}{R} = \frac{1}{R_1} + \frac{1}{R_2} + \ldots + \frac{1}{R_n} \qquad (C.3)$$

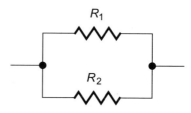

○ **Figure C.3**   Circuit diagram of two resistors in parallel.

For the case of two resistors in parallel, this reduces to

$$R = \frac{R_1 R_2}{R_1 + R_2} \qquad (C.4)$$

EXAMPLE C.2

a. What is the total resistance of two 10 Ω resistors in series? In parallel?

See answer C.2a on page 280.

b. What is the total resistance of a 10 Ω resistor and a 1 Ω resistor in series? In parallel?

See answer C.2b on page 280.

CAPACITORS

A capacitor is a device that stores electric charge; in our analogy to fluid flow, a capacitor is equivalent to a tank or a bucket that holds water. Its behavior is very different from that of a resistor and is not discussed in detail here. The important thing about capacitance is that it is a common physical property that we often must account for. It typically attenuates the voltage that we try to measure, and its effects can be noticeable on certain data. For example, high-speed devices such as telephones and computer networks have very thin cables because the capacitance of thicker cables would essentially absorb the small amounts of electricity being sent through them. This is analogous to the fact that a garden hose holds water: Water does not come out of the hose for a few seconds after the faucet is turned on because the water must first fill the hose to capacity. It is for this reason that some accelerometers have extremely thin cables. Similarly, EMG electrodes are preamplified to provide a stronger source of electricity that can overcome the capacitance of the wires.

Note that capacitance is not a *bad* factor, but simply a factor that must be taken into account. We in fact exploit the behavior of capacitors so that radios can be tuned to different stations. Capacitors can also be used to filter signals in the same way as the digital filters introduced in chapter 1 and detailed in chapter 11. Readers interested in relevant examples may again refer to any linear-circuits text.

Along with capacitors, impedance is also important. Impedance, denoted Z, is a more general term for all of the factors that limit electrical flow through a circuit. Thus, impedance includes the net effects of all resistors and capacitors in the circuit.

The symbol for a capacitor—two lines—represents the two plates that hold the electric charge. Sometimes the plates are drawn parallel, but at other times, the plate with the lower voltage is denoted with a curved shape.

OHM'S LAW

Ohm's law is perhaps the most fundamental law in all of electronics. It states that the voltage across a resistor equals the product of its resistance and the current flowing through it:

$$V = IR \qquad (C.5)$$

where V is the voltage across the resistor, I is the current through the resistor, and R is the magnitude of the resistance. There are different ways to express this law. If we increase the voltage in a circuit, the current increases in proportion, and if we increase the size of a resistor, the current decreases proportionately. When plotted, this function is a straight line, as shown in figure C.4. This is a linear function. It remains true regardless of the magnitude of the current or how it changes over time. We express this mathematically as

$$V(t) = I(t)\,R \qquad (C.6)$$

Because of this linearity, resistor circuits are the most straightforward to analyze, although they, too, can get complicated.

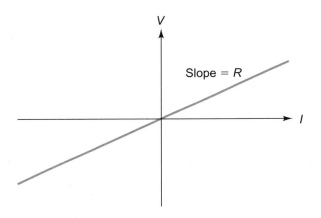

Figure C.4   Graphical representation of Ohm's law.

Ohm's law is analogous to fluid flow. The electrical resistance R corresponds to the resistance of the piping, the current I corresponds to the volume rate of fluid flow, and the voltage V corresponds to pressure. If we increase the voltage, more current flows, in the same way that if we increase the pressure of water in a pipe, more water flows. If we increase the resistance in a circuit, less current flows—again, just like water.

EXAMPLE C.3

a. What is the current flowing in this circuit?

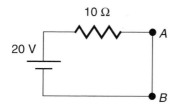

20 V   10 Ω

• A

• B

See answer C.3a on page 280.

b. In this circuit, what is the size of the resistor R?

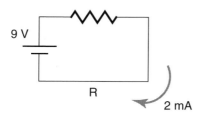

9 V

R

2 mA

See answer C.3b on page 280.

c. Suppose a 110 V house outlet is wired to a 15 A circuit breaker. What is the minimum resistance that can be applied to this outlet?

See answer C.3c on page 280.

In practice, we are not concerned with currents flowing through loops, as these examples have illustrated. Instead, we often speak of the voltage drop across a resistor, which has to do with the manner in which you measure voltage. Since voltage is a potential difference between two points, a voltmeter measures the difference with two probes. For example, in the circuit in example 2, if we placed one probe before the 9 V battery and one probe after it, the voltmeter would register a voltage gain of 9 V. If we placed the probes across the resistor, the voltmeter would measure a voltage drop of 9 V (i.e., it would read –9 V). If you placed the probes across points A and B, the voltmeter would register 0 V because there is no resistance between these points. In measuring any electrical device, there is a specific component *across which* the changes in voltage are measured.

Earlier in this appendix, we discussed impedance. When measuring impedance, the formula is analogous to Ohm's law:

$$Z = \frac{V}{I} \qquad (C.7)$$

where Z is the impedance, V is the voltage across the circuit, and I is the current through it. If a circuit is made up entirely of resistors, then the impedance is equal to the resistance. However, for reasons beyond the scope of our discussion, if the voltage varies with time, we will observe the effects of the capacitance of the circuit.

POWER LAWS

Sliding friction between two objects generates heat. In a similar manner, electrical resistance generates heat. This is simply a matter of energetics: If the electrical potential between sites is different and current flows between them, the energy must be dissipated in some manner, whether through a motor, light bulb, or heating element. The power dissipated by a resistor is given by

$$P = IV \qquad (C.8)$$

where P is the power dissipated by the resistor, I is the current through it, and V is the voltage across it. That is, the power dissipated by a resistor as heat is given by the product of the current flowing though and the voltage across it. Power, as in mechanical applications, has units of watts (W). With Ohm's law, we can also derive two other forms of the power law,

$$P = I^2 R \quad \text{and} \quad P = \frac{V^2}{R} \qquad (C.9)$$

where R is the magnitude of the resistance. These equations demonstrate that heating devices such as ovens and hair dryers work by having low resistances. A heating element (a coil) is simply a resistor; as current flows through it, energy dissipates as heat. Using the equation on the left, we see that the power increases as the square of the current. Therefore, a decrease in the resistance of the heating element causes a proportional increase in the current.

EXAMPLE C.4

a. What is the resistance of the heating coils of a 1200 W toaster that runs on 110 V house circuitry?

See answer C.4a on page 280.

b. In an earlier example, we had a 110 V house outlet on a 15 A circuit breaker. What is the maximum wattage appliance you can plug into this outlet?

See answer C.4b on page 281.

MEASUREMENT OF PHYSICAL SYSTEMS

Having discussed the basic behaviors of simple circuit components, we now turn to the way we use these components in the laboratory. We begin with a discussion of how we convert human movements into electrical signals that our computers can measure.

TRANSDUCERS

In the vast majority of cases in which we measure physical quantities electrically, we measure changes in voltage. This is a fundamental principle that cannot be overemphasized. A *0* or a *1* in a computer is represented by a voltage of 0 or 5 V, respectively. When sound is transmitted through a wire to a speaker, the changes in voltage are interpreted as sound. When radio signals are transmitted to a satellite, these, too, are registered by the voltages they impart on the receiver. This is also true for the measurement of EMG activity, force, and even the reflections of body markers to a camera's lens.

The process of converting a physical dimension into a voltage is called transduction. A device that performs this function is a transducer. Some of the types of transducers are force, pressure, linear displacement, rotary displacement, and acceleration transducers. The common principle in all of these devices is that the quantity being measured causes the resistance of the transducer to change. For example, a force transducer (used in a force platform) has tiny resistors that deform slightly when force is applied. An electrogoniometer has a rotary resistor that changes as it is rotated. When these resistances change, then, in accordance with Ohm's law, a constant current through a transducer causes the voltage to change proportionately.

EXAMPLE C.5

Suppose we have a blood-pressure transducer connected to a 10 mA current supply. As the pressure changes from 80 to 120 mmHg, the transducer's resistance changes from 1000 to 1200 Ω. What will the voltage outputs be at these two pressures?

See answer C.5 on page 281.

VOLTAGE DIVIDERS

How would we measure a sensor with a variable resistance? This is slightly more complicated than the blood-pressure example, because most electrical supplies have a constant voltage, not a constant cur-rent. Suppose we have a variable resistor and connect a voltage source across it as shown in figure C.5. The standard nomenclature is to label the source voltage V_{in} and the measured voltage V_{out}. For a simple circuit like this, no matter how much the variable resistance R_V changes, V_{out} will always equal V_{in}, so this circuit is useless for measuring changes in R_V.

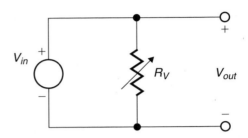

○ **Figure C.5** Circuit diagram of a variable resistor connected to a voltage source.

In a modification of this circuit (figure C.6), a resistor R is in series with the variable resistance. We want to know the voltage V_{out}. To do this, we can determine using Ohm's law that the current, I, is

$$I = \frac{V_{in}}{R + R_V} \qquad (C.10)$$

Because current flows through both resistors, we can substitute it in Ohm's law for R_V and V_{out}:

$$V_{out} = \frac{V_{in} R_V}{R + R_V} \qquad (C.11)$$

This circuit is referred to as a voltage divider. V_{out} for this circuit varies over an easily measurable range when R and R_V are of similar magnitudes. The circuit is commonly used for a simple potentiometer.

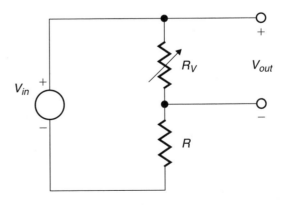

○ **Figure C.6** Circuit diagram of a resistor in series with a variable resistance.

EXAMPLE C.6

In the voltage divider, what is V_{out} for the following cases?

1. $V_{in} = 15$ V, $R_V = 100\ \Omega$, and $R = 100\ \Omega$
2. $V_{in} = 15$V, $R_V = 110\ \Omega$, and $R = 100\ \Omega$
3. $V_{in} = 15$V, $R_V = 100\ \Omega$, and $R = 10\ \Omega$
4. $V_{in} = 15$V, $R_V = 110\ \Omega$, and $R = 10\ \Omega$

See answer C.6 on page 281.

WHEATSTONE BRIDGES

Voltage dividers have two problems. In many sensors, the variability of resistance is small, often less than 5%. Also, we often "zero" a sensor rather than subtracting a constant voltage to establish zero for the quantity we are measuring. These difficulties are overcome with a Wheatstone bridge (figure C.7), a circuit of two parallel voltage dividers. In its neutral state, it has four equivalent resistances. When the variable resistance changes, we can compare the amount by which the variable resistance has changed from its neutral state by measuring V_{out}.

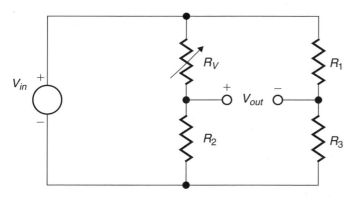

° **Figure C.7**   Circuit diagram of a Wheatstone bridge.

EXAMPLE C.7

In the Wheatstone bridge, what is the formula for V_{out}? Let $R_1 = R_2 = R_3$.

See answer C.7 on page 281.

COMMON LABORATORY INSTRUMENTS

Many common laboratory instruments employ the principles discussed in this appendix. These instruments include (1) the linear variable differential transducer (LVDT), (2) the electrogoniometer, (3) strain-gauge force transducers, and (4) amplifiers.

LINEAR VARIABLE DIFFERENTIAL TRANSDUCER

The LVDT (figure C.8) is a common instrument that measures a linear movement over a short range of motion, typically less than 30 cm. Its main cylinder contains a finely manufactured and calibrated linear potentiometer. Therefore, its resistance changes linearly as it is moved. LVDTs can measure to within fractions of a millimeter—a computer-controlled milling machine, for instance, measures to within 2.5 μm. Common lab applications include treadmill inclination adjustments, digital calipers, footwear impact testers, and knee arthrometers (for measuring joint laxity or stiffness).

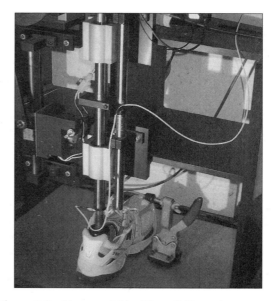

° **Figure C.8**   Linear variable differential transducer as used in a footwear impact tester.

ELECTROGONIOMETER

An electrogoniometer, as its name suggests, measures joint angles electronically. Its basic component is a rotary potentiometer. Its internal structure is diagrammed in figure C.9. The end terminals (A and C) are connected to the ends of the resistive material. The middle terminal (B) is connected to a rotating slider. As the knob of this slider is turned, the middle contact moves across the resistive material. Because resistance is a function of material length, we observe the change in resistance. For example, if we have a 10 kΩ potentiometer, the resistance from A to C will measure 10 kΩ. As the rotating slider is moved from A to C, we will measure a resistance across A and B that changes from 0 to 10 kΩ, while the resistance from B to C changes from 10 kΩ to 0 (figure C.10).

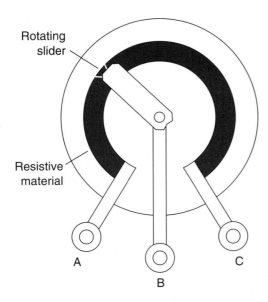

Rotating slider

Resistive material

A

B

C

○ **Figure C.9**   Schematic of a rotary potentiometer.

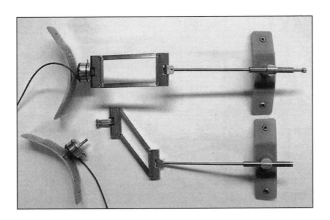

○ **Figure C.10**   Two electrogoniometers with four-bar linkages.

STRAIN-GAUGE FORCE TRANSDUCERS

When force is applied to a material, it deforms. This is called mechanical strain. Because resistance is a function of material length, we observe a change in resistance when a material is deformed. This is the basic principle of a strain gauge. If we have a resistor with a precisely known resistance glued to a deformable object, we can measure the object's change in resistance as it deforms. The gauges themselves are usually much smaller in area than a postage stamp, but equally as thin (see figure C.11). Once glued to the surface of a structure, they bend with the material without altering the structural properties. Strain gauges are usually placed in a Wheatstone bridge circuit.

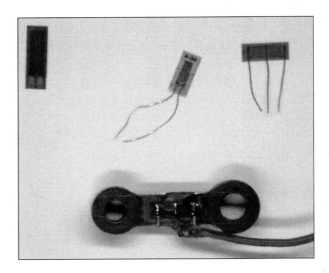

○ **Figure C.11**   Three types of strain gauges and a strain-gauged link for measuring axial loads.

Strain gauges are commonly used to measure forces in human movement with such devices as floor-mounted force plates, tension transducers, pressure transducers, and even accelerometers. It is also common in biomedical research to mount gauges to orthoses and prostheses, as well as to cadaver samples of bone, cartilage, and tendon.

AMPLIFIERS

An amplifier is a device that increases the voltage of a signal. Figure C.12 shows how the idealized amplifier is designated in circuit diagrams.

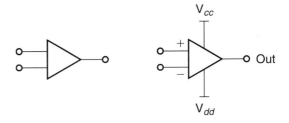

○ **Figure C.12**   Symbols for an operational amplifier. The detailed form, on the right, labels the inputs and the power supply for the amplifier V_{cc} and V_{dd}.

The most common type of amplifier is the operational amplifier. They are commonly installed on silicon chips. Unlike resistors and capacitors, *op-amps* are active circuit elements and therefore require current to power them. Op-amps have many different uses and implementations. Two common connections are the inverting and noninverting configurations (figure C.13). The noninverting op-amp circuit

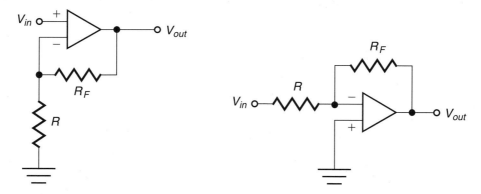

○ Figure C.13   Circuit diagrams for noninverting (left) and inverting (right) operational amplifiers.

increases the magnitude of the voltage it measures. The inverting op-amp circuit increases the incoming voltage and inverts it (i.e., takes the negative of it).

The performance of an amplifier is called the gain. Gain is the ratio of the incoming voltage to the amplified voltage, $\frac{V_{out}}{V_{in}}$. In op-amp circuits, gain is controlled by altering the ratio of the resistors R_F and R. For noninverting configurations, the gain is given by $1 + \frac{R_F}{R}$, and for inverting configurations the gain is $-\frac{R_F}{R}$. For a variable gain, a potentiometer may be substituted for either R_F or R.

Applications for amplifiers are too numerous to mention. Their most common usage is for turning weak electromagnetic waves into audible sound in radios and cell phones. In human movement science, we use them to measure tiny EMG, electrocardiographic, and electroencephalographic signals and in force plates and other force transducers and in accelerometers. They can also be used to construct analog filters, integrators, and differentiators.

An important characteristic of op-amps and other active circuits is their input impedance. This is a measure of the sensitivity of the op-amp: High input impedance means, in effect, that the op-amp needs to draw very little current from the measured quantity to function. This is very important in human movement, because most biophysical signals are

extremely small. Ideally, the input impedance of an EMG amplifier, for example, would be infinite. Typically, amplifiers have input impedances of 1 MΩ, but EMG or bioamplifiers have input impedances of 10 MΩ or more. EMG amplifiers require higher impedances because skin can have resistances of about 20 to 100 kΩ, and when unprepared, as high as 2 MΩ or more.

SUGGESTED READINGS

Bobrow, L.S. 1987. *Elementary Linear Circuit Analysis.* 2nd ed. Oxford: Oxford University Press.

Cathey, J.J. 2002. *Schaum's Outline of Electronic Devices and Circuits.* 2nd ed. New York: McGraw-Hill.

Cobbold, R.S.C. 1974. *Transducers for Biomedical Measurements: Principles and Applications.* Toronto: John Wiley & Sons.

Horowitz, P., and W. Hill. 1989. *The Art of Electronics.* 2nd ed. Cambridge: Cambridge University Press.

Ohanian, H.C. 1994. Electric force and electric charge. In *Principles of Physics.* 2nd ed. New York: Norton.

O'Malley, J. 1992. *Schaum's Outline of Basic Circuit Analysis.* 2nd ed. New York: McGraw-Hill.

Winter, D.A., and A.E. Patla. 1997. *Signal Processing and Linear Systems for Movement Sciences.* Waterloo, ON: Waterloo Biomechanics.

Vector Operations

To illustrate the following operations, two vectors, $\vec{A} = (1, 5, 2)$ and $\vec{B} = (6, 1, 3)$, will be used in the examples in this appendix.

MAGNITUDE OR NORM OF A VECTOR

The magnitude or norm of a vector represents the length of the vector. For a vector $\vec{A}$, with components A_x, A_y, and A_z representing distances from the Y-Z, X-Z, and X-Y planes, respectively, the norm of the vector is calculated using the Pythagorean relationship as follows:

$$\left\|\vec{A}\right\| = \sqrt{A_x^2 + A_y^2 + A_z^2} \qquad \text{(D.1)}$$

EXAMPLE D.1

For $\vec{A} = (1, 5, 2)$, the norm or the magnitude of $\vec{A}$ is:

$$\left\|\vec{A}\right\| = \sqrt{1^2 + 5^2 + 2^2} \qquad \text{(D.3)}$$

$$\left\|\vec{A}\right\| = \sqrt{30} = 5.48 \qquad \text{(D.4)}$$

ADDITION OF VECTORS

If vectors $\vec{A}$ and $\vec{B}$ are designated as $\vec{A} = (A_x, A_y, A_z)$ and $\vec{B} = (B_x, B_y, B_z)$ then:

$$\vec{A} + \vec{B} = (A_x + B_x, A_y + B_y, A_z + B_z) \qquad \text{(D.2)}$$

Using unit vector notation, the sum of A and B is:

$$\vec{A} + \vec{B} = (A_x + B_x)\vec{i} + (A_y + B_y)\vec{j} + (A_z + B_z)\vec{k} \qquad \text{(D.3)}$$

EXAMPLE D.2

For $\vec{A} = (1, 5, 2)$ and $\vec{B} = (6, 1, 3)$, the vector $\vec{A} + \vec{B}$ is:

$$\vec{A} + \vec{B} = (1 + 6)\vec{i} + (5 + 1)\vec{j} + (2 + 3)\vec{k}$$
$$\vec{A} + \vec{B} = 7\vec{i} + 6\vec{j} + 5\vec{k} = (7, 6, 5)$$

SUBTRACTION OF VECTORS

Technically, there is no operation in which vectors subtract; however, the negative of one vector may be added to another vector. Thus, if we wished to find the difference between vectors $\vec{A}$ and $\vec{B}$, then $\vec{A} - \vec{B}$ is really $\vec{A} + (-\vec{B})$. As a result, if vectors $\vec{A} = A_x, A_y, A_z$ and $\vec{B} = B_x, B_y, B_z$ then the difference between two vectors A and B is:

$$\vec{A} + (-\vec{B}) = (A_x - B_x)\vec{i} + (A_y - B_y)\vec{j} + (A_z - B_z)\vec{k} \qquad \text{(D.4)}$$

EXAMPLE D.3

For $\vec{A} = (1, 5, 2)$ and $\vec{B} = (6, 1, 3)$, the vector $\vec{A} - \vec{B}$ is:

$$\vec{A} - \vec{B} = (1 - 6)\vec{i} + (5 - 1)\vec{j} + (2 - 3)\vec{k}$$
$$\vec{A} - \vec{B} = -5\vec{i} + 4\vec{j} - 1\vec{k} \text{ or } (-5, 4, -1)$$

MULTIPLICATION OF A VECTOR BY A SCALAR

Vectors can be multiplied by a scalar by multiplying each component by the scalar. Thus, for a vector $\vec{A}$ and scalar c:

$$c\vec{A} = cA_x\vec{i} + cA_y\vec{j} + cA_z\vec{k} \qquad \text{(D.5)}$$

EXAMPLE D.4

For $\vec{A} = (1, 5, 2)$ and scalar $c = 2$, the vector $c\vec{A}$ is:

$$c\vec{A} = 2(1)\,\vec{i} + 2(5)\,\vec{j} + 2(2)\vec{k}$$

$$c\vec{A} = 2\vec{i} + 10\vec{j} + 4\vec{k} = (2, 10, 4)$$

DOT OR SCALAR PRODUCT

Given two vectors $\vec{A}$ and $\vec{B}$, whose components are A_x, A_y, A_z and B_x, B_y, B_z, respectively, the dot or scalar product, $\vec{A} \cdot \vec{B}$, is defined by the equation:

$$\vec{A} \cdot \vec{B} = A_x B_x + A_y B_y + A_z B_z \qquad (\text{D.6})$$

The result of the dot product calculation is always a scalar and hence the alternative name "scalar product." Since the result is a quantity with only magnitude and no direction, we can state that the dot product operation is commutative (i.e., $\vec{A} \cdot \vec{B} = \vec{B} \cdot \vec{A}$).

EXAMPLE D.5

For $\vec{A} = (1, 5, 2)$ and $\vec{B} = (6, 1, 3)$, $\vec{A} \cdot \vec{B}$ is:

$$\vec{A} \cdot \vec{B} = (1 \times 6) + (5 \times 1) + (2 \times 3)$$

$$\vec{A} \cdot \vec{B} = 6 + 5 + 6$$

$$\vec{A} \cdot \vec{B} = 17$$

In analytical geometry, the cosine of the angle between two line segments is given by the formula:

$$\cos\theta = \frac{A_x B_x + A_y B_y + A_z B_z}{|\vec{A}||\vec{B}|} \qquad (\text{D.7})$$

We can rewrite this equation using our previous definition of the dot product to be:

$$\cos\theta = \frac{\vec{A} \cdot \vec{B}}{|\vec{A}||\vec{B}|} \qquad (\text{D.8})$$

or

$$\vec{A} \cdot \vec{B} = |\vec{A}||\vec{B}|\cos\theta \qquad (\text{D.9})$$

Geometrically, $\vec{A} \cdot \vec{B}$ is equal to the length of the projection of $\vec{A}$ on $\vec{B}$, times the magnitude of $\vec{B}$. If the angle between these two vectors is 90° and the cosine of the angle is zero, then the dot product is equal to zero. That is, $\vec{A}$ is perpendicular to $\vec{B}$ provided that neither is equal to (0,0,0). From the definitions of unit vector coordinates, $\vec{i}$, $\vec{j}$, and $\vec{k}$, the following relationships hold:

$$\vec{i} \cdot \vec{i} = 1 \qquad (\text{D.10})$$

$$\vec{j} \cdot \vec{j} = 1 \qquad (\text{D.11})$$

$$\vec{k} \cdot \vec{k} = 1 \qquad (\text{D.12})$$

because the angle and the cosine of the angle between the pairs of coordinates is zero. Since the unit vector coordinates are perpendicular to each other, the cosine of 90° is zero. Thus:

$$\vec{i} \cdot \vec{j} = 0 \qquad (\text{D.13})$$

$$\vec{i} \cdot \vec{k} = 0 \qquad (\text{D.14})$$

$$\vec{j} \cdot \vec{k} = 0 \qquad (\text{D.15})$$

When the vectors $\vec{A}$ and $\vec{B}$ are unit vectors, then the dot product of the two vectors is the cosine of the angle between them:

$$\vec{A} \cdot \vec{B} = \cos\theta \qquad (\text{D.16})$$

CROSS OR VECTOR PRODUCT

For two vectors, $\vec{A}$ and $\vec{B}$, the cross product, $\vec{A} \times \vec{B}$, is defined by the following equation:

$$\vec{A} \times \vec{B} = [A_y B_z - A_z B_y,\ A_z B_x - A_x B_z,\ A_x B_y - A_y B_x] \qquad (\text{D.17})$$

The result of this calculation is always a vector, hence the alternative name, the vector product. Geometrically, the resulting vector, $\vec{C}$, of the cross product of two vectors, $\vec{A}$ and $\vec{B}$ is perpendicular to the plane formed by both $\vec{A}$ and $\vec{B}$ (figure D.1). The direction of the vector $\vec{C}$ is determined by the right-hand rule. That is, if the fingers of the right hand are placed

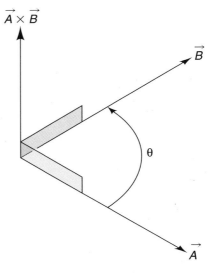

○ Figure D.1 Geometric representation of the cross or vector product of $\vec{A} \times \vec{B}$.

along the vector $\vec{A}$ and rotated to the vector $\vec{B}$, the right thumb would point in the direction of the resulting vector $\vec{C}$. It should be intuitive that:

$$\vec{A} \times \vec{B} \neq \vec{B} \times \vec{A} \qquad (D.18)$$

However,

$$\vec{A} \times \vec{B} = -\vec{B} \times \vec{A} \qquad (D.19)$$

Therefore, we say that the cross-product operation is not commutative.

It is clear that the vector $\vec{C} = \vec{A} \times \vec{B}$ is determined from a right-handed triad of the three vectors $\vec{A}$, $\vec{B}$, and $\vec{C}$. Therefore, since $\vec{C}$ is perpendicular to the plane of $\vec{A}$ and $\vec{B}$, the magnitude of the cross product can be written as:

$$\left|\vec{A} \times \vec{B}\right| = \left|\vec{A}\right|\left|\vec{B}\right| \sin \theta \qquad (D.20)$$

where $|\vec{A}|$ and $|\vec{B}|$ are the norms of the two vectors and θ is the angle between $\vec{A}$ and $\vec{B}$. In terms of unit vector coordinates, it can be shown that:

$$\vec{i} \times \vec{i} = 0 \qquad (D.21)$$

$$\vec{j} \times \vec{j} = 0 \qquad (D.22)$$

$$\vec{k} \times \vec{k} = 0 \qquad (D.23)$$

because the angle between each of these pairs of unit vectors is zero and the sine of $0°$ is zero. Also,

$$\vec{i} \times \vec{j} = \vec{k} \qquad (D.24)$$

$$\vec{j} \times \vec{k} = \vec{i} \qquad (D.25)$$

$$\vec{k} \times \vec{i} = \vec{j} \qquad (D.26)$$

Rather than trying to remember the formula for a cross product, it is probably easier to determine the result as the expansion of a *determinant* (see appendix E for more details about computing determinants). If a 2×2 matrix, [M], is defined as:

$$[M] = \begin{bmatrix} A & B \\ C & D \end{bmatrix} \qquad (D.27)$$

then the determinant of this matrix, |M|, is:

$$|M| = AD - BC \qquad (D.28)$$

We can use this calculation to determine the resulting vector from a cross product. The cross product can be written as follows:

$$\vec{A} \times \vec{B} = \begin{vmatrix} \vec{i} & \vec{j} & \vec{k} \\ A_x & A_y & A_z \\ B_x & B_y & B_z \end{vmatrix} \qquad (D.29)$$

Using the concept of the determinant and that $\vec{i} \times \vec{j} = \vec{k}$, $\vec{j} \times \vec{k} = \vec{i}$, and $\vec{k} \times \vec{i} = \vec{j}$, we can reconfigure the structure above to be a series of determinants as follows:

$$\vec{A} \times \vec{B} = \begin{vmatrix} A_y & A_z \\ B_y & B_z \end{vmatrix} \vec{i} + \begin{vmatrix} A_z & A_x \\ B_z & B_x \end{vmatrix} \vec{j} + \begin{vmatrix} A_x & A_y \\ B_x & B_y \end{vmatrix} \vec{k} \quad (D.30)$$

Calculating the determinant of each 2×2 matrix, gives us:

$$\vec{A} \times \vec{B} = (A_y B_z - A_z B_y)\vec{i} + (A_z B_x - A_x B_z)\vec{j} + (A_x B_y - A_y B_x)\vec{k} \qquad (D.31)$$

or

$$\vec{A} \times \vec{B} = [A_y B_z - A_z B_y, A_z B_x - A_x B_z, A_x B_y - A_y B_x] \qquad (D.32)$$

or

$$\vec{A} \times \vec{B} = [A_y B_z - A_z B_y, -(A_x B_z - A_z B_x), A_x B_y - A_y B_x] \quad (D.33)$$

EXAMPLE D.6

For $\vec{A} = (1, 5, 2)$ and $\vec{B} = (6, 1, 3)$, $\vec{A} \times \vec{B}$ is:

$$\vec{A} \times \vec{B} = \begin{vmatrix} \vec{i} & \vec{j} & \vec{k} \\ 1 & 5 & 2 \\ 6 & 1 & 3 \end{vmatrix}$$

$$\vec{A} \times \vec{B} = \begin{vmatrix} 5 & 2 \\ 1 & 3 \end{vmatrix} \vec{i} + \begin{vmatrix} 2 & 1 \\ 3 & 6 \end{vmatrix} \vec{j} + \begin{vmatrix} 1 & 5 \\ 6 & 1 \end{vmatrix} \vec{k}$$

$$\vec{A} \times \vec{B} = [(5 \times 3) - (2 \times 1)]\vec{i} + [(2 \times 6) - (1 \times 6)]\vec{j} + [(1 \times 1) - (5 \times 6)]\vec{k}$$

$$\vec{A} \times \vec{B} = [15 - 2]\vec{i} + [12 - 6]\vec{j} + [1 - 30]\vec{k}$$

$$\vec{A} \times \vec{B} = 13\vec{i} + 6\vec{j} - 29\vec{k}$$

Matrix Operations

To illustrate the following operations, a matrix [A] will be used throughout the examples.

$$[A] = \begin{bmatrix} 6 & 1 & 3 \\ -1 & 1 & 2 \\ 4 & 1 & 3 \end{bmatrix} \qquad (E.1)$$

DETERMINANT OF A 3 × 3 MATRIX

In appendix D, we demonstrated the method of calculating the determinant of a 2 × 2 matrix. We will now show how to calculate the determinant of a 3 × 3 matrix. Given the matrix [A], we copy the first two columns to the right of the matrix. We first draw three arrows to the right beginning with the column furthest to the left. The elements of each arrow are multiplied together. For example, for the first arrow, the product of the elements is $(a_{11} \times a_{22} \times a_{33})$. We then draw three arrows to the left beginning with the column furthest to the right. The elements of each arrow are also multiplied together. The determinant is calculated by adding the products of the arrows to the right and subtracting each of the products of the arrows to the left. This technique is illustrated here:

EXAMPLE E.1

For $[A] = \begin{bmatrix} 6 & 1 & 3 \\ -1 & 1 & 2 \\ 4 & 1 & 3 \end{bmatrix}$, the determinant of [A] is:

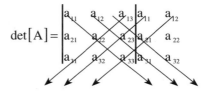

$$\det[A] = (6 \times 1 \times 3) + (1 \times 2 \times 4) + (3 \times -1 \times 1)$$
$$- (1 \times -1 \times 3) - (6 \times 2 \times 1) - (3 \times 1 \times 4)$$
$$\det[A] = 18 + 8 - 3 - (-3) - 12 - 12$$
$$\det[A] = 2$$

COFACTOR MATRIX

A further matrix operation that is important in 3-D kinematics is the calculation of the *cofactor matrix*. This operation is used in the calculation of the inverse of a matrix. The formula for calculating the cofactor matrix (A^c) of a 3 × 3 matrix [A] is:

$$\det[A] = \begin{vmatrix} a_{11} & a_{12} & a_{13} \\ a_{21} & a_{22} & a_{23} \\ a_{31} & a_{32} & a_{33} \end{vmatrix} \begin{matrix} a_{11} & a_{12} \\ a_{21} & a_{22} \\ a_{31} & a_{32} \end{matrix} \qquad (E.2)$$

$$= (a_{11} \times a_{22} \times a_{33}) + (a_{12} \times a_{23} \times a_{31}) + (a_{13} \times a_{21} \times a_{32})$$
$$- (a_{12} \times a_{21} \times a_{33}) - (a_{11} \times a_{23} \times a_{32}) - (a_{13} \times a_{22} \times a_{31})$$

$$A^c = \begin{vmatrix} (a_{22} \times a_{33} - a_{23} \times a_{32}) & -(a_{21} \times a_{33} - a_{23} \times a_{31}) & (a_{21} \times a_{32} - a_{22} \times a_{31}) \\ -(a_{12} \times a_{33} - a_{13} \times a_{32}) & (a_{11} \times a_{33} - a_{13} \times a_{31}) & -(a_{11} \times a_{32} - a_{12} \times a_{31}) \\ (a_{12} \times a_{23} - a_{13} \times a_{22}) & -(a_{11} \times a_{23} - a_{13} \times a_{21}) & (a_{11} \times a_{22} - a_{12} \times a_{21}) \end{vmatrix} \qquad (E.3)$$

EXAMPLE E.2

For $[A] = \begin{bmatrix} 6 & 1 & 3 \\ -1 & 1 & 2 \\ 4 & 1 & 3 \end{bmatrix}$, A^c is:

$$A^c = \begin{vmatrix} (1 \times 3 - 2 \times 1) & -((-1) \times 3 - 2 \times 4) & (-1 \times 1 - 1 \times 4) \\ -(1 \times 3 - 3 \times 1) & (6 \times 3 - 3 \times 4) & -(6 \times 1 - 1 \times 4) \\ (1 \times 2 - 3 \times 1) & -(6 \times 2 - 3 \times (-1)) & (6 \times 1 - 1 \times (-1)) \end{vmatrix}$$

$$A^c = \begin{vmatrix} 1 & -11 & -5 \\ 0 & 6 & -2 \\ -1 & -15 & 7 \end{vmatrix}$$

INVERTING A MATRIX

The *inverse* of a matrix D is designated as $[D]^{-1}$ and is one that satisfies the expression

$$[A]^{-1}[A] = [I] \qquad (E.4)$$

where $[A] \neq 0$ (i.e., not singular) and $[I]$ is the identity matrix.

The inverse of a matrix is calculated in several steps: (1) calculate the determinant of the matrix; (2) calculate the cofactor matrix; (3) take the transpose of the cofactor matrix; and (4) multiply the transpose of the cofactor matrix by the inverse of the determinant. To invert matrix $[A]$, the formula then is

$$[A]^{-1} = \frac{1}{\det[A]} [A^c]^T \qquad (E.5)$$

EXAMPLE E.3

For $[A] = \begin{bmatrix} 6 & 1 & 3 \\ -1 & 1 & 2 \\ 4 & 1 & 3 \end{bmatrix}$, the inverse matrix of $[A]$ is:

$$[A]^{-1} = \frac{1}{\det[A]} \text{cofactor} \begin{bmatrix} 1 & -11 & -5 \\ 0 & 6 & -2 \\ -1 & -15 & 7 \end{bmatrix}^T$$

$$[A]^{-1} = \frac{1}{2} \text{cofactor} \begin{bmatrix} 1 & 0 & -1 \\ -11 & 6 & -15 \\ -5 & -2 & 7 \end{bmatrix}$$

$$[A]^{-1} = \begin{vmatrix} 0.5 & 0 & -0.5 \\ -5.5 & 3 & -7.5 \\ -2.5 & -1 & 3.5 \end{vmatrix}$$

To understand 3-D kinematics, it is important to be fully cognizant of the mathematical manipulations that must be performed with vectors and matrices. For a more in-depth exposure to matrices and matrix algebra, we suggest that you refer to a textbook that deals particularly with vectors and matrices.

Numerical Integration of Double Pendulum Equations

A fourth-order Runge-Kutta method is more difficult to implement in computer code than an Euler method, and it is described here in a format that can be readily adapted to a computer language. The first step is to implement functions that return the angular accelerations of each segment given the angular positions and velocities of each segment, that is,

$$\alpha_T = f(\theta_T, \theta_L, \omega_T, \omega_L) \text{ and } \alpha_L = f(\theta_T, \theta_L, \omega_T, \omega_L)$$

These functions implement the result of simultaneously solving equations G.1 and G.2 (equations of motion for the leg and thigh) for α_T and α_L, that is,

$$\alpha_T = \frac{C_1 I_2 - C_2 A}{B}$$

$$\alpha_L = \frac{C_2 \left(I_T + m_L L_T^2\right) - C_1 A}{B}$$

where

$$A = m_L L_T d_L \cos(\theta_L - \theta_T)$$

$$B = I_L (I_T + m_L L_T^2) - A^2$$

$$C_1 = m_L L_T d_L \omega_L^2 \sin(\theta_L - \theta_T) +$$
$$(a_{Hx}\cos\theta_T + (a_{Hy} + g)\sin\theta_T)(m_T d_T + m_L L_T)$$

$$C_2 = - m_L L_T d_L \omega_T^2 \sin(\theta_L - \theta_T) +$$
$$(a_{Hx} \cos\theta_L + (a_{Hy} + g) \sin\theta_L) m_L d_L$$

The next step is to implement functions that compute the change in velocity,

$$\Delta\omega_T = f(\theta_T, \theta_L, \omega_T, \omega_L) \text{ and } \Delta\omega_L = f(\theta_T, \theta_L, \omega_T, \omega_L)$$

which are implemented by using a Euler method applied to equations A1 and A2:

$$\Delta\omega_T = \alpha_T (\theta_T, \theta_L, \omega_T, \omega_L) \Delta t \text{ and}$$
$$\Delta\omega_L = \alpha_L (\theta_T, \theta_L, \omega_T, \omega_L) \Delta t$$

where Δt is the time increment of the numerical integration procedure.

Two functions are needed to compute the change in angular positions using the Euler method:

$$\Delta\theta_T = \omega_T \Delta t + \frac{1}{2} \alpha_T(\theta_T, \theta_L, \omega_T, \omega_L)\Delta t^2$$

and

$$\Delta\theta_L = \omega_L \Delta t + \frac{1}{2} \alpha_L(\theta_T, \theta_L, \omega_T, \omega_L)\Delta t^2$$

In the functions α, $\Delta\omega$, and $\Delta\theta$, the values for θ_T, θ_L, ω_T, and ω_L are temporary function parameters and therefore should be given different names when used in computer code. Using these functions, a fourth-order Runge-Kutta method can be implemented. Given values for θ_T, θ_L, ω_T, and ω_L at an instant in time, the values of these four parameters at the next instant in time are computed using the following procedure:

$$\delta\omega_{T1} = \Delta\omega_T\,(\theta_T,\,\theta_L,\,\omega_T,\,\omega_L) \qquad \delta\theta_{T1} = \Delta\theta_T\,(\theta_T,\,\theta_L,\,\omega_T,\,\omega_L)$$

$$\delta\omega_{L1} = \Delta\omega_L\,(\theta_T,\,\theta_L,\,\omega_T,\,\omega_L) \qquad \delta\theta_{L1} = \Delta\theta_L\,(\theta_T,\,\theta_L,\,\omega_T,\,\omega_L)$$

$$\delta\omega_{T2} = \Delta\omega_T\left(\theta_T + \frac{\partial\theta_{1T}}{2},\,\theta_L + \frac{\partial\theta_{1L}}{2},\,\omega_T + \frac{\partial\omega_{1T}}{2},\,\omega_L + \frac{\partial\omega_{1L}}{2}\right)$$

$$\delta\omega_{L2} = \Delta\omega_L\left(\theta_T + \frac{\partial\theta_{1T}}{2},\,\theta_L + \frac{\partial\theta_{1L}}{2},\,\omega_T + \frac{\partial\omega_{1T}}{2},\,\omega_L + \frac{\partial\omega_{1L}}{2}\right)$$

$$\delta\theta_{T2} = \Delta\theta_T\left(\theta_T + \frac{\partial\theta_{1T}}{2},\,\theta_L + \frac{\partial\theta_{1L}}{2},\,\omega_T + \frac{\partial\omega_{1T}}{2},\,\omega_L + \frac{\partial\omega_{1L}}{2}\right)$$

$$\delta\theta_{L2} = \Delta\theta_L\left(\theta_T + \frac{\partial\theta_{1T}}{2},\,\theta_L + \frac{\partial\theta_{1L}}{2},\,\omega_T + \frac{\partial\omega_{1T}}{2},\,\omega_L + \frac{\partial\omega_{1L}}{2}\right)$$

$$\delta\omega_{T3} = \Delta\omega_T\left(\theta_T + \frac{\partial\theta_{2T}}{2},\,\theta_L + \frac{\partial\theta_{2L}}{2},\,\omega_T + \frac{\partial\omega_{2T}}{2},\,\omega_L + \frac{\partial\omega_{2L}}{2}\right)$$

$$\delta\omega_{L3} = \Delta\omega_L\left(\theta_T + \frac{\partial\theta_{2T}}{2},\,\theta_L + \frac{\partial\theta_{2L}}{2},\,\omega_T + \frac{\partial\omega_{2T}}{2},\,\omega_L + \frac{\partial\omega_{2L}}{2}\right)$$

$$\delta\theta_{T3} = \Delta\theta_T\left(\theta_T + \frac{\partial\theta_{2T}}{2},\,\theta_L + \frac{\partial\theta_{2L}}{2},\,\omega_T + \frac{\partial\omega_{2T}}{2},\,\omega_L + \frac{\partial\omega_{2L}}{2}\right)$$

$$\delta\theta_{L3} = \Delta\theta_L\left(\theta_T + \frac{\partial\theta_{2T}}{2},\,\theta_L + \frac{\partial\theta_{2L}}{2},\,\omega_T + \frac{\partial\omega_{2T}}{2},\,\omega_L + \frac{\partial\omega_{2L}}{2}\right)$$

$$\delta\omega_{T4} = \Delta\omega_T\,(\theta_T + \delta\theta_{3T},\,\theta_L + \delta\theta_{3L},\,\omega_T + \delta\omega_{3T},\,\omega_L + \delta\omega_{3L})$$

$$\delta\omega_{L4} = \Delta\omega_L\,(\theta_T + \delta\theta_{3T},\,\theta_L + \delta\theta_{3L},\,\omega_T + \delta\omega_{3T},\,\omega_L + \delta\omega_{3L})$$

$$\delta\theta_{T4} = \Delta\theta_T\,(\theta_T + \delta\theta_{3T},\,\theta_L + \delta\theta_{3L},\,\omega_T + \delta\omega_{3T},\,\omega_L + \delta\omega_{3L})$$

$$\delta\theta_{L4} = \Delta\theta_L\,(\theta_T + \delta\theta_{3T},\,\theta_L + \delta\theta_{3L},\,\omega_T + \delta\omega_{3T},\,\omega_L + \delta\omega_{3L})$$

Next value for $\omega_T = \omega_T + \dfrac{1}{6}\left(\partial\omega_{T1} + \partial\omega_{T4}\right) + \dfrac{1}{3}\left(\partial\omega_{T2} + \partial\omega_{T3}\right)$

Next value for $\omega_L = \omega_L + \dfrac{1}{6}\left(\partial\omega_{L1} + \partial\omega_{L4}\right) + \dfrac{1}{3}\left(\partial\omega_{L2} + \partial\omega_{L3}\right)$

Next value for $\theta_T = \theta_T + \dfrac{1}{6}\left(\partial\theta_{T1} + \partial\theta_{T4}\right) + \dfrac{1}{3}\left(\partial\theta_{T2} + \partial\theta_{T3}\right)$

Next value for $\theta_L = \theta_L + \dfrac{1}{6}\left(\partial\theta_{L1} + \partial\theta_{L4}\right) + \dfrac{1}{3}\left(\partial\theta_{L2} + \partial\theta_{L3}\right)$

The first step of this Runge-Kutta procedure is the Euler method. The remaining steps compute three more Euler-method estimates. The final values are weighted averages of the four estimates. This procedure repeats until the desired end conditions—such as an instant in time or a leg angle—are reached.

Derivation of Double Pendulum Equations

Table G.1 Notation

m	Segment mass
I_{cm}	Segment moment of inertia about mass center
I	Segment moment of inertia about proximal end
L	Segment length
d	Distance from segment mass center to proximal end
θ, ω, and α	Segment angular position, velocity, and acceleration
a_H	Acceleration of the hip
g	Acceleration of gravity
GRF	Ground reaction force
CP	Center of pressure of ground reaction force

The subscripts T and L denote the thigh and leg, respectively; the subscripts x and y denote A/P and vertical coordinate system directions, respectively.

The kinetic energy of the double pendulum is derived from the first principle that the kinetic energy of an object in general planar motion equals the sum of its rotational and translational kinetic energies. The rotational kinetic energy is calculated using the moment of inertia about the mass center, and the translational velocity equals the linear velocity of the mass center.

$$KE = \frac{1}{2}\left(I_{Tcm}\omega_T^2 + I_{Lcm}\omega_L^2 + m_T v_T^2 + m_L v_L^2\right)$$

where v_T is the linear velocity of the thigh center of mass, which is the vector sum of the velocities of the hip and the thigh mass center relative to the hip, that is,

$$v_L^2 = (v_{Hx} + \omega_T d_T \cos\theta_T)^2 + (v_{Hy} + \omega_T d_T \sin\theta_T)^2$$

where v_L is the linear velocity of the leg center of mass, which is the vector sum of the velocities of the hip, the knee relative to the hip, and the leg mass center relative to the knee, that is,

$$v_L^2 = (v_{Hx} + \omega_T L_T \cos\theta_T + \omega_L d_L \cos\theta_L)^2 + (v_{Hy} + \omega_T L_T \sin\theta_T + \omega_L d_L \sin\theta_L)^2$$

The potential energy of the double pendulum is derived by determining the effects of the three degrees of freedom, namely the position of the hip and the angular positions of the thigh and leg, on the segment mass centers. As shown in figure G.1, the hip position changes both mass centers. The thigh angular position affects the locations of both thigh and leg mass centers, whereas the leg angular position affects only the leg center of mass.

Direct analysis of this geometry yields

$$PE = g((m_T + m_L) y_H + (m_T d_T + m_L L_T)(1 - \cos\theta_T) + m_L d_L (1 - \cos\theta_L))$$

Lagrange's equation of motion,

$$\frac{d}{dt}\frac{\partial KE}{\partial \omega_i} - \frac{\partial KE}{\partial \theta_i} + \frac{\partial PE}{\partial \theta_i} = M_i$$

is then implemented as follows.

Thigh

$$\frac{\partial KE}{\partial \omega_T} = I_T \omega_T + m_T d_T \left(v_{Hx}\cos\theta_T + v_{Hy}\sin\theta_T\right) +$$

$$m_L \left((v_{Hx} + \omega_T L_T \cos\theta_T + \omega_L d_L \cos\theta_L)(L_T \cos\theta_T) + (v_{Hy} + \omega_T L_T \sin\theta_T + \omega_L d_L \sin\theta_L)(L_T \sin\theta_T)\right)$$

$$\frac{d}{dt}\frac{\partial KE}{\partial \omega_T} = I_T \alpha_T + m_T d_T \left(a_{Hx}\cos\theta_T - v_{Hx}\omega_T\sin\theta_T + a_{Hy}\sin\theta_T + v_{Hy}\omega_T\cos\theta_T\right)$$

$$+ m_L \left((v_{Hx} + \omega_T L_T \cos\theta_T + \omega_L d_L \cos\theta_L)(-L_T \omega_T \sin\theta_T)\right)$$

$$+ (a_{Hx} + \alpha_T L_T \cos\theta_T - \omega_T^2 L_T \sin\theta_T + \alpha_L d_L \cos\theta_L - \omega_L^2 d_L \sin\theta_L)(L_T \cos\theta_T)$$

$$+ (v_{Hy} + \omega_T L_T \sin\theta_T + \omega_L d_L \sin\theta_L)(L_T \omega_T \cos\theta_T)$$

$$+ (a_{Hy} + \alpha_T L_T \sin\theta_T + \omega_T^2 L_T \cos\theta_T + \alpha_L d_L \sin\theta_L + \omega_L^2 d_L \cos\theta_L)(L_T \sin\theta_T))$$

$$= I_T \alpha_T + m_T d_T \left((a_{Hx} + v_{Hy}\omega_T)\cos\theta_T + (a_{Hy} - v_{Hx}\omega_T)\sin\theta_T\right)$$

$$+ m_L (L_T^2 \alpha_T + (v_{Hx} + \omega_L d_L \cos\theta_L)(-L_T \omega_T \sin\theta_T)$$

$$+ (a_{Hx} + \alpha_L d_L \cos\theta_L - \omega_L^2 d_L \sin\theta_L)(L_T \cos\theta_T)$$

$$+ (v_{Hy} + \omega_L^2 d_L \sin\theta_L)(L_T \omega_T \cos\theta_T)$$

$$+ (a_{Hy} + \alpha_L d_L \sin\theta_L + \omega_L^2 d_L \cos\theta_L)(L_T \sin\theta_T))$$

$$= (I_T + m_L L_T^2) \alpha_T + m_T d_T \left((a_{Hx} + v_{Hy}\omega_T)\cos\theta_T + (a_{Hy} - v_{Hx}\omega_T)\sin\theta_T\right)$$

$$+ m_L (L_T \omega_T (v_{Hy}\cos\theta_T - v_{Hx}\sin\theta_T) + L_T (a_{Hx}\cos\theta_T + a_{Hy}\sin\theta_T)$$

$$+ L_T d_L \omega_T \omega_L \sin(\theta_L - \theta_T) + L_T d_L \alpha_L \cos(\theta_L - \theta_T)$$

$$- L_T d_L \omega_L^2 \sin(\theta_L - \theta_T))$$

$$= (I_T + m_L L_T^2) \alpha_T + m_L L_T d_L (\alpha_L \cos(\theta_L - \theta_T) - \omega_T^2 \sin(\theta_L - \theta_T))$$

$$+ \omega_T \omega_L \sin(\theta_L - \theta_T))$$

$$+ (m_T d_T + m_L L_T) \left((a_{Hx} + v_{Hy}\omega_T)\cos\theta_T + (a_{Hy} - v_{Hx}\omega_T)\sin\theta_T\right)$$

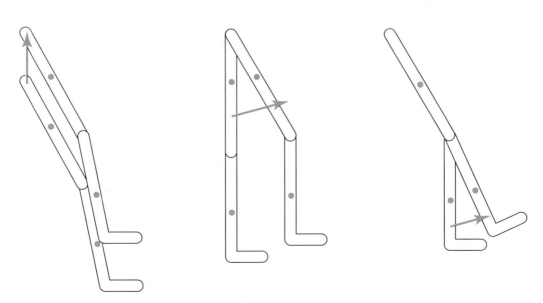

° **Figure G.1**   The hip position, thigh position, and leg position each have different effects on the mass centers.

$$\frac{\partial KE}{\partial \theta_T} = m_T d_T \omega_T \left(-v_{Hx} \sin\theta_T + v_{Hy}\cos\theta_T\right)$$

$$+ m_L L_T \left((v_{Hx} + \omega_T L_T \cos\theta_T + \omega_L d_L \cos\theta_L)(-\omega_T L_T \sin\theta_T)\right.$$

$$+ (v_{Hy} + \omega_T L_T \sin\theta_T + \omega_L d_L \sin\theta_L)(\omega_T L_T \cos\theta_T))$$

$$= m_T d_T \omega_T (v_{Hy}\cos\theta_T - v_{Hx}\sin\theta_T)$$

$$+ m_L L_T \omega_T (v_{Hy}\cos\theta_T - v_{Hx}\sin\theta_T)$$

$$+ m_L L_T d_L \omega_T \omega_L (\cos\theta_T \sin\theta_L - \sin\theta_T \cos\theta_L)$$

$$= \omega_T (m_T d_T + m_L L_T)(v_{Hy}\cos\theta_T - v_{Hx}\sin\theta_T)$$

$$+ m_L L_T d_L \omega_T \omega_L \sin(\theta_L - \theta_T)$$

$$\frac{\partial PE}{\partial \theta_T} = g(m_T d_T + m_L L_T)\sin\theta_T$$

Leg/Foot

$$\frac{\partial KE}{\partial \omega_L} = I_{Leg}\omega_L + m_L \left((v_{Hx} + \omega_T L_T \cos\theta_T + \omega_L d_L \cos\theta_L)\right.$$

$$(d_L \cos\theta_L)$$

$$+ (v_{Hy} + \omega_T L_T \sin\theta_T + \omega_L d_L \sin\theta_L)(d_L \sin\theta_L))$$

$$\frac{d}{dt}\frac{\partial KE}{\partial \omega_L} = I_L \alpha_L + m_L \left((v_{Hx} + \omega_T L_T \cos\theta_T + \omega_L d_L \cos\theta_L)\right.$$

$$(-d_L \omega_L \sin\theta_L)$$

$$+ (a_{Hx} + \alpha_T L_T \cos\theta_T - \omega_T^2 L_T \sin\theta_T$$

$$+ \alpha_L d_L \cos\theta_L - \omega_L^2 d_L \sin\theta_L)(d_L \cos\theta_L)$$

$$+ (v_{Hy} + \omega_T L_T \sin\theta_T + \omega_L d_L \sin\theta_L)(d_L \omega_L \cos\theta_L)$$

$$+ (a_{Hy} + \alpha_T L_T \sin\theta_T + \omega_T^2 L_T \cos\theta_T$$

$$+ \alpha_L d_L \sin\theta_L + \omega_L^2 d_L \cos\theta_L)(d_L \sin\theta_L))$$

$$= I_{Leg}\alpha_L + m_L d_L ((a_{Hx} + v_{Hy}\omega_L)\cos\theta_L$$

$$+ (a_{Hy} - v_{Hx}\omega_L)\sin\theta_L)$$

$$+ m_L (\alpha_T L_T d_L \cos(\theta_L - \theta_T) + \omega_T^2 L_T d_L \sin(\theta_L - \theta_T)$$

$$+ \alpha_L d_L^2 - L_T d_L \omega_T \omega_L \sin(\theta_L - \theta_T))$$

$$= I_L \alpha_L + m_L d_L ((a_{Hx} + v_{Hy}\omega_L)\cos\theta_L$$

$$+ (a_{Hy} - v_{Hx}\omega_L)\sin\theta_L)$$

$$+ m_L L_T d_L (\alpha_T \cos(\theta_L - \theta_T) + \omega_T^2 \sin(\theta_L - \theta_T)$$

$$- \omega_T \omega_L \sin(\theta_L - \theta_T))$$

$$\frac{\partial KE}{\partial \theta_L} = m_L \left((v_{Hx} + \omega_T L_T \cos\theta_T + \omega_L d_L \cos\theta_L)\right.$$

$$(-\omega_L d_L \sin\theta_L)$$

$$+ (v_{Hy} + \omega_T L_T \sin\theta_T + \omega_L d_L \sin\theta_L)(\omega_L d_L \cos\theta_L))$$

$$= m_L (d_L \omega_L (v_{Hy}\cos\theta_L - v_{Hx}\sin\theta_L)$$

$$+ L_T d_L \omega_T \omega_L (\sin\theta_T \cos\theta_L - \cos\theta_T \sin\theta_L))$$

$$= m_L (d_L \omega_L (v_{Hy}\cos\theta_L - v_{Hx}\sin\theta_L)$$

$$- L_T d_L \omega_T \omega_L \sin(\theta_T - \theta_L))$$

$$\frac{\partial PE}{\partial \theta_L} = g\, m_L d_L \sin\theta_L$$

Thus, applying Lagrange's equation yields:

Thigh

$$(I_T + m_L L_T^2)\alpha_T + m_L L_T d_L (\alpha_L \cos(\theta_L - \theta_T)$$

$$- \omega_L^2 \sin(\theta_L - \theta_T)$$

$$+ \omega_T \omega_L \sin(\theta_L - \theta_T))$$

$$+ (m_T d_T + m_L L_T)((a_{Hx} + v_{Hy}\omega_T)\cos\theta_T$$

$$+ (a_{Hy} - v_{Hx}\omega_T)\sin\theta_T))$$

$$- (\omega_T (m_T d_T + m_L L_T)(v_{Hy}\cos\theta_T - v_{Hx}\sin\theta_T)$$

$$+ m_L L_T d_L \omega_T \omega_L \sin(\theta_L - \theta_T))$$

$$+ g(m_T d_T + m_L L_T)\sin\theta_T$$

$$\Rightarrow (I_T + m_L L_T^2)\alpha_T + m_L L_T d_L (\alpha_L \cos(\theta_L - \theta_T)$$

$$- \omega_L^2 \sin(\theta_L - \theta_T))$$

$$+ (m_T d_T + m_L L_T)(a_{Hx}\cos\theta_T$$

$$+ (a_{Hy} + g)\sin\theta_T) = 0 \qquad (G.1)$$

Leg/Foot

$$I_L \alpha_L + m_L d_L ((a_{Hx} + v_{Hy}\omega_L)\cos\theta_L$$

$$+ (a_{Hy} - v_{Hx}\omega_L)\sin\theta_L)$$

$$+ m_L L_T d_L (\alpha_T \cos(\theta_L - \theta_T) + \omega_T^2 \sin(\theta_L - \theta_T)$$

$$- \omega_T \omega_L \sin(\theta_L - \theta_T))$$

$$- m_L (d_L \omega_L (v_{Hy}\cos\theta_L - v_{Hx}\sin\theta_L)$$

$$+ L_T d_L \omega_T \omega_L \sin(\theta_T - \theta_L))$$

$$+ g\, m_L d_L \sin\theta_L = 0$$

$$\Rightarrow I_L \alpha_L + m_L d_L (L_T \alpha_T \cos(\theta_L - \theta_T) + L_T \omega_T^2 \sin(\theta_L - \theta_T)$$

$$+ a_{Hx}\cos\theta_L + (a_{Hy} + g)\sin\theta_L) = 0 \qquad (G.2)$$

Discrete Fourier Transform Subroutine

This is Visual Basic code to calculate the power spectrum. The method is slow but simple.

```
DIM s(numpnts), c(numpnts), h(numpnts), power(numpnts)

w = (2 * pi) / numpnts

m = numpnts / 2 + 1

FOR k = 1 TO m

k1w = (k - 1) * w

FOR j = 1 TO numpnts

alpha = k1w * (j - 1)

s(k) = s(k) + h(j) * SIN(alpha)

c(k) = c(k) + h(j) * COS(alpha)

NEXT j

s(k) = 2 * s(k)

c(k) = 2 * c(k)

power(k) = s(k)^2 + c(k)^2

NEXT k
```

Shannon's Reconstruction Subroutine

This is Visual Basic code to implement Shannon's formula for reconstructing data sampled above the Nyquist rate.

```
'olddelta = original sampling rate

'newdelta = new sampling rate

'samptime = duration of the trial

'fc = Nyquist frequency

'newpoints = number of reconstructed data points

'oldpoints = original number of points

'd!(n) = nth point of the original signal

's!(i) = ith point in the reconstructed signal

pi = 3.14159265

samptime = (oldpoints - 1) * olddelta

newpoints = samptime/newdelta

fc = 1/(2 * olddelta)

fc2 = 2 * fc

For i = 1 To newpoints

t = (i - 1) * newdelta

For n = 1 To oldpoints

If (t - (n - 1) * olddelta) <> 0 Then

m = Sin(fc2 * pi * (t - (n - 1) * olddelta))/(pi * t - (n - 1) * olddelta))

Else

m = 1/olddelta

End If

newdata(i) = newdata(i) + olddata(n) * m

Next n

newdata(i) = newdata(i) * olddelta

Next I
```

EXAMPLE ANSWERS

CHAPTER 3

EXAMPLE 3.1

$$m_{thigh} = P_{thigh}\, m_{total} = 0.100 \times 180.0 = 18.00 \text{ kg}$$

EXAMPLE 3.2

$$x_{cg} = -12.80 + 0.433\,(7.3 - (-12.80)) = -4.10 \text{ cm}$$

$$y_{cg} = 83.3 + 0.433\,(46.8 - 83.3) = 67.5 \text{ cm}$$

Notice that the coordinates of the thigh's center of gravity (−4.10, 67.5) must fall between the two endpoints. Carefully preserve the signs of the coordinates during the computations. These coordinates were taken from frame 10 of table 1.3.

EXAMPLE 3.3

$$l_{thigh} = \sqrt{\left(7.3 - (-12.80)\right)^2 + \left(46.8 - 83.3\right)^2} = 41.67 \text{ cm}$$

$$k_{cg} = K_{cg(thigh)} \times l_{thigh} = 0.323 \times 41.67$$
$$= 13.46 \text{ cm} = 0.1346 \text{ m}$$

In example 3.1, the thigh mass was calculated to be 18.00 kg.

$$I_{cg} = mk^2 = 18.00 \times 0.1346^2 = 0.326 \text{ kg·m}^2$$

Notice that the radius of gyration (k_{cg}) was calculated before the moment of inertia could be computed. This in turn required calculating the thigh segment's length (l_{thigh}). Also notice that the units of the radius of gyration were converted to meters before squaring.

To compute the moment of inertia about the proximal end of this thigh, we must apply the parallel axis theorem after first computing the distance from the thigh center to the proximal end. This distance is called $r_{proximal}$.

$$r_{proximal} = 0.433 \times l_{thigh} = 0.433 \times 41.67$$
$$= 18.04 \text{ cm} = 0.1804 \text{ m}$$

$$I_{proximal} = I_{cg} + m_{thigh}\, r_{proximal}^2$$
$$= 0.326 + 18.00 \times 0.1804^2 = 0.912 \text{ kg·m}^2$$

Notice that the moment of inertia about the proximal end is larger than that about the center of gravity. The moment of inertia is always smallest about an axis through the center of gravity.

CHAPTER 4

EXAMPLE 4.1

The issue with drawing the wind is that we are unable to locate its center of pressure. This is the same problem with FBDs of swimmers and cyclists.

CHAPTER 5

EXAMPLE 5.1a

The FBD for this example is almost the same as figure 5.10d, except that in the FBD for this example we included the forces of gravity and the vertical GRF. Note that X- and Y-axes have been drawn to indicate positive direction. For the sake of example, note that the horizontal acceleration (−64 m/s²) is drawn with an arrow in the negative horizontal direction (to the left) with a positive 64 m/s². Similarly note that the angular acceleration is drawn in a negative (clockwise) direction with a positive 28 rad/s² value. In these two cases, it is also acceptable to draw them pointing in the opposite directions with negative values.

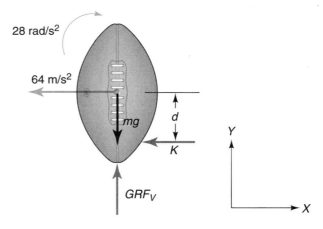

As indicated by the grey arrows, the unknowns in this diagram are the vertical GRF, the force, K, and the distance, d, between K's line of action and the center of the ball. We start by solving for the forces in the vertical direction because they are the simplest in this case.

$$\Sigma F_y = ma_y:$$

$$GRF_y - mg = ma_y$$

$$\Rightarrow GRF_y - (0.25 \text{ kg})(9.81 \text{ m/s}^2) = 0$$

$$\Rightarrow GRF_y = 2.45 \text{ N}$$

This is the expected result when the ground is simply supporting the weight of the ball.

When solving for the horizontal force K, note that the FBD is used to determine that K should have a minus sign because it points to the left (−x). That we calculated a positive K means that the force acts in the direction drawn in the FBD.

$$\Sigma F_x = ma_x:$$

$$-K = ma_x$$

$$\Rightarrow -K = (0.25 \text{ kg})(-64 \text{ m/s}^2)$$

$$\Rightarrow K = 16.0 \text{ N}$$

As is standard in human-movement problems, we calculate moments about the ball's mass center. The FBD shows that there is only one force, K, that does not act through the mass center. Its moment equals the product Kd. We write this term down and then put a negative sign in front of it because this moment causes a *clockwise* (negative) effect. This somewhat tricky part requires visualization. Imagine the mass center as a fixed point and observe that the force K turns about the mass center in a clockwise fashion.

$$\Sigma M = I\alpha:$$

$$-K(d) = I\alpha$$

$$\Rightarrow d = \frac{(0.04 \text{ kg}\cdot\text{m}^2)(-28 \text{ rad/s}^2)}{16 \text{ N}} = -0.070 \text{ m}$$

The distance, d, to the force is negative, indicating that the ball was kicked below the mass center.

EXAMPLE 5.1b
The FBD is almost the same as in the previous example, except for the addition of the horizontal tee force. Note that the tee force has a positive sign:

$$\Sigma F_x = ma_x:$$

$$-K + F_T = ma_x$$

$$\Rightarrow -K + 4 \text{ N} = (0.25 \text{ kg})(-64 \text{ m/s}^2) = -16 \text{ N}$$

$$\Rightarrow K = 20.0 \text{ N}$$

This is the logical result: The kicking force was the sum of the tee force and the ball's reaction.

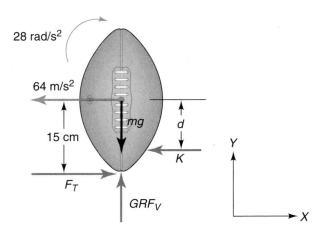

In summing moments about the mass center, note that the moment of the force K is negative, as before, but that the moment of the force F_T is positive:

$$\Sigma M = I\alpha:$$

$$-K(d) + F_T(0.15 \text{ m}) = I\alpha$$

$$\Rightarrow d = \frac{(0.04 \text{ kg}\cdot\text{m}^2)(-28 \text{ rad/s}^2) - 4(0.15 \text{ m})}{16 \text{ N}}$$

$$= -0.1075 \text{ m}$$

Thus, the force was lower on the ball in this instance.

EXAMPLE 5.2
This is an apparently complex situation resolvable with an FBD; in fact, it is very similar to our football example. To draw this correctly, it helps to first identify the forces involved: body weight (gravity), the unknown GRFs, and the reaction of a body mass being accelerated. There is also an unknown ankle moment, M_A. In the FBD, they look like this:

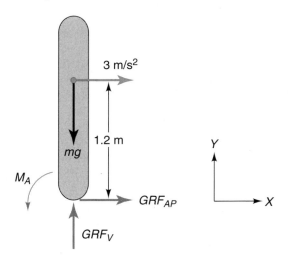

Because the commuter maintains a still posture, her body's angular acceleration is 0 rad/s. The unknown reactions GRF_x, GRF_y, and M_A are at the floor; they are drawn in positive coordinate system directions and determined with our three equations of motion:

$$\Sigma F_x = ma_x:$$

$$GRF_x = ma_x$$

$$\Rightarrow GRF_x = (60 \text{ kg}) (3 \text{ m/s}^2) = 180.0 \text{ N}$$

$$\Sigma F_y = ma_y:$$

$$GRF_y - mg = ma_y$$

$$\Rightarrow GRF_y - (60 \text{ kg}) (9.81 \text{ m/s}^2) = 0$$

$$\Rightarrow GRF_y = 589 \text{ N}$$

Both of these reactions are positive, indicating that they act in the directions drawn.

We will again sum moments about the mass center. Note that we have drawn the unknown moment M_A in a counterclockwise direction, so it is positive in the first equation below. The moment of the vertical GRF_y is zero and the moment of the horizontal GRF_x is positive because it points counterclockwise about the mass center.

$$\Sigma M = I\alpha:$$

$$M_A + GRF_x(1.2 \text{ m}) = I\alpha$$

$$\Rightarrow M_A + (180 \text{ N}) (1.2 \text{ m}) = (130 \text{ kg·m}^2) (0 \text{ rad/s}^2)$$

$$\Rightarrow M_A = -216 \text{ N·m}$$

This is a large ankle moment, which is why sudden subway starts usually cause people to either take a step or grab a handle.

EXAMPLE 5.3

The FBD of the racket is quite simple. The only tricky part is that the hand has an unknown action. Therefore, we must draw unknown actions in each coordinate system direction:

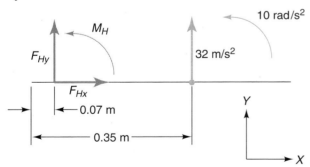

Solving our three equations of motion:

$$\Sigma F_x = ma_x:$$

$$F_{Hx} = (0.5 \text{ kg}) (0 \text{ m/s}^2)$$

$$\Rightarrow F_{Hx} = 0 \text{ N}.$$

$$\Sigma F_y = ma_y:$$

$$F_{Hy} = (0.5 \text{ kg}) (32 \text{ m/s}^2) = 16 \text{ N}.$$

$$\Sigma M = I\alpha:$$

$$M_H - F_{Hy}(0.35 \text{ m} - 0.07 \text{ m}) = (0.1 \text{ kg·m}^2) (10 \text{ rad/s}^2)$$

$$\Rightarrow M_H = (16 \text{ N}) (0.35 \text{ m} - 0.07 \text{ m})$$
$$+ (0.1 \text{ kg·m}^2) (10 \text{ rad/s}^2) = 5.5 \text{ N·m}$$

A moment is the result of a force couple, that is, two equal, opposite, noncollinear forces. We can assume that the hand area by the forefinger is pushing the racket handle forward and the area of the little finger is pulling it backward:

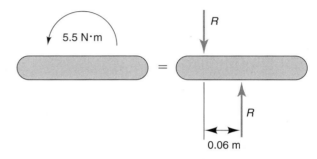

If these points are about 6 cm apart, we can estimate that each member of the force couple is about 91.7 N.

EXAMPLE 5.4

We have measured the endpoints of the object with our camera and determined where the object's mass center is located. We start by redrawing our FBD with the distances in the X and Y directions between the forces and the mass center, about which we are going to calculate the moments of force:

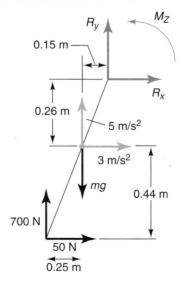

We then apply the three equations of 2-D motion:

$$\Sigma F_x = ma_x:$$

$$R_x + 50 \text{ N} = (8 \text{ kg}) (3 \text{ m/s}^2)$$

$$\Rightarrow R_x = -26.0 \text{ N}$$

Note that R_x is negative. This means that the force points in the direction opposite to that shown in the FBD.

$$\Sigma F_y = ma_y:$$

$$R_y + 700 \text{ N} - mg = (8 \text{ kg})(5 \text{ m/s}^2)$$

$$\Rightarrow R_y = -700 \text{ N} + (8 \text{ kg})(9.81 \text{ m/s}^2) + (8 \text{ kg})(5 \text{ m/s}^2) = -581.5 \text{ N}$$

As with R_x, our calculated value for R_y is negative, indicating that it points downward, the opposite of what we drew in the FBD.

In writing our moment equation, we first write down all the terms and then decide if each term is positive or negative according to the right-hand rule:

$$\Sigma M_z = I\alpha:$$

$$M_z + (50 \text{ N})(0.44 \text{ m}) - (700 \text{ N})(0.25 \text{ m}) - R_x(0.26 \text{ m}) + R_y(0.15 \text{ m})$$

$$= (0.2 \text{ kg·m}^2)(10 \text{ rad/s}^2)$$

We then substitute the numerical values for R_x and R_y that we calculated previously. Note that we put these in parentheses with their negative signs to preserve the signs that we had determined for the signs of the moments:

$$\Rightarrow M_z + (50 \text{ N})(0.44 \text{ m}) - (700 \text{ N})(0.25 \text{ m}) - (-26 \text{ N})(0.26 \text{ m}) + (-582 \text{ N})(0.15 \text{ m})$$

$$= (0.2 \text{ kg·m}^2)(10 \text{ rad/s}^2)$$

We then bring all terms but our unknown moment M_z to the right-hand side of the equation and solve:

$$\Rightarrow M_z = -(50 \text{ N})(0.44 \text{ m}) + (700 \text{ N})(0.25 \text{ m}) + (-26 \text{ N})(0.26 \text{ m})$$

$$- (-582 \text{ N})(0.15 \text{ m}) + (0.2 \text{ kg·m}^2)(10 \text{ rad/s}^2)$$

$$= 235 \text{ N·m}$$

EXAMPLE 5.5

The FBD is:

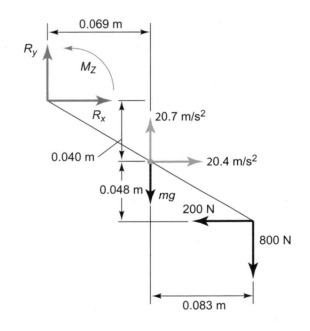

$\Sigma F_x = ma_x:$

$$R_x - 200 \text{ N} = (0.1 \text{ kg})(-0.4 \text{ m/s}^2)$$

$$\Rightarrow R_x = 200 \text{ N}$$

The exact value is 199.96 N, but it rounds to 200 N. The crank is so light that its mass-acceleration product is negligible. The same thing happens in the y-direction:

$$\Sigma F_y = ma_y:$$

$$R_y - 800 \text{ N} - (0.1 \text{ kg})(9.81 \text{ m/s}^2)$$

$$= (0.1 \text{ kg})(-0.7 \text{ m/s}^2)$$

$$\Rightarrow R_y = 801 \text{ N}$$

Again, it is helpful when calculating moments to redraw the FBD with distances between points:

$\Sigma M = I\alpha:$

$$M_z - R_x(0.040 \text{ m}) - R_y(0.069 \text{ m}) - (200 \text{ N})(0.048 \text{ m}) - (800 \text{ N})(0.083 \text{ m})$$

$$= (0.003 \text{ kg·m}^2)(10 \text{ rad/s}^2)$$

$$\Rightarrow M_z = (200 \text{ N})(0.040 \text{ m}) + (801 \text{ N})(0.069 \text{ m}) + (200 \text{ N})(0.048 \text{ m})$$

$$+ (800 \text{ N})(0.083 \text{ m}) + (0.1 \text{ kg·m}^2)(10 \text{ rad/s}^2)$$

$$\Rightarrow M_z = 139.3 \text{ N·m}$$

EXAMPLE 5.6

In this example we need only to section the arm at the shoulder and analyze that one piece. Because this is a static case, the ma and $I\alpha$ terms are zero. To construct the FDB, we section the arm and draw the three unknowns at the shoulder joint. We also draw the weight of the upper arm, forearm, and hand:

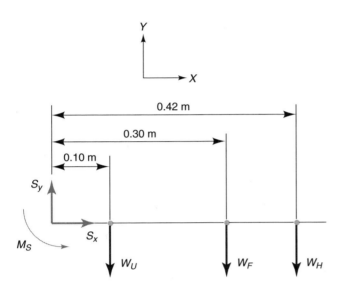

To solve for the three unknowns:

$$\Sigma F_x = ma_x:$$

$$S_x = 0 \text{ N}$$

$$\Sigma F_y = ma_y:$$

$$S_y - W_U - W_F - W_H = 0$$

$$\therefore S_y = W_U + W_F + W_H$$

$$\therefore S_y = (4 \text{ kg})(9.81 \text{ m/s}^2) + (3 \text{ kg})(9.81 \text{ m/s}^2) + (1 \text{ kg})(9.81 \text{ m/s}^2) = 78.5 \text{ N}$$

In this example, it is convenient to calculate moments about the shoulder. Once again, the procedure is first to write down the products of forces and distances and then establish the sign of each moment.

$$\Sigma M_z = 0:$$

$$M_S - W_U(0.10 \text{ m}) - W_F(0.30 \text{ m}) - W_H(0.42 \text{ m}) = 0$$

$$\therefore M_S = W_U(0.10 \text{ m}) + W_F(0.30 \text{ m}) + W_H(0.42 \text{ m})$$

$$\therefore M_S = (4 \text{ kg})(9.81 \text{ m/s}^2)(0.10 \text{ m}) + (3 \text{ kg})(9.81 \text{ m/s}^2)(0.30 \text{ m})$$

$$+ (1 \text{ kg})(9.81 \text{ m/s}^2)(0.42 \text{ m}) = 16.9 \text{ N·m}$$

EXAMPLE 5.7

The FBD is almost identical to the previous example, except for the added weight W_1:

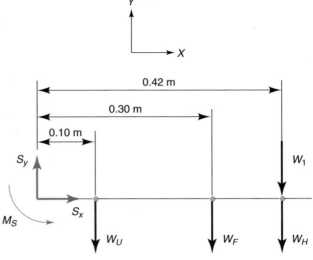

Solving for the three unknowns:

$$\Sigma F_x = ma_x:$$

$$S_x = 0 \text{ N}$$

$$\Sigma F_y = ma_y:$$

$$S_y - W_U - W_F - W_H - W_1 = 0$$

$$\therefore S_y = W_U + W_F + W_H + W_1$$

$$\therefore S_y = (4 \text{ kg})(9.81 \text{ m/s}^2) + (3 \text{ kg})(9.81 \text{ m/s}^2)$$

$$+ (1 \text{ kg})(9.81 \text{ m/s}^2)$$

$$+ (2 \text{ kg})(9.81 \text{ m/s}^2) = 98.1 \text{ N}$$

This is the expected change (compared to example 5.6) of about 20 N.

Once again, to sum the moments, we first write down each moment, then establish its sign:

$$\Sigma M = 0:$$

$$M_S - W_U(0.10 \text{ m}) - W_F(0.30 \text{ m})$$
$$- (W_H + W_1)(0.42 \text{ m}) = 0$$

$$\therefore M_S = W_U(0.10 \text{ m}) + W_F(0.30 \text{ m})$$
$$+ (W_H + W_1)(0.42 \text{ m})$$

$$\therefore M_S = (4 \text{ kg})(9.81 \text{ m/s}^2)(0.10 \text{ m})$$
$$+ 3 \text{ kg} (9.81 \text{ m/s}^2)(0.30 \text{ m})$$

$$+ (1 \text{ kg} + 2 \text{ kg})(9.81 \text{ m/s}^2)(0.42 \text{ m}) = 25.1 \text{ N·m}$$

EXAMPLE 5.8

The FBD of the forearm is again similar to the previous example. We draw the new unknowns at the elbow joint in a positive sense and calculate the distances from the forces to the elbow:

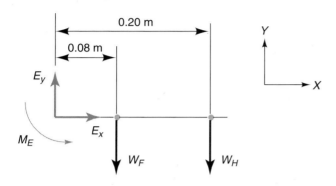

Solving for the three unknowns:

$$\Sigma F_x = ma_x:$$

$$E_x = 0 \text{ N}$$

$$\Sigma F_y = ma_y:$$

$$E_y - W_F - W_H = 0$$

$$\Rightarrow E_y = W_F + W_H$$

$$\Rightarrow E_y = (3 \text{ kg})(9.81 \text{ m/s}^2) + (1 \text{ kg})(9.81 \text{ m/s}^2)$$
$$= 39.2 \text{ N}$$

It is convenient to calculate moments about the elbow:

$$\Sigma M = 0:$$

$$M_E - W_F(0.08 \text{ m}) - W_H(0.20 \text{ m}) = 0$$

$$\Rightarrow M_E = W_F(0.08 \text{ m}) + W_H(0.20 \text{ m})$$

$$\Rightarrow M_E = (3 \text{ kg})(9.81 \text{ m/s}^2)(0.08 \text{ m})$$
$$+ (1 \text{ kg})(9.81 \text{ m/s}^2)(0.20 \text{ m}) = 4.3 \text{ N·m}$$

To solve for the upper arm, the method of sections requires that we draw the upper arm with forces and moments at the elbow having directions equal but opposite to those on the forearm:

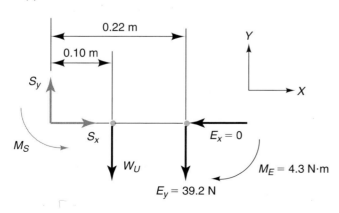

$$\Sigma F_x = ma:$$

$$S_x - E_x = 0$$

$$\Rightarrow S_x = 0 \text{ N}$$

$$\Sigma F_Y = ma:$$

$$S_y - W_U - E_y = 0$$

$$\Rightarrow S_Y = W_U + E_y$$

$$\Rightarrow S_Y = (4 \text{ kg})(9.81 \text{ m/s}^2) + 39.2 \text{ N} = 78.4 \text{ N}$$

This is the same value we had calculated earlier. Summing the moments about the shoulder:

$$\Sigma M = 0:$$

$$M_S - M_E - W_U(0.10 \text{ m}) - E_y(0.22 \text{ m}) = 0$$

$$\Rightarrow M_S = M_E + W_U(0.10 \text{ m}) + E_y(0.22 \text{ m})$$

$$\Rightarrow M_S = 4.32 \text{ N·m} + (4 \text{ kg})(9.81 \text{ m/s}^2)(0.10 \text{ m})$$
$$+ (39.2 \text{ N})(0.22 \text{ m}) = 16.9 \text{ N·m}$$

This is also the same value that we had calculated earlier.

EXAMPLE 5.9

The FBD of the foot:

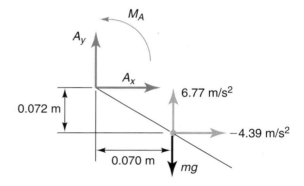

Solving for the reaction forces:

$$\Sigma F_x = ma_x:$$

$$A_x = m_f a_f$$

$$\Rightarrow A_x = (1.2 \text{ kg})(-4.39 \text{ m/s}^2) = -5.27 \text{ N}$$

$$\Sigma F_y = ma_y:$$

$$A_y - m_f g = m_f a_f$$

$$\Rightarrow A_y = m_f g + m_f a_f$$

$$\Rightarrow A_y = (1.2 \text{ kg})(9.81 \text{ m/s}^2)$$
$$+ (1.2 \text{ kg})(6.77 \text{ m/s}^2) = 19.9 \text{ N}$$

Because the foot mass is small, the reactions are small.

Solving for the ankle joint moment, we sum moments about the mass center. There are three moments; of these, the ankle moment M_A is positive because that is how it was drawn. The moments of the reaction forces are both negative because they turn clockwise about the mass center:

$$\Sigma M = I\alpha:$$

$$M_A - A_x(0.072 \text{ m}) - A_y(0.070 \text{ m}) = I_f\alpha_f$$

$$\Rightarrow M_A = A_x(0.072 \text{ m}) + A_y(0.070 \text{ m}) + I_f\alpha_f$$

We now substitute the numerical values of the reaction forces A_x and A_y in parentheses so that we preserve their signs:

$$\Rightarrow M_A = (-5.27 \text{ N})(0.072 \text{ m}) + (19.9 \text{ N})(0.070 \text{ m})$$
$$+ (0.011 \text{ kg·m}^2)(5.12 \text{ rad/s}^2)$$
$$\Rightarrow M_A = 1.1 \text{ N·m}$$

This is a very small joint moment. Although it is essentially zero, because it is positive it is a dorsiflexor moment, assuming the person is facing the right.

The FBD of the leg is as follows. Note that we placed the numerical values for the ankle reactions into this diagram and retained their original signs.

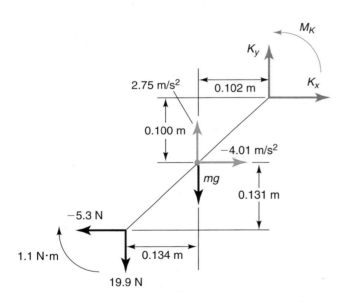

$$\Sigma F_x = ma_x:$$

$$K_x - A_x = m_l a_l$$

$$\Rightarrow K_x = A_x + m_l a_l$$

$$\Rightarrow K_x = -5.3 \text{ N} + (2.4 \text{ kg})(-4.01 \text{ m/s}^2) = -14.9 \text{ N}$$

$$\Sigma F_y = ma_y:$$

$$K_y - A_y - m_l g = m_l a_l$$

$$\Rightarrow K_y = A_y + m_l g + m_l a_l$$

$$\Rightarrow K_y = 19.9 \text{ N} + (2.4 \text{ kg})(9.81 \text{ m/s}^2)$$
$$+ (2.4 \text{ kg})(2.75 \text{ m/s}^2) = 50.0 \text{ N}$$

When summing leg moments, note that the moments of the vertical (y) forces are positive, whereas the moments of the horizontal (x) forces are negative.

$$\Sigma M = I\alpha:$$

$$M_K - M_A - K_x(0.100 \text{ m}) + K_y(0.102 \text{ m})$$
$$- A_x(0.131 \text{ m}) + A_y(0.134 \text{ m}) = I_l\alpha_l$$

$$\Rightarrow M_K = M_A + K_x(0.100 \text{ m}) - K_y(0.102 \text{ m})$$
$$+ A_x(0.131 \text{ m})$$

$$- A_y(0.134 \text{ m}) + I_l\alpha_l$$

Note how we again substitute for the reaction forces the numerical values in parentheses to preserve their signs:

$$\Rightarrow M_K = 1.1 \text{ N·m} + (-14.9 \text{ N})(0.100 \text{ m})$$
$$- (50.0 \text{ N})(0.102 \text{ m}) + (-5.3 \text{ N})(0.131 \text{ m})$$
$$- (19.90 \text{ N})(0.134 \text{ m}) + (0.064 \text{ kg·m}^2)(-3.08 \text{ rad/s}^2)$$
$$\Rightarrow M_K = -9.1 \text{ N·m}$$

This is a small knee flexor moment, assuming the person is facing to the right.

This is the FBD of the thigh. Again note that we placed the numerical values for the knee reactions into this diagram with the same signs that they were calculated as having.

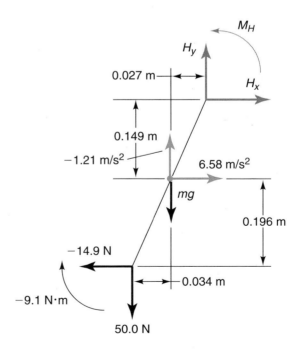

$$\Sigma F_x = ma_x:$$

$$H_x - K_x = m_t a_t$$

$$\Rightarrow H_x = K_x + m_t a_t$$

$$\Rightarrow H_x = -14.9 \text{ N} + (6.0 \text{ kg})(6.58 \text{ m/s}^2) = 24.6 \text{ N}$$

$$\Sigma F_y = ma_y:$$

$$H_y - K_y - m_t g = m_t a_t$$

$$\Rightarrow H_y = K_y + m_t g + m_t a_t$$

$$\Rightarrow H_y = 50.0 \text{ N} + (6.0 \text{ kg})(9.81 \text{ m/s}^2)$$
$$+ (6.0 \text{ kg})(-1.21 \text{ m/s}^2) = 101.6 \text{ N}$$

$$\Sigma M = I\alpha:$$

$$M_H - M_K - H_x(0.149 \text{ m}) + H_y(0.027 \text{ m})$$
$$- K_x(0.196 \text{ m}) + K_y(0.034 \text{ m}) = I_t(\alpha_t)$$

$$\Rightarrow M_H = M_K + H_x(0.149 \text{ m}) - H_y(0.027 \text{ m})$$
$$+ K_x(0.196 \text{ m})$$

$$- K_y(0.034 \text{ m}) + I_t(\alpha_t)$$

$$\Rightarrow M_H = -9.1 \text{ N·m} + (24.6 \text{ N})(0.149 \text{ m})$$
$$- (101.6 \text{ N})(0.027 \text{ m})$$

$$+ (-14.9 \text{ N})(0.196 \text{ m}) - (50.0 \text{ N})(0.034 \text{ m})$$
$$+ (0.130 \text{ kg·m}^2)(8.62 \text{ rad/s}^2)$$

$$\Rightarrow M_H = -11.7 \text{ N·m}$$

Because this is a negative result, it is a hip extensor moment, assuming the person is facing to the right.

EXAMPLE 5.10

The free-body diagram of the foot is the same as for the swing phase except for the ground reaction forces. With these we must be careful to place them in the proper location. It is also helpful to draw them in the positive direction and put their values on, positive or negative.

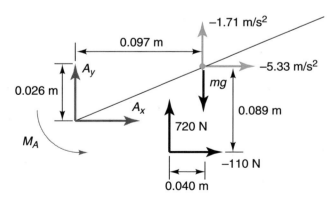

Solving for the reaction forces:

$$\Sigma F_x = ma_x:$$

$$A_x + GRF_x = m_f a_f$$

$$\Rightarrow A_x = -GRF_x + m_f a_f$$

$$\Rightarrow A_x = -(-110 \text{ N}) + 1.2 \text{ kg}(-5.33 \text{ m/s}^2)$$

$$\Rightarrow A_x = 103.6 \text{ N}$$

$$\Sigma F_y = ma_y:$$

$$A_y + GRF_y - m_f g = m_f a_f$$

$$\Rightarrow A_y = -GRF_y + m_f g + m_f a_f$$

$$\Rightarrow A_y = -(720.0 \text{ N}) + 1.2 \text{ kg}(9.81 \text{ m/s}^2)$$
$$+ 1.2 \text{ kg}(-1.71 \text{ m/s}^2)$$

$$\Rightarrow A_y = -710.3 \text{ N}$$

Because the foot mass is small, the ankle joint reaction forces are nearly equal and opposite to the ground reaction forces. Note that the vertical reaction A_y is negative; this means that the actual force is pushing down on the ankle joint, which is a logical result given that this joint is bearing body weight.

Solving for the ankle joint moment, we sum moments about the mass center. There are five moments; of these the ankle moment M_A is positive because of how it is drawn. The moment of the horizontal joint reaction force is positive because it runs counterclockwise around the mass center; however, the moment of vertical joint reaction force is negative because it turns clockwise about the mass center. Similarly, the moment of the horizontal ground reaction force is positive, and the moment of the vertical ground reaction force is negative. Note that the center-of-pressure location is critical in establishing the moment of the vertical GRF; also note that the moment arm of the moment of the horizontal GRF is the vertical position of the mass center because that force is located on the ground (i.e., at y = 0.0 m):

$$\Sigma M = I\alpha:$$

$$M_A - A_x(0.026 \text{ m}) - A_y(0.097 \text{ m}) + GRF_x(0.089 \text{ m})$$
$$- GRF_y(0.040 \text{ m}) = I_f \alpha_f$$

$$\Rightarrow M_A = A_x(0.026 \text{ m}) + A_y(0.097 \text{ m}) - GRF_x(0.089 \text{ m})$$
$$+ GRF_y(0.040 \text{ m}) + I_f \alpha_f$$

$$\Rightarrow M_A = (103.6 \text{ N})(0.026 \text{ m}) + (-710.3 \text{ N})(0.097 \text{ m})$$
$$- (-110 \text{ N})(0.089 \text{ m})$$

$$+ (720 \text{ N})(0.040 \text{ m}) + (0.011 \text{ kg·m}^2)(-20.2 \text{ rad/s}^2)$$

$$\Rightarrow M_A = -33.2 \text{ N·m}$$

This is a plantarflexor action. The free-body diagram of the leg is shown here.

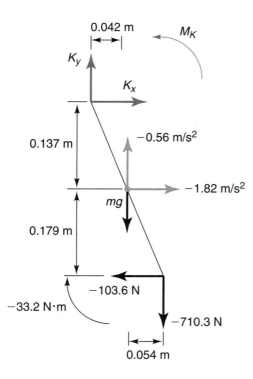

$$\Sigma F_x = m a_x:$$

$$K_x - A_x = m_l\, a_l$$

$$\Rightarrow K_x = A_x + m_l\, a_l$$

$$\Rightarrow K_x = 103.6\ \mathrm{N} + 2.4\ \mathrm{kg}\,(-1.82\ \mathrm{m/s^2})$$

$$\Rightarrow K_x = 99.2\ \mathrm{N}$$

$$\Sigma F_y = m a_y:$$

$$K_y - A_y - m_l\, g = m_l\, a_l$$

$$\Rightarrow K_y = A_y + m_l\, g + m_l\, a_l$$

$$\Rightarrow K_y = -710.3\ \mathrm{N} + 2.4\ \mathrm{kg}\,(9.81\ \mathrm{m/s^2})$$
$$+ 2.4\ \mathrm{kg}\,(-0.56\ \mathrm{m/s^2})$$

$$\Rightarrow K_y = -688.1\ \mathrm{N}$$

When summing leg moments, note that the moments of both the vertical and horizontal forces are negative.

$$\Sigma M = I\alpha:$$

$$M_K - M_A - K_x\,(0.137\ \mathrm{m}) - K_y\,(0.042\ \mathrm{m})$$
$$- A_x\,(0.179\ \mathrm{m}) - A_y\,(0.054\ \mathrm{m}) = I_l\,\alpha_l$$

$$\Rightarrow M_K = M_A + K_x\,(0.137\ \mathrm{m}) + K_y\,(0.042\ \mathrm{m})$$
$$+ A_x\,(0.179\ \mathrm{m})$$

$$+ A_y\,(0.054\ \mathrm{m}) + I_l\,\alpha_l$$

$$\Rightarrow M_K = -33.2\ \mathrm{N\cdot m} + (99.2\ \mathrm{N})\,(0.137\ \mathrm{m})$$
$$+ (-688.1\ \mathrm{N})\,(0.042\ \mathrm{m})$$

$$+ (103.6\ \mathrm{N})\,(0.179\ \mathrm{m}) + (-710.3\ \mathrm{N})\,(0.054\ \mathrm{m})$$
$$+ (0.064\ \mathrm{kg\cdot m^2})\,(-22.4\ \mathrm{rad/s^2})$$

$$\Rightarrow M_K = -69.8\ \mathrm{N\cdot m}$$

This is a knee flexor moment.

The free-body diagram of the thigh is as follows. Again note that we have placed the numerical values for the knee reactions into this diagram with the same signs that they were calculated as having:

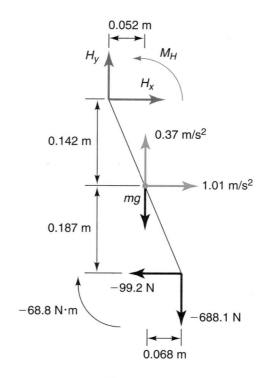

$$\Sigma F_x = m a_x:$$

$$H_x - K_x = m_l\, a_l$$

$$\Rightarrow H_x = K_x + m_l\, a_l$$

$$\Rightarrow H_x = 99.2\ \mathrm{N} + 6.0\ \mathrm{kg}\,(1.01\ \mathrm{m/s^2})$$

$$\Rightarrow H_x = 105.3\ \mathrm{N}$$

$$\Sigma F_y = m a_y:$$

$$H_y - K_y - m_l\, g = m_l\, a_l$$

$$\Rightarrow H_y = K_y + m_l\, g + m_l\, a_l$$

$$\Rightarrow H_y = -688.1\ \mathrm{N} + 6.0\ \mathrm{kg}\,(9.81\ \mathrm{m/s^2})$$
$$+ 6.0\ \mathrm{kg}\,(0.37\ \mathrm{m/s^2})$$

$$\Rightarrow H_y = -627.0\ \mathrm{N}$$

As when we summed the leg moments, note that the moments of both the vertical and horizontal forces are negative:

$$\Sigma M = I\alpha:$$

$$M_H - M_K - H_x\,(0.142\ \mathrm{m}) - H_y\,(0.052\ \mathrm{m})$$
$$- K_x\,(0.187\ \mathrm{m}) - K_y\,(0.068\ \mathrm{m}) = I_t\,\alpha_t$$

$$\Rightarrow M_H = M_K + H_x\,(0.142\ \mathrm{m}) + H_y\,(0.052\ \mathrm{m})$$
$$+ K_x\,(0.187\ \mathrm{m})$$

$$+ K_y \, (0.068 \text{ m}) + I_t \alpha_t$$

$$\square \; M_H = -69.8 \text{ N·m} + (105.3 \text{ N})(0.142 \text{ m})$$
$$+ (-627.0 \text{ N})(0.052 \text{ m})$$

$$+ (99.2 \text{ N})(0.187 \text{ m}) + (-688.1 \text{ N})(0.068 \text{ m})$$
$$+ (0.130 \text{ kg·m}^2)(8.6 \text{ rad/s}^2)$$

$$\square \; M_H = -114.6 \text{ N·m}$$

This is a hip extensor moment. Note how all three of these lower-extremity joint moments are much larger than their swing-phase counterparts.

CHAPTER 6

EXAMPLE 6.1

First, convert the workload to newtons:

$$2.5 \times 9.81 = 24.53 \text{ N}$$

Second, calculate the number of revolutions of the crank:

$$R = 60 \times 20 = 1200$$

Finally, calculate the work:

$$W = 24.53 \times 1200 \times 6 = 176\,616 \text{ J} = 176.6 \text{ kJ}$$

Notice, that the final answer was converted to kilojoules (kJ).

EXAMPLE 6.2

$$W = \Delta E = E_{final} - E_{initial}$$

Assuming that the only change in energy results from a change in speed:

$$\text{work} = 1/2 \; m \, v^2 - 0 = 1/2 \times 180 \times 6^2 = 3240 \text{ J}$$

$$\text{power} = \text{work/duration} = 3260/4 = 810 \text{ W}$$

EXAMPLE 6.3

First, compute the potential energy:

$$E_{gpe} = 18.0 \times 9.81 \times 1.20 = 212.9 \text{ J}$$

Second, calculate the translational kinetic energy:

$$E_{tke} = 1/2 \times 18.0 \times 8^2 = 576 \text{ J}$$

Third, calculate the rotational kinetic energy:

$$E_{rke} = 1/2 \times 0.50 \times 20.0^2 = 100.0 \text{ J}$$

Last, sum to obtain the total energy:

$$E_{tme} = 212.9 + 576 + 100 = 889 \text{ J}$$

APPENDIX C

EXAMPLE C.1

Total resistance *(R)* is the product of the resistivity and the length:

$$R = (10^{-4} \; \Omega/\text{m})(10^{-2} \text{ m}), \text{ which equals } 10^{-6} \; \Omega$$

EXAMPLE C.2a

In series, the total resistance is 10 Ω + 10 Ω, which equals 20 Ω. In parallel, the total resistance is

$$\frac{(10\Omega)(10\Omega)}{10\Omega + 10\Omega}$$

which equals 5 Ω.

EXAMPLE C.2b

In series, the total resistance is 10 Ω + 1 Ω, which equals 11 Ω. In parallel, the total resistance is

$$\frac{(10\Omega)(1\Omega)}{10\Omega + 1\Omega}$$

which equals 0.909 Ω.

Note that when resistors are in series, the total resistance must be greater than the largest resistor in the circuit. However, when resistors are connected in parallel, the total resistance must be less than the smallest resistor in the circuit.

EXAMPLE C.3a

By Ohm's law,

$$I = \frac{V}{R} \Rightarrow I = \frac{20 \text{ V}}{10\Omega} \Rightarrow I = 2 \text{ A}$$

EXAMPLE C.3b

By Ohm's law,

$$R = \frac{V}{I} \Rightarrow R = \frac{9 \text{ V}}{0.002 \text{ A}} \Rightarrow R = 4.5 \text{ k}\Omega$$

EXAMPLE C.3c

By Ohm's law,

$$R = \frac{110 \text{ V}}{15 \text{ A}} \Rightarrow R = 7.3\Omega$$

EXAMPLE C.4a

Rearranging the power law,

$$R = \frac{V^2}{P} \Rightarrow R = 10.1\Omega$$

EXAMPLE C.4b

Using the power law,

$$P = (15 \text{ A})(110 \text{ V}) \quad P = 1650 \text{ W}$$

EXAMPLE C.5

The change in voltage will be given by Ohm's law:

$$V_{80} = 1000 \ \Omega \ (10 \text{ mA}) = 10.0 \text{ V}$$

$$V_{120} = 1200 \ \Omega \ (10 \text{ mA}) = 12.0 \text{ V}$$

This change is easily measured, because typical voltmeters measure to units at least as small as millivolts.

EXAMPLE C.6

1. $V_{out} = \dfrac{15 \text{V}(100 \Omega)}{100 \Omega + 100 \Omega} = 7.5 \text{ V}$

2. $V_{out} = \dfrac{15 \text{V}(110 \Omega)}{100 \Omega + 100 \Omega} = 8.25 \text{ V}$

3. $V_{out} = \dfrac{15 \text{V}(100 \Omega)}{10 \Omega + 100 \Omega} = 13.63 \text{ V}$

4. $V_{out} = \dfrac{15 \text{V}(110 \Omega)}{10 \Omega + 110 \Omega} = 13.75 \text{ V}$

Voltage dividers are most sensitive to changes in one resistor when the two resistances are of similar magnitude. Note that between cases 1 and 2 and between cases 3 and 4, R_v changed by 10%. When R was 100 Ω, there was a 0.75 V (5%) change in V_{out} when R_v changed by 10%. However, when R was 10 Ω, the 10% change in R_v yielded only a 0.12 V (0.8%) change in V_{out}.

EXAMPLE C.7

V_{in} is applied across each of two voltage dividers. From the previous example, we can write a formula for the voltage across R_2 as

$$V_2 = R_2 \frac{V_{in}}{R_v + R_2}$$

and we can write a formula for the voltage across R_3 as

$$V_3 = R_3 \frac{V_{in}}{R_1 + R_3}$$

V_{out} represents the difference in voltage between these two points, that is,

$$V_{out} = V_2 - V_3 = R_2 \frac{V_{in}}{R_v + R_2} - R_3 \frac{V_{in}}{R_1 + R_3}$$

Letting $R_1 = R_2 = R_3$ and doing some algebra yields

$$V_{out} = V_{in} \frac{R_1 - R_v}{2(R_1 + R_v)}$$

This is the classic formula for the output voltage of a Wheatstone bridge.

GLOSSARY

acceleration—A rate of change of velocity and second time derivative of displacement; symbol is *a*.

accelerometer—An instrument that directly measures acceleration; used in impact testing and car crash studies.

accelerometry—Measurement of acceleration by an accelerometer.

accuracy—The ability of an instrument or transducer to obtain the true measure of a quantity; same as *validity* in statistics.

action (force)—A force created when a reaction force is possible; see *law of reaction*.

ammeter—A device for measuring current or amperage.

ampere—An SI unit of electrical current; equals 1 coulomb of charge per second; symbol is A.

amplifier—A device for increasing the magnitude of an analog signal.

amplify—To increase a signal's magnitude; opposite of *attenuate*.

amplitude—Half of the peak-to-peak magnitude in a sinusoidal signal. The value *a* in the sinusoidal function, $W = a \sin (2\pi t + \phi)$. See also *frequency* and *phase angle*.

analog—*(a)* A voltage-varying signal; *(b)* a continuous signal; *(c)* the opposite of digital, as in analog vs. digital signal, timepiece, or computer.

analog signal—A continuous electrical signal that has the same characteristics as another physical signal (for example, force, pressure, acceleration).

analog-to-digital (A/D)—A process by which an analog signal is converted to a digital signal that is suitable to input into a digital computer or device.

analogue—A variant of *analog*.

angular impulse—The time integral of a resultant moment of force acting on a body.

angular momentum—A product of mass moment of inertia and angular velocity.

anthropometry—The measurement of human physical dimensions and the relationships these measurements have with performance.

area moment of inertia—The second moment of a geometrical body.

attenuate—To reduce signal magnitude; opposite of *amplify*.

axis of rotation—The axis about which a body appears to rotate.

biomechanics—The science that studies the influence of forces on living bodies.

body segment parameters—The inertial or physical properties of body segments, especially mass, density, locations of center of mass and center of gravity, and moment of inertia.

capacitor—An electrical component that stores electrical charge; symbol is *C*.

center of gravity—The same as the *center of mass* when the body is near a large astronomical body (Earth).

center of mass—The point at which any line passing through it divides the body's mass in half.

center of percussion—The point on a body at which a collision causes no pressure at the suspension point.

center of pressure—The point at which an equivalent single force causes the same affect on a rigid body as a distributed force.

central force—A force that is always directed at a single point in space; a force that acts through the center of gravity of a body.

centrifugal force—A pseudo-force that is "felt" when a body is following a curved path; the negative of the *centripetal force*.

centripetal force—A force that causes a body to follow a curved path; always directed toward the center of curvature of the path.

centroidal moment of inertia—The moment of inertia of a body about an axis through its center of mass.

cinematography—*(a)* Recording images on film; *(b)* the study of the factors that influence the quality of recording of film images

circuit diagram—A formal means of representing an electric circuit, in which standard icons and straight lines are employed.

coefficient of kinetic friction—The ratio of kinetic friction to normal force, $\mu_{kinetic}$.

coefficient of restitution—The ratio of changes in velocity of two undeformed bodies after and before colliding with each other.

coefficient of static friction—The ratio of maximum static friction to normal force, μ_{static}.

concentric contraction—A contraction in which the muscle force is directed toward the center of the muscle, that is, the muscle shortens while contracting.

contraction—In muscle, the state it is in when it has been induced internally (neurally) or externally (electrically) to shorten; requires neural or electrical stimulation and chemical energy, in the form of adenosine triphosphate or creatine phosphate, and produces an electromyographic signal and force in the tendon. Note that muscle can shorten (concentric contraction), lengthen (eccentric contraction), or remain the same length (isometric contraction) and still be in a state of contraction; however, a muscle can produce a force passively without being in contraction.

Coriolis force—A pseudo-force that appears to exist when a body is moving within a moving frame of reference; for example, an airplane flying due north appears to follow a curved path with respect to the rotating Earth.

coulomb—An SI unit of charge, corresponding to about 6.25×10^{18} electrons; abbreviation is C.

couple—See *force couple.*

current—The rate of flow of electrons or electricity; measured in amps.

decibel—One-hundredth of a *bel*, a unit for describing the ratio of two powers or intensities or for comparison to a reference power or intensity; abbreviated *dB*. Used in electronics and acoustics. For intensities, n dB = 20 $\log_{10}(I_1/I_2)$. For powers, n dB = 10 $\log_{10}(P_1/P_2)$. For example, an amplifier gain (intensity) of 1000 = 20 $\log_{10}(1000/1)$ = 60 dB; a power gain of 20 dB is a gain of 100.

deformable body—A body that can deform under the influence of forces; an elastic body.

diagonal matrix—A matrix whose elements are all zero except elements that have the same row and column numbers.

digital—A numeric; can be represented by a number suitable for use by digital computers.

digital-to-analog (D/A)—A process by which a digital signal can be converted to an equivalent analog (voltage-varying) signal.

digitizer—A device for converting positional information to digital form, usually used to quantify motion from filmed or videotaped images.

direct dynamics—Derivation of kinematics from forces and moments of force.

displacement—The vector that quantifies change of linear position of a particle or a body's center of gravity; sometimes called *linear displacement.*

distortion—Any error introduced to a signal; see *amplitude, frequency,* and *phase distortion.*

dynamics—The mechanics of bodies in motion; see *direct dynamics* and *inverse dynamics.*

dynamometry—The measurement of mechanical forces, moments of force, and power.

eccentric contraction—A contraction in which the muscle force is directed away from the center of the muscle, that is, the muscle lengthens while contracting.

eccentric force or thrust—An impulsive force with a line of action that does not pass through the center of gravity of the body, causing angular acceleration.

electrocardiography (ECG)—A recording of the electrical potentials produced by the cardiac muscles.

electrogoniometer (elgon)—A goniometer that measures joint angles electronically; often consists of a potentiometer with two armatures.

electromyogram (EMG)—A recording from an electromyograph.

electromyograph—A device for measuring the electrical potentials produced by skeletal muscles; usually consists of a differential amplifier with high input impedance (10 $M\Omega$) and high common mode rejection (>80 decibels).

electromyography—The recording of electrical potentials produced by skeletal muscles.

energy—The ability to do work; can be potential or kinetic.

ensemble—A group of related (especially temporally related) data; a history; a digitized signal.

entropy—The loss of usable energy after any transformation of energy from one form to another.

ergometer—A device for measuring mechanical work or permitting human exercise (an exercise machine), such as bicycle or rowing ergometers.

ergometry—The measurement of mechanical work.

ergonomics—*(a)* The study of factors influencing human work, especially in occupational settings; *(b)* literally, work economics; *(c)* "fitting the task to the worker."

event—A unique instant in time, such as the heel-strike in walking, ball contact in striking or impacting activities, and the "catch" in rowing.

external force—Any environmental force that acts on a body.

force—The action of one body on another.

force couple—The turning effect of two parallel forces of equal magnitude but opposite direction; a free moment.

force platform (or plate)—An instrumented, rigid plate capable of quantifying forces applied to its surface.

free-body diagram (FBD)—A diagram of a body free from its environment, but including all the external forces it experiences.

free moment—A moment of force caused by forces, especially force couples, at which the location of the axis of rotation is arbitrary.

frequency—The cyclic rate of a periodic signal in cycles per second or hertz (Hz). Can also mean an angular frequency (ω) in radians per second where $\omega = 2\pi f$. The value f in the sinusoidal function, $W = a \sin(2\pi ft + \phi)$. See also *amplitude* and *phase angle.*

friction—A force caused by the transverse (that is, sheering) interactions of two surfaces.

gain—In an amplifier, the ratio of the original voltage to the amplified voltage.

g-force—A pseudo-force that occurs when a body is rapidly accelerated; results from the inertia of the body, that is, its reluctance to accelerate in response to the applied force.

goniometer—A device for measuring joint angles; see *electrogoniometer.*

ground—The electrical reference point assigned zero voltage.

ground reaction force (GRF)—A single equivalent force equal to the sum of a distribution of forces applied to a surface.

Hall-effect (transducer)—An effect due to the movement of electrons perpendicularly to a permanent magnet; used to quantify force.

hysteresis—The maximum difference between the loading and unloading curves of a *transducer.*

impedance—The sum of all effects on current flow, including resistance and capacitance.

impulse—The time integral of the resultant force acting on a body.

inertia—The reluctance of a body to change its state of rest or motion along a straight line; measured by mass and moment of inertia.

inertial force—A force equal to the negative of the resultant force; used with d'Alembert's principle (the sum of the resultant force and inertial force equals zero).

input impedance—The resistance between the input of a circuit and its ground.

internal force—A force whose action and reaction occur within the same body; a muscle force whose origin and insertion act within the same free body.

inverse dynamics—Computation of forces or moments of force from a body's kinematics and inertial properties.

invert—In electronics, to take the negative of a voltage.

isokinetic contraction—Contraction of a joint at which the joint angular speed is constant (compare *isovelocity contraction*).

isometric contraction—A constant-length contraction, meaning a muscle contraction in which the muscle has no appreciable change in its length.

isotonic contraction—*(a)* A contraction of an excised *(in vitro)* muscle in which the muscle contracts against a constant load; *(b)* weightlifting, that is, a whole muscle (and joint) contraction against a constant load, such as a weight, barbell, or dumbbell; *(c)* a contraction in which a muscle contracts against an artificially produced constant load.

isovelocity contraction—A muscle contraction in which the muscle shortens or lengthens at a constant velocity or speed or when a joint's angular velocity is constant (compare with *isokinetic contraction*).

jerk—The rate of change of acceleration; sometimes called *jolt;* symbol is *j.* Measured in m/s^3 or $g/s.$

kilopond—A force equal to a one-kilogram mass on earth; equal to 9.81 newtons.

kinematics—The study of motion without regard to its causes or quantities of motion, such as velocity, speed, acceleration, angular displacement, and so on.

kinesiology—The science that studies the causes of human motion and the factors that influence human motion.

kinetic friction—Dry *friction* that occurs when two contacting surfaces are in motion.

kinetics—The study of the causes of motion; the study of forces and moments of force and their characteristics, such as work, energy, impulse, momentum, power, and so on.

law of acceleration—Newton's second law; the acceleration of an object is proportional to the sum of the external forces, $\Sigma \vec{F} = m\vec{a}.$

law of gravitation—See *universal law of gravitation.*

law of inertia—Newton's first law; in the absence of an external force, an object remains motionless or in constant speed along a straight line.

law of reaction—Newton's third law; for a force to exist, there must be a reaction force equal in magnitude but opposite in direction.

limiting static friction force—The maximum static friction force before two contacting surfaces "slip."

linearity—The ability of a transducer to produce an output signal that is directly proportional to the input amplitude; the closeness of the relationship between a transducer's input and output signals to a straight line as measured statically or at low frequency. Measured by Pearson's product-moment correlation coefficient (r), which is the same as a linear least-squares curve fit.

linear potentiometer—A resistor with a fixed connection on each end and a sliding connection between the two ends so that translation of the sliding connection alters the resistance between it and each end.

local angular momentum—The product of a body's centroidal mass moment of inertia and its angular velocity.

mass moment of inertia—The second moment of mass of a body about a particular axis.

matrix—Any rectangular array of numbers in which each number in the array is an element.

mechanics—A science that studies the influence of forces on bodies.

moment (arm)—*(a)* The perpendicular distance from a point to a line or surface; radius; the moment arm in the moment of force, moment of momentum, and moment of inertia; *(b)* short for moment of force.

moment of force—The turning effect of a force on a body.

moment of inertia—The reluctance of a body to change its rotational state.

moment of inertia tensor—A 3×3 matrix corresponding to the three-dimensional moment of inertia of a body.

moment of momentum—Vector product of position and linear momentum ($\vec{r} \times m\vec{v}$). Same as remote angular momentum.

momentum—A product of mass and linear velocity; Newton's quantity of motion.

motion-analysis system—A system for collecting and processing the motion of sensors or markers attached to a body.

net force—A force equivalent to the sum of all forces acting across a joint.

net moment of force—A moment of force equivalent to the sum of all moments of force acting across a joint.

noise—Any unwanted random or systematic component in a waveform. Random noise that is uncorrelated with the true signal can be reduced by filtering or averaging. Systematic noise is caused by interference produced by external sources and can be reduced in a variety of ways, including removal of the source of the interference or shielding the electronics.

normal—(a) Perpendicular to a surface or line; (b) perpendicular to tangential.

normal force—The force component that is perpendicular to a surface (compare *tangential force*).

normalize—To perform a form of scaling involving division of a set of numbers by a factor such as body weight, cycle time, or maximum force.

ohm—An SI unit of electrical resistance; symbol is Ω.

Ohm's law—A law stating the linear relation between voltage and current in a linear circuit, $V = IR$.

operational amplifier—A specific type of electronic component that amplifies a voltage.

oscillograph—A device for recording signals on paper, such as a strip chart, pen recorder, or ultraviolet (UV) recorder.

oscilloscope—A device for recording signals on cathode ray tubes (CRTs).

parallel—In electronics, a connection scheme in which the corresponding ends of two or more devices are connected so that the electrical current branches through one or another of the devices.

parallelogram law—A law that defines the addition of vectors; the resultant of two vectors is the diagonal of a parallelogram formed from the two vectors.

phase—A period of time, such as a swing phase or recovery phase.

phase angle—The amount of lead or lag of a sinusoidal waveform compared to a second sinusoidal waveform of the same frequency, used in a Fourier series; measured in degrees or radians. The value ϕ in the sinusoidal function, $W = a \sin(2\pi f t + \phi)$. See also frequency and phase angle.

piezoelectric effect—Occurs when certain crystals, such as quartz, are mechanically stressed causing a voltage.

piezoelectric (transducer)—An effect from pressure exerted on certain crystals that causes them to produce a voltage; used in force and acceleration transducers and precision timepieces.

potentiometer—An electronic device that permits variable resistance; used in electrogoniometers, amplifier controls, volume controls, and other similar devices.

precision—The ability of a device to produce the same measurement repeatedly; the same as *statistical reliability* or *repeatability*.

pressure—The force exerted over an area; symbol is P; units are pascals (Pa).

principal distance—The camera's distance setting, which should be set equal to the distance between the lens and the object being filmed.

principal mass moments of inertia—The diagonal elements of an inertia tensor.

products of inertia—The off-diagonal elements of an inertia tensor.

pseudo-force—An apparent force that exists as the result of a moving frame of reference; includes *centrifugal forces*, *g-forces*, and *Coriolis forces*.

radial—In a direction away from the center of the radius of curvature of a path; the direction perpendicular to transverse.

radius of gyration—The radius of a point mass that has the same mass moment of inertia as a particular body.

reaction (force)—A force that occurs whenever an action force is created (see *law of reaction*).

refine—To scale or otherwise transform digitized motion-picture data to real units in a known frame of reference.

remote angular momentum—The same as *moment of momentum*.

resistance—In electronics, the effect of a particular device or component that is directly proportional to the voltage applied across it.

resistivity—A material property that expresses the electrical resistance per unit length.

resistor—A device that limits current flow in direct proportion to the voltage across it.

resultant—The vector sum of two or more vectors; see *parallelogram law*.

resultant force—The sum of all forces acting on a body; also called *external force*.

Riemann integration—Integration by adding the elements and multiplying by the duration of the sample.

rigid body—A group of particles occupying fixed positions with respect to each other; a theoretical body that is not deformable and has fixed inertial properties.

rotary potentiometer—A resistor with a fixed connection on each end and a sliding connection between the two ends so that rotating the control alters the resistance between the sliding connection and each end.

scalar—A quantity that can be characterized by a magnitude alone (e.g., mass, distance, speed, time, energy and power [compare *vector*]).

scale—To alter the magnitude of a digital signal; to multiply by a constant; compare *normalize, refine.*

sensitivity—For transducers, this is the ratio of the input signal divided by the output signal; e.g., 100 newtons per volt.

series—In electronics, a connection between two or more components in which one follows the other so that all current passes through each component.

signal—The information content of a waveform; the opposite of noise.

Simpson's rule integration—A method of integration based on Simpson's rule approximation of a series of data.

spatial synchronization—The synchronization of two or more data sets in space.

static friction—Dry friction when there is no relative motion of two contacting surfaces.

statics—The mechanics of bodies at rest or in uniform (constant linear) motion.

strain—The change in length divided by resting length; normalized deformation.

strain gauge—A resistor-based device designed to be attached to the surface of a material so that its resistance changes as the material deforms.

stress—The loading force per cross-sectional area; measured in kilopascals (kPa); the normalized load force.

tangent—(a) A line that is perpendicular to the normal surface or curve; (b) the slope of a line or rise/run; (c) the tangent of an angle (tan θ); the ratio of opposite to adjacent sides of a right triangle.

tangential—The direction that is parallel to the tangent line of a curved path; perpendicular to the normal.

tangential force—The force component that is perpendicular to a *normal force* and parallel to a surface.

telemetry—The transmission of a signal over a distance, usually by radio transmission.

temporal—Relating to time; in the time domain.

temporal synchronization—Synchronization of two or more data sets in time.

tensor—A mathematical or physical quantity possessing a specified system of components for every coordinate system; a generalized vector with more than three components, each of which is a function of the coordinates of an arbitrary point in space of an appropriate number of dimensions.

thermodynamics—The branch of science concerned with transduction of heat and energy.

torque—The same as *moment of force*, especially when the moment of force is about the longitudinal axis of a body.

transducer—A device that is actuated by power from one system and supplies power, usually in another form, to a second system; a device that changes one form of energy to another. An input transducer converts a physical signal, such as force, temperature, or power, into an electrical signal, usually voltage. An output transducer converts an electrical signal into a physical quantity, as, for example, loudspeakers, oscillographs, and multimeters do.

transduction—The process of converting a physical dimension into a voltage.

trapezoidal integration—A method of integrating a series of numbers by adding adjacent trapezoids.

universal law of gravitation—A law stating that two objects have a force of attraction proportional to the product of their masses divided by the square of the distance between their centers of mass; $F_g = \dfrac{Gm_1 m_2}{r^2}$

vector—A mathematical expression possessing magnitude and direction that adds according to the *parallelogram law*; examples are force, acceleration, and displacement, but not finite rotations; compare *scalar.*

velocity—A vector rate of change of displacement that includes the direction of motion; symbol is v.

voltage—An SI unit of electrical potential, equal to 1 joule of energy per coulomb of charge; symbol is V or E.

voltage divider—A circuit of two or more resistors in series, with their object being to employ or measure voltage at an intermediate point.

voltage drop—The voltage across a resistor.

voltmeter—A device that measures voltage.

waveform—Any continuously varying quantity consisting of signal or noise components.

weight—A force due to the gravitational attraction of a massive body such as the Earth or Moon.

Wheatstone bridge—An electrical circuit including two parallel pairs of series-connected resistors.

work—The change in energy of a body.

REFERENCES

Abdel-Aziz, Y.I., and H.M. Karara. 1971. Direct linear transformation from comparator coordinates into object space coordinates in close-range photogrammetry. In *American Society for Photogrammetry Symposium on Close Range Photogrammetry,* Falls Church, VA: American Society for Photogrammetry.

Albert, W.J., and D.I. Miller. 1996. Takeoff characteristics of single and double axel figure skating jumps. *Journal of Applied Biomechanics* 12:72-87.

Aleshinsky, S.Y. 1986a. An energy "sources" and "fractions" approach to the mechanical energy expenditure problem. I: Basic concepts, description of the model, analysis of a one-link system movement. *Journal of Biomechanics* 19:287-94.

Aleshinsky, S.Y. 1986b. An energy "sources" and "fractions" approach to the mechanical energy expenditure problem. IV: Criticism of the concept of "energy transfers within and between links." *Journal of Biomechanics* 19: 307-9.

Aleshinsky, S.Y. 1986c. An energy "sources" and "fractions" approach to the mechanical enegy expenditure problem. V: The mechanical energy expenditure deduction during motion of a multi-link system. *Journal of Biomechanics* 19:310-5.

Alexander, R.M. 1990. Optimum take-off techniques for high and long jumps. *Philosophical Transactions of the Royal Society of London B* 329:3-10.

Alexander, R.M. 1992. Simple models of walking and jumping. *Human Movement Science* 11:3-9.

Alexander, R.M., M.B. Bennett, and R.F. Ker. 1986. Mechanical properties and function of the paw pads of some mammals. *Journal of Zoology (London)* 209: 405-19.

Allard, P., R. Lachance, R. Aissaoui, and M. Duhaime. 1996. Simultaneous bilateral able-bodied gait. *Human Movement Science* 15:327-46.

Ambroz, C., A. Scott, A. Ambroz, and E.O. Talbott. 2000. Chronic low back pain assessment using surface electromyography. *Journal of Occupational and Environmental Medicine* 42:660-9.

American Society for Photogrammetry and Remote Sensing. 1980. Manual of Photogrammetry, 4th ed. Falls Church, VA: American Society for Photogrammetry and Remote Sensing.

An, K.-N., and E.Y.S. Chao. 1991. Kinematic analysis. In *Biomechanics of the Wrist Joint,* ed. K.-N. An, R.A. Berger, and W.P. Cooney. New York: Springer-Verlag.

Anderson, F.C., and M.G. Pandy. 2001. Contributions of individual muscles to support during normal gait. *Gait and Posture* 13:292-3.

Andreassen, S., and L. Arendt-Nielsen. 1987. Muscle fibre conduction velocity in motor units of the human anterior tibial muscle: A new size principle parameter. *Journal of Physiology* 391:561-71.

Andreassen, S., and A. Rosenfalck. 1978. Recording from a single motor unit during strong effort. *IEEE Transactions on Biomedical Engineering* 25:501-8.

Andriacchi, T.P., B.J. Andersson, R.W. Fermier, D. Stern, and J.O. Galante. 1980. A study of lower limb mechanics during stair climbing. *Journal of Bone and Joint Surgery* 62:749-57.

Angeloni, C., A. Cappozzo, F. Catani, and A. Leardini. 1993. Quantification of relative displacement of skin- and plate-mounted markers with respect to bones. *Journal of Biomechanics* 26:864.

Aoki, F., H. Nagasaki, and R. Nakamura. 1986. The relation of integrated EMG of the triceps brachii to force in rapid elbow extension. *Tohoku Journal of Experimental Medicine* 149:287-91.

Areblad, M., B.M. Nigg, J. Ekstrand, K.O. Olsson, and H. Ekstrom. 1990. Three-dimensional measurement of rearfoot motion during running. *Journal of Biomechanics* 23:933-40.

Arun, K., T. Huang, and S. Blostein. 1987. Least squares fitting of two 3D point sets. *IEEE Transactions on Pattern Analysis and Machine Intelligence* 5:698-700.

Audu, M.L., and D.T. Davy. 1988. A comparison of optimal control algorithms for complex bioengineering studies. *Optimal Control Applications & Methods* 9:101-6.

Bach, T.M. 1995. Optimizing mass and mass distribution in lower limb prostheses. *Prosthetics & Orthotics Australia* 10:29-34.

Bahler, A.S. 1967. Series elastic component of mammalian skeletal muscle. *American Journal of Physiology* 213: 1560-4.

Baildon, R., and A.E. Chapman. 1983. A new approach to the human muscle model. *Journal of Biomechanics* 16:803-9.

Barkhaus, P.E., and S.D. Nandedkar. 1994. Recording characteristics of the surface EMG electrodes. *Muscle and Nerve* 17:1317-23.

Barter, J.T. 1957. *Estimation of the Mass of Body Segments.* WADC Technical Report 57-260. Wright-Patterson Air Force Base, OH.

Basmajian, J.V. 1989. *Biofeedback: Principles and Practices for Clinicians*. Baltimore: Lippincott, Williams & Wilkins.

Basmajian, J.V., and G. Stecko. 1962. A new biopolar electrode for electromyography. *Journal of Applied Physiology* 17:849.

Bellemare, F., and N. Garzaniti. 1988. Failure of neuromuscular propagation during human maximal voluntary contraction. *Journal of Applied Physiology* 64:1084-93.

Bernstein, N. 1967. *Co-ordination and Regulation of Movements*. London: Pergamon Press.

Betts, B., and J.L. Smith. 1979. Period-amplitude analysis of EMG from slow and fast extensors of cat during locomotion and jumping. *Electroencephalography and Clinical Neurophysiology* 47:571-81.

Biewener, A.A., D.D. Konieczynski, and R.V. Baudinette. 1998. *In vivo* muscle force-length behavior during steady-speed hopping in tammar wallabies. *Journal of Experimental Biology* 201:1681-94.

Biewener, A.A., and T.J. Roberts. 2000. Muscle and tendon contributions to force, work and elastic energy savings: A comparative perspective. *Exercise and Sport Science Reviews* 28:99-107.

Bigland-Ritchie, B., R. Johansson, O.C.J. Lippold, S. Smith, and J.J. Woods. 1983. Changes in motoneurone firing rates during sustained maximal voluntary contractions. *Journal of Physiology* 340:335-46.

Bilodeau, M., A.B. Arsenault, D. Gravel, and D. Bourbonnais. 1990. The influence of an increase in the level of force on the EMG power spectrum of elbow extensors. *European Journal of Applied Physiology* 61:461-6.

Bobbert, M.F., P.A. Huijing, and G.J. van Ingen Schenau. 1986. A model of the human triceps surae muscle-tendon complex applied to jumping. *Journal of Biomechanics* 19:887-98.

Bobbert, M.F., and G.J. van Ingen Schenau. 1990. Isokinetic plantar flexion: Experimental results and model calculations. *Journal of Biomechanics* 23:105-19.

Bobbert, M.F., and J.P. van Zandwijk. 1999. Dynamics of force and muscle stimulation in human vertical jumping. *Medicine and Science in Sports and Exercise* 31:303-10.

Boer, R.W. de, J. Cabri, W. Vaes, J.P. Clarijs, A.P. Hollander, G. de Groot, and G.J. van Ingen Schenau. 1987. Moments of force, power, and muscle coordination in speed-skating. *International Journal of Sports Medicine* 8:371-8.

Bogert, A.J. van den, K.G.M. Gerritsen, and G.K. Cole. 1998. Human muscle modeling from a user's perspective. *Journal of Electromyography and Kinesiology* 8:119-24.

Bouisset, S., and B. Maton. 1972. Quantitative relationship between surface EMG and intramuscular electromyographic activity in voluntary movement. *American Journal of Physical Medicine* 51:285-95.

Braune, W., and O. Fischer. 1889. Über den Scherpunkt des menschlichen Körpers, mit Rücksicht auf die Aüsrustung des deutschen Infanteristen [The center of gravity of the human body as related to the equipment of the German Infantry]. *Abhandlungen der mathematisch-physischen Klasse der Königlich-Sächsischen Gesellschaft der Wissenschaften* 26:561-672.

Braune, W., and O. Fischer. 1895-1904. *Der Gang des Menschen*. Berlin: B.G. Teubner. Trans. P. Maquet and R. Furlong, *The Human Gait*. Berlin: Springer-Verlag, 1987.

Bresler, B., and F.R. Berry. 1951. *Energy and Power in the Leg During Normal Level Walking*. Prosthetic Devices Research Project, Ser. II, Iss. 15. Berkeley: University of California.

Bresler, B., and J.P. Frankel. 1950. The forces and moments in the leg during level walking. *Transactions of the American Society of Mechanical Engineers* 72:27-36.

Brismar, T., and L. Ekenvall. 1992. Nerve conduction in the hands of vibration exposed workers. *Electroencephalography and Clinical Neurophysiology* 85:173-6.

Broker, J.P., and R.J. Gregor. 1990. A dual piezoelectric force pedal for kinetic analysis of cycling. *International Journal of Sports Biomechanics* 6:394-403.

Brooks, C.B., and A.M. Jacobs. 1975. The gamma mass scanning technique for inertial anthropometric measurement. *Medicine and Science in Sports* 7:290-4.

Brown, T.D., L. Sigal, G.O. Njus, N.M. Njus, R.J. Singerman, and R.A. Brand. 1986. Dynamic performance characteristics of the liquid metal strain gage. *Journal of Biomechanics* 19:165-73.

Buchanan, T.S., D.J. Almdale, J.L. Lewis, and W.Z. Rymer. 1986. Characteristics of synergic relations during isometric contractions of human elbow muscles. *Journal of Neurophysiology* 56:1225-41.

Buchanan, T.S., G.P. Rovai, and W.Z. Rymer. 1989. Strategies for muscle activation during isometric torque generation at the human elbow. *Journal of Neurophysiology* 62:1201-12.

Buchthal, F., and P. Rosenfalck. 1958. Rate of impulse conduction in denervated human muscle. *Electroencephalography and Clinical Neurophysiology* 10:521-6.

Burgar, C.G., F.J. Valero-Cuevas, and V.R. Hentz. 1997. Fine-wire electromyographic recording during force generation: Application to index finger kinesiologic studies. *American Journal of Physical Medicine* 76:494-501.

Caldwell, G.E. 1995. Tendon elasticity and relative length: Effects on the Hill two-component muscle model. *Journal of Applied Biomechanics* 11:1-24.

Caldwell, G.E., W.B. Adams, and M.L. Whetstone. 1993. Torque/velocity properties of human knee muscles: Peak and angle-specific estimates. *Canadian Journal of Applied Physiology* 18(3): 274-90.

Caldwell G.E., and A.E. Chapman. 1989. Applied muscle modeling: Implementation of muscle-specific models. *Computers in Biology and Medicine* 19:417-34.

Caldwell G.E., and A.E. Chapman. 1991. The general distribution problem: A physiological solution which includes antagonism. *Human Movement Science* 10:355-92.

Caldwell, G.E., and L.W. Forrester. 1992. Estimates of mechanical work and energy transfers: Demonstration of a rigid body power model of the recovery leg in gait. *Medicine and Science in Sports and Exercise* 24:1396-412.

Caldwell, G.E., L. Li, S.D. McCole, and J.M. Hagberg. 1998. Pedal and crank kinetics in uphill cycling. *Journal of Applied Biomechanics* 14:245-59.

Cappozzo, A., F. Figura, M. Marchetti, and A. Pedotti. 1976. The interplay of muscular and external forces in human ambulation. *Journal of Biomechanics* 9:35-43.

Cappozzo, A., T. Leo, and A. Pedotti. 1975. A general computing method for the analysis of human locomotion. *Journal of Biomechanics* 8:307-20.

Carrière, L., and A. Beuter. 1990. Phase plane analysis of biarticular muscles in stepping. *Human Movement Science* 9:23-5.

Cavagna, G.A., N.C. Heglund, and C.R. Taylor. 1977. Mechanical work in terrestrial locomotion: Two basic mechanisms for minimizing energy expenditure. *American Journal of Physiology* 233:R243-61.

Cavagna, G.A., and M. Kaneko. 1977. Mechanical work and efficiency in level walking and running. *Journal of Physiology* 268:467-481.

Cavagna, G.A., L. Komarek, and S. Mazzoleni. 1971. The mechanics of sprint running. *Journal of Physiology* 217: 709-21.

Cavagna, G.A., F.P. Saibene, and R. Margaria. 1963. External work in walking. *Journal of Applied Physiology* 18:1-9.

Cavagna, G.A., F.P. Saibene, and R. Margaria. 1964. Mechanical work in running. *Journal of Applied Physiology* 19:249-56.

Cavanagh, P.R. 1978. A technique for averaging center of pressure paths from a force platform. *Journal of Biomechanics* 11:487-91.

Chandler, R.F., C.E. Clauser, J.T. McConville, H.M. Reynolds, and J.W. Young. 1975. *Investigation of Inertial Properties of the Human Body*. AMRL Technical Report 74-137. Wright-Patterson Air Force Base, OH.

Chao, E.Y.S. 1980. Justification for a triaxial goniometer for the measurement of joint rotation. *Journal of Biomechanics* 13:989-1006.

Chapman, A.E. 1985. The mechanical properties of human muscle. *Exercise and Sports Science Reviews* 13:443-501.

Chapman, A., G. Caldwell, R. Herring, R. Lonergan, and S. Selbie. 1987. Mechanical energy and the preferred style of running. In *Biomechanics X-B*, ed. B. Jonsson, 875-9. International Series on Biomechanics. Champaign, IL: Human Kinetics.

Chapman, A.E., G.E. Caldwell, and W.S. Selbie. 1985. Mechanical output following muscle stretch in forearm supination against inertial loads. *Journal of Applied Physiology* 59(1):78-86.

Chen, J.J., T.Y. Sun, T.H. Lin, and T.S. Lin. 1997. Spatiotemporal representation of multichannel EMG firing patterns and its clinical applications. *Medical Engineering and Physics* 19:420-30.

Clamann, H.P., and K.T. Broecker. 1979. Relation between force and fatigability of red and pale skeletal muscles in man. *American Journal of Physical Medicine* 58:70-85.

Clamann, H.P., and T.B. Schelhorn. 1988. Nonlinear force addition of newly recruited motor units in the cat hindlimb. *Muscle and Nerve* 11:1079-89.

Clauser, C.E., J.T. McConville, and J.W. Young. 1969. *Weight, Volume and Center of Mass of Segments of the Human Body*. AMRL Technical Report 60-70. Wright-Patterson Air Force Base, OH.

Cole, G.K., B.M. Nigg, J.L. Ronsky, and M.R. Yeadon. 1993. Application of the joint coordinate system to three-dimensional joint attitude and movement representation: A standardization proposal. *Journal of Biomechanical Engineering* 115:344-9.

Connolly, S., D.G. Smith, D. Doyle, and C.J. Fowler. 1993. Chronic fatigue: Electromyographic and neuropathological evaluation. *Journal of Neurology* 240:435-8.

Contini, R. 1972. Body segment parameters, part II. *Artificial Limbs* 16:1-19.

Cram, J.R., G.S. Kasman, and J. Holtz. 1998. *Introduction to Surface Electromyography*. Gaithersburg, MD: Aspen.

Cronin, A., and D.G.E. Robertson. 2000. Two-segment model of the foot: Reduction of power imbalance. *Archives of Physiology & Biochemistry* 108:120.

Crowninshield, R.D., and R.A. Brand. 1981. A physiologically based criterion of muscle force prediction in locomotion. *Journal of Biomechanics* 14:783-801.

Cuturic, M., and S. Palliyath. 2000. Motor unit number estimate (MUNE) testing in male patients with mild to moderate carpal tunnel syndrome. *Electromyography and Clinical Neurophysiology* 40:67-72.

Cywinska-Wasilewska, G., J.J. Ober, and J. Koczocik-Przedpelska. 1998. Power spectrum of the surface EMG in post-polio syndrome. *Electromyography and Clinical Neurophysiology* 38:463-6.

Dainis, A. 1980. Whole body and segment center of mass determination from kinematic data. *Journal of Biomechanics* 13:647-51.

Dapena, J. 1978. A method to determine the angular momentum of a human body about three orthogonal axes passing through its center of gravity. *Journal of Biomechanics* 11:251-6.

Dapena, J., E.A. Harman, and J.A. Miller. 1982. Three-dimensional cinematography with control object of unknown shape. *Journal of Biomechanics* 15:11-19.

D'Apuzzo, N. 2001. Motion capture from multi image video sequences. In *Proceedings of the XVIIIth Congress of the International Society of Biomechanics, July 8-13, Zurich, Switzerland*. CD-ROM, paper 0106.

Davis, R.B., S. Ounpuu, D. Tyburski, and J.R. Gage. 1991. A gait analysis data collection and reduction technique. *Human Movement Science* 10:575-87.

de Leva, P. 1996. Adjustments to Zatsiorsky-Seluyanov's segment inertia parameters. *Journal of Biomechanics* 29: 1223-30.

DeLisa, J.A., and K. Mackenzie. 1982. *Manual of Nerve Conduction Velocity Techniques.* New York: Raven Press.

Delp, S.L., A.S. Arnold, and S.J. Piazza. 1998. Graphics-based modeling and analysis of gait abnormalities. *Bio-medical Materials and Engineering* 8:227-40.

Delp, S.L., J.P. Loan, M.G. Hoy, F.E. Zajac, E.L. Topp, and J.M. Rosen. 1990. An interactive graphics-based model of the lower extremity to study orthopaedic surgical procedures. *IEEE Transactions on Biomedical Engineering* 37:757-67.

Deluzio, K.J., U.P. Wyss, P.A. Costigan, C. Sorbie, and B. Zee. 1999. Gait assessment in unicompartmental knee arthroplasty patients: Principle component modeling of gait waveforms and clinical status. *Human Movement Science* 18:701-11.

Dempster, W.T. 1955. *Space Requirements of the Seated Operator: Geometrical, Kinematic, and Mechanical Aspects of the Body with Special Reference to the Limbs.* WADC Technical Report 55-159. Wright-Patterson Air Force Base, OH.

Derrick, T.R. 1998. Circular continuity of non-periodic data. *Proceedings of the Third North American Congress on Biomechanics.* Waterloo: American and Canadian Societies of Biomechanics, 313-4.

Derrick, T.R., G.E. Caldwell, and J. Hamill. 2000. Modeling the stiffness characteristics of the human body while running with various stride lengths. *Journal of Applied Biomechanics* 16:36-51.

Dewald, J.P., P.S. Pope, J.D. Given, T.S. Buchanan, and W.Z. Rymer. 1995. Abnormal muscle coactivation patterns during isometric torque generation at the elbow and shoulder in hemiparetic subjects. *Brain* 118:495-510.

Donelan, J.M., R. Kram, and A.D. Kuo. 2002. Simultaneous positive and negative external mechanical work in human walking. *Journal of Biomechanics* 35:117-24.

Dowling, J.J. 1997. The use of electromyography for the noninvasive prediction of muscle forces: Current issues. *Sports Medicine* 24:82-96.

Drillis, R., R. Contini, and M. Bluestein. 1964. Body segment parameters: A survey of measurement techniques. *Artificial Limbs* 8:44-66.

Duchateau, J., S. Le Bozec, and K. Hainaut. 1986. Contributions of slow and fast muscles of triceps surae to a cyclic movement. *European Journal of Applied Physiology* 55:476-81.

Dumitru, D., and J.C. King. 1999. Motor unit action potential duration and muscle length. *Muscle and Nerve* 22:1188-95.

Dumitru, D., J.C. King, and S.D. Nandedkar. 1997. Motor unit action potential duration recorded by monopolar and concentric needle electrodes: Physiologic implications. *American Journal of Physical Medicine* 76:488-93.

Durkin, J.L., and J. Dowling. 2003. Analysis of body segment parameter differences between four human populations and the estimation of errors of four popular mathematical models. *Journal of Biomechanical Engineering* 125:515-22.

Durkin, J.L., J. Dowling, and D.M. Andrews. 2002. The measurement of body segment inertial parameters using dual energy X-ray absorptiometry. *Journal of Biomechanics* 35:1575-80.

Eberstein, A., and B. Beattie. 1985. Simultaneous measurement of muscle conduction velocity and EMG power spectrum changes during fatigue. *Muscle and Nerve* 8:768-73.

Edman, K.A.P. 1975. Mechanical deactivation induced by active shortening in isolated muscle fibres of the frog. *Journal of Physiology* 246:255-75.

Edman, K.A.P. 1988. Double-hyperbolic force-velocity relation in frog muscle fibres. *Journal of Physiology* 404:301-21.

Edman, K.A.P., G. Elzinga, and M.I.M. Noble. 1978. Enhancement of mechanical performance by stretch during tetanic contractions of vertebrate skeletal muscle fibres. *Journal of Physiology* 466:535-52.

Elftman, H. 1934. A cinematic study of the distribution of pressure in the human foot. *Anatomical Record* 59:481-91.

Elftman, H. 1939a. Forces and energy changes in the leg during walking. *American Journal of Physiology* 125:339-56.

Elftman, H. 1939b. The function of muscles in locomotion. *American Journal of Physiology* 12:357-66.

Elftman, H. 1940. The work done by muscle in running. *American Journal of Physiology* 129:672-84.

Engsberg, J., S.K. Grimston, and J. Wackwitz. 1988. Predicting talocalcaneal joint attitudes from talocalcaneal/talocrural joint attitudes. *Journal of Orthopedic Research* 6:749-57.

Etnyre, B.R., and L.D. Abraham. 1988. Antagonist muscle activity during stretching: A paradox re-assessed. *Medicine and Science in Sports and Exercise* 20:285-9.

Feldman, R.G., P.H. Travers, J. Chirico-Post, and W.M. Keyserling. 1987. Risk assessment in electronic assembly workers: Carpal tunnel syndrome. *Journal of Hand Surgery—American Volume* 12:849-55.

Fenn, W.O. 1929. Work against gravity and work due to velocity changes in running. *American Journal of Physiology* 262:639-57.

Fenn, W.O. 1930. Frictional and kinetic factors in the work of sprint running. *American Journal of Physiology* 92:583-611.

Filligoi, G., and F. Felici. 1999. Detection of hidden rhythms in surface EMG signals with a non-linear time-series tool. *Medical Engineering and Physics* 21:439-48.

Finer, J.T., R.M. Simmons, and J.A. Spudich. 1994. Single myosin molecule mechanics: Piconewton forces and nanometre steps. *Nature* 368:113-9.

Finucane, S.D., T. Rafeei, J. Kues, R.L. Lamb, and T.P. Mayhew. 1998. Reproducibility of electromyographic recordings of submaximal concentric and eccentric muscle contractions in humans. *Electroencephalography and Clinical Neurophysiology* 109:290-6.

Fischer, O. 1906. *Theoretische Grundlagen fur eine Mechanik der Lebenden Körper mit Speziellen Anwendungen auf den Menschen, sowie auf einige Bewegungs-vorange an Maschine* [Theoretical fundamentals for a mechanics of living bodies, with special applications to man, as well as to some processes of motion in machines]. Leipzig: B.G. Tuebner.

Froese, E.A., and M.E. Houston. 1985. Torque-velocity characteristics and muscle fiber type in human vastus lateralis. *Journal of Applied Physiology* 59: 309-14.

Fuglevand, A.J., D.A. Winter, A.E. Patla, and D. Stashuk. 1992. Detection of motor unit action potentials with surface electrodes: Influence of electrode size and spacing. *Biological Cybernetics* 67:143-53.

Fuller, J., L.-J. Lui, M.C. Murphy, and R.W. Mann. 1997. A comparison of lower-extremity skeletal kinematics measured using skin- and pin-mounted markers. *Human Movement Science* 16:219-42.

Gerard, M.J., T.J. Armstrong, A. Franzblau, B.J. Martin, and D.M. Rempel. 1999. The effects of keyswitch stiffness on typing force, finger electromyography, and subjective discomfort. *American Industrial Hygiene Association Journal* 60:762-9.

Gerber, H., and E. Stuessi. 1987. A measuring system to assess and compute the double stride. *Biomechanics X-B*, ed. B. Jonsson, 1055-8. International Series on Biomechanics. Champaign, IL: Human Kinetics.

Gerilovsky, L., P. Tsvetinov, and G. Trenkova. 1986. H-reflex potentials shape and amplitude changes at different lengths of relaxed soleus muscle. *Electromyography and Clinical Neurophysiology* 26:641-53.

Gerilovsky, L., P. Tsvetinov, and G. Trenkova. 1989. Peripheral effects on the amplitude of monopolar and bipolar H-reflex potentials from the soleus muscle. *Experimental Brain Research* 76:173-81.

Gerleman, D.G., and T.M. Cook. 1992. Instrumentation. In *Selected Topics in Surface Electromyography for Use in the Occupational Setting: Expert Perspectives*, ed. G.L. Soderberg, 44-68. Washington, DC: National Institute for Occupational Safety and Health.

Gervais, P., and F. Tally. 1993. The beat swing and mechanical descriptors of three horizontal bar release-regrasp skills. *Journal of Applied Biomechanics* 9:66-83.

Gilchrist, L.A., and D.A. Winter. 1997. A multisegment computer simulation of normal human gait. *IEEE Transactions on Rehabilitation Engineering* 5:290-9.

Gitter, J.A., and M.J. Czerniecki. 1995. Fractal analysis of the electromyographic interference pattern. *Journal of Neuroscience Methods* 58:103-8.

Gollhofer, A., G.A. Horstmann, D. Schmidtbleicher, and D. Schonthal. 1990. Reproducibility of electromyographic patterns in stretch-shortening type contractions. *European Journal of Applied Physiology* 60:7-14.

Gordon, A.M., A.F. Huxley, and F.J. Julian. 1966. The variation in isometric tension with sarcomere length in vertebrate muscle fibres. *Journal of Physiology* 184: 170-92.

Gregor, R.J., and T.A. Abelew. 1994. Tendon force measurements and movement control: A review. *Medicine and Science in Sports and Exercise* 26:1359-72.

Gregor, R.J., P.V. Komi, R.C. Browning, and M. Jarvinen. 1991. A comparison of the triceps surae and residual muscle moments at the ankle during cycling. *Journal of Biomechanics* 24:287-97.

Gregor, R.J., P.V. Komi, and M. Jarvinen. 1987. Achilles tendon forces during cycling. *International Journal of Sports Medicine* 8:9-14.

Grieve, D.W., S. Pheasant, and P.R. Cavanagh. 1978. Prediction of gastrocnemius length from knee and ankle joint posture. In *Biomechanics VI-A*, ed. E. Asmussen and K. Jorgensen, 405-12. Baltimore: University Park Press.

Griffiths, R.I. 1991. Shortening of muscle fibres during stretch of the active cat medial gastrocnemius muscle: The role of tendon compliance. *Journal of Physiology* 436: 219-36.

Grood, E.S., and W.J. Suntay. 1983. A joint coordination system for the clinical description of three-dimensional motions: Application to the knee. *Journal of Biomedical Engineering* 105:136-44.

Hagberg, M., and B.-E. Ericson. 1982. Myoelectric power spectrum dependence on muscular contraction level of elbow flexors. *European Journal of Applied Physiology* 48:147-56.

Hagemann, B., G. Luhede, and H. Luczak. 1985. Improved "active" electrodes for recording bioelectric signals in work physiology. *European Journal of Applied Physiology* 54:95-8.

Hägg, G. 1981. Electromyographic fatigue analysis based on the number of zero crossings. *Pflugers Archiv European Journal of Physiology* 391:78-80.

Hamill, J., G.E. Caldwell, and T.R. Derrick. 1997. A method for reconstructing digital signals using Shannon's sampling theorem. *Journal of Applied Biomechanics* 13: 226-38.

Hammelsbeck, M., and W. Rathmayer. 1989. Intracellular Na^+, K^+ and Cl^- activity in tonic and phasic muscle fibers of the crab *Eriphia*. *Pflugers Archiv (European Journal of Physiology)* 413:487-92.

Hanavan, E.P. 1964. *A Mathematical Model of the Human Body*. AMRL Technical Report 64-102. Wright-Patterson Air Force Base, OH.

Hannaford, B., and S. Lehman. 1986. Short time Fourier analysis of the electromyogram: Fast movements and constant contraction. *IEEE Transactions on Biomedical Engineering* 12:1173-81.

Hannah, R., S. Cousins, and J. Foort. 1978. The CARS-UBC electrogoniometer, a clinically viable tool. In *7th Canadian Medical and Biological Engineering Conference*, 133-4.

Hannerz, J. 1974. An electrode for recording single motor unit activity during strong muscle contractions. *Electroencephalography and Clinical Neurophysiology* 37:179-81.

Harless, E. 1860. The static moments of the component masses of the human body. *Treatises of the Mathematics-Physics Class, Royal Bavarian Academy of Sciences.* 8:69-96, 257-94. Trans. FTD Technical Report 61-295. Wright-Patterson Air Force Base, OH, 1962.

Harris, G.F., and J.J. Wertsch. 1994. Procedures for gait analysis. *Archives of Physical Medicine and Rehabilitation* 75:216-25.

Harvey, R., and E. Peper. 1997. Surface electromyography and mouse use position. *Ergonomics* 40:781-9.

Hashimoto, S., J. Kawamura, Y. Segawa, Y. Harada, T. Hanakawa, and Y. Osaki. 1994. Waveform changes of compound muscle action potential (CMAP) with muscle length. *Journal of the Neurological Sciences* 124: 21-4.

Hatze, H. 1975. A new method for the simultaneous measurement of the moment on inertia, the damping coefficient and the location of the centre of mass of a body segment *in situ. European Journal of Applied Physiology* 34:217-26.

Hatze, H. 1979. A model for the computational determination of parameter values of anthropometric segments. NRIMS Technical Report TWISK 79. Pretoria, South Africa.

Hatze, H. 1980. A mathematical model for the computational determination of parameter values of anthropometric segments. *Journal of Biomechanics* 13:833-43.

Hatze, H. 1981. The use of optimally regularised Fourier series for estimating higher-order derivatives of noisy biomechanical data. *Journal of Biomechanics* 14:13-8.

Hatze, H. 1993. The relationship between the coefficient of restitution and energy losses in tennis rackets. *Journal of Applied Biomechanics* 9:124-42.

Hatze, H. 1998. Validity and reliability of methods for testing vertical jumping performance. *Journal of Applied Biomechanics* 14:127-40.

Hatze, H. 2002. The fundamental problem of myoskeletal inverse dynamics and its application. *Journal of Biomechanics* 35:109-16.

Hay, J.G. 1973. The center of gravity of the human body. *Kinesiology III* 20-44.

Hay, J.G. 1974. Moment of inertia of the human body. *Kinesiology IV* 43-52.

Hayward, M. 1983. Quantification of interference patterns. In *Computer-Aided Electromyography,* ed. J.E. Desmedt, 128-49. New York: Karger.

Hermens, H.J., T.A.M. van Bruggen, C.T.M. Baten, W.L.C. Rutten, and H.B.K. Boom. 1992. The median frequency of the surface EMG power spectrum in relation to motor unit firing and action potential properties. *Journal of Electromyography and Kinesiology* 2:15-25.

Herzog, W. 1988. The relation between the resultant moments at a joint and the moments measured by an isokinetic dynamometer. *Journal of Biomechanics* 21: 5-12.

Herzog, W., and T.R. Leonard. 1991. Validation of optimization models that estimate the forces exerted by synergistic muscles. *Journal of Biomechanics* 24:31-9.

Herzog, W., and T.R. Leonard. 1991. Validation of optimization models that estimate the forces exerted by synergistic muscles. *Journal of Biomechanics* 24:31-9.

Herzog, W., T.R. Leonard, and J.Z. Wu. 2000. The relationship between force depression following shortening and mechanical work in skeletal muscle. *Journal of Biomechanics* 33:659-68.

Hill, A.V. 1938. The heat of shortening and the dynamic constants of muscle. *Proceedings of the Royal Society B* 126: 136-95.

Hinrichs, R.N. 1985. Regression equations to predict segmental moments of inertia from anthropometric measurements: An extension of the data of Chandler et al. (1975). *Journal of Biomechanics* 18:621-4.

Hodges, P.W., and B.H. Bui. 1996. A comparison of computer-based methods for the determination of onset of muscle contraction using electromyography. *Electroencephalography and Clinical Neurophysiology* 101:511-9.

Holt, K.G., J. Hamill, and R.O. Andres. 1990. The force-driven harmonic oscillator as a model for human locomotion. *Human Movement Science* 9:55-68.

Huijing, P.A. 1998. Muscle, the motor of movement: Properties in function, experiment and modeling. *Journal of Electromyography and Kinesiology* 8:61-77.

Huijing, P.A. 1999. Muscle as a collagen fiber reinforced composite: A review of force transmission in muscle and whole limb. *Journal of Biomechanics* 32:329-45.

Huxley, A.F. 1957. Muscle structure and theories of contraction. *Progress in Biophysics and Biophysical Chemistry* 7:255-318.

Huxley, A.F., and R.M. Simmons. 1971. Proposed mechanism of force generation in striated muscle. *Nature* 233: 533-8.

Ikegawa, S., M. Shinohara, T. Fukunaga, J.P. Zbilut, and C.L.J. Webber. 2000. Nonlinear time-course of lumbar muscle fatigue using recurrence quantifications. *Biological Cybernetics* 82:373-82.

Inbar, G.F., J. Allin, O. Paiss, and H. Kranz. 1986. Monitoring surface EMG spectral changes by the zero crossing rate. *Medical and Biological Engineering and Computing* 24:10-8.

Ingen Schenau, G.J. van. 1989. From rotation to translation: Constraints on multi-joint movements and the unique action of bi-articular muscles. *Human Movement Science* 8:301-37.

Ingen Schenau, G.J. van, and P.R. Cavanagh. 1990. Power equations in endurance sports. *Journal of Biomechanics* 23:865-81.

Ingen Schenau, G.J. van, W.L.M. van Woensel, P.J.M. Boots, R.W. Snackers, and G. de Groot. 1990. Determination and interpretation of mechanical power in human movement: Application to ergometer cycling. *European Journal of Applied Physiology* 61:11-19.

Ito, M., Y. Kawakami, Y. Ichinose, S. Fukashiro, and T. Fukunaga. 1998. Nonisometric behavior of fascicles during isometric contractions of a human muscle. *Journal of Applied Physiology* 85:1230-5.

Jackson, K.M. 1979. Fitting of mathematical functions to biomechanical data. *IEEE Transactions on Biomedical Engineering* 26(2): 122-4.

Jacobs, R., and G.J. van Ingen Schenau. 1992. Control of an external force in leg extensions in humans. *Journal of Physiology* 457:611-26.

Jensen, B.R., B. Schibye, K. Sogaard, E.B. Simonsen, and G. Sjøgaard. 1993. Shoulder muscle load and muscle fatigue among industrial sewing-machine operators. *European Journal of Applied Physiology* 67:467-75.

Jensen, R.K. 1976. Model for body segment parameters. In *Biomechanics V-B*, ed. P.V. Komi, 380-6. Baltimore: University Park Press.

Jensen, R.K. 1978. Estimation of the biomechanical properties of three body types using a photogrammetric method. *Journal of Biomechanics* 11:349-58.

Jensen, R.K. 1986. Body segment mass, radius and radius of gyration proportions of children. *Journal of Biomechanics* 19:359-68.

Jensen, R.K. 1989. Changes in segment inertial proportions between 4 and 20 years. *Journal of Biomechanics* 22:529-36.

Jensen, R.K., T. Treitz, and S. Doucet. 1996. Prediction of human segment inertias during pregnancy. *Journal of Applied Biomechanics* 12:15-30.

Johnson, S.W., P.A. Lynn, J.S.G. Miller, and G.L. Reed. 1977. Miniature skin-mounted preamplifier for measurement of surface electromyographic potentials. *Medical and Biological Engineering and Computing* 15:710-1.

Jonas, D., C. Bischoff, and B. Conrad. 1999. Influence of different types of surface electrodes on amplitude, area and duration of the compound muscle action potential. *Clinical Neurophysiology* 110:2171-5.

Juel, C. 1988. Muscle action potential propagation velocity changes during activity. *Muscle and Nerve* 11:714-9.

Kamen, G., S.V. Sison, C.C. Du, and C. Patten. 1995. Motor unit discharge behavior in older adults during maximal-effort contractions. *Journal of Applied Physiology* 79:1908-13.

Kameyama, O., R. Ogawa, T. Okamoto, and M. Kumamoto. 1990. Electric discharge patterns of ankle muscles during the normal gait cycle. *Archives of Physical Medicine and Rehabilitation* 71:969-74.

Kang, W.J., J.R. Shiu, C.K. Cheng, J.S. Lai, H.W. Tsao, and T.S. Kuo. 1995. The application of cepstral coefficients and maximum likelihood method in EMG pattern recognition. *IEEE Transactions on Biomedical Engineering* 42:777-85.

Karlsson, D., and R. Tranberg. 1999. On skin movement artifact: Resonant frequencies of skin markers attached to the leg. *Human Movement Science* 18:627-35.

Karlsson, S., J. Yu, and M. Akay. 1999. Enhancement of spectral analysis of myoelectric signals during static contractions using wavelet methods. *IEEE Transactions on Biomedical Engineering* 46:670-84.

Karlsson, S., J. Yu, and M. Akay. 2000. Time-frequency analysis of myoelectric signals during dynamic contractions: A comparative study. *IEEE Transactions on Biomedical Engineering* 47:228-38.

Kawakami, Y., and R.L. Lieber. 2000. Interaction between series compliance and sarcomere kinetics determines internal sarcomere shortening during fixed-end contraction. *Journal of Biomechanics* 33:1249-55.

Kilbom, A., G.M. Hägg, and C. Kall. 1992. One-handed load carrying: Cardiovascular, muscular and subjective indices of endurance and fatigue. *European Journal of Applied Physiology* 65:52-8.

Kim, M.J., W.S. Druz, and J.T. Sharp. 1985. Effect of muscle length on electromyogram in a canine diaphragm strip preparation. *Journal of Applied Physiology* 58:1602-7.

Kirkwood, R.N., E.G. Culham, and P. Costigan. 1999. Radiographic and non-invasive determination of the hip joint center location: Effect on hip joint moments. *Clinical Biomechanics* 14:227-35.

Knaflitz, M., and P. Bonato. 1999. Time-frequency methods applied to muscle fatigue assessment during dynamic contractions. *Journal of Electromyography and Kinesiology* 9:337-50.

Koh, T.J., and M.D. Grabiner. 1992. Cross talk in surface electromyograms of human hamstring muscles. *Journal of Orthopedic Research* 10:701-9.

Komi, P.V. 1990. Relevance of *in vivo* force measurements to human biomechanics. *Journal of Biomechanics* 23 (Suppl. no. 1): 23-34.

Komi, P.V., A. Belli, V. Huttunen, R. Bonnefoy, A. Geyssant, and J.R. Lacour. 1996. Optic fibre as a transducer of tendomuscular forces. *European Journal of Applied Physiology and Occupational Physiology* 72:278-80.

Koning, J.J., G. de Groot, and G.J. van Ingen Schenau. 1991. Speed skating the curves: A study of muscle coordination and power production. *International Journal of Sport Biomechanics* 7:344-58.

Krogh-Lund, C. 1993. Myo-electric fatigue and force failure from submaximal static elbow flexion sustained to exhaustion. *European Journal of Applied Physiology* 67:389-401.

Kumar, S., and A. Mital, eds. 1996. *Electromyography in Ergonomics*. London: Taylor & Francis.

Kwon, Y.-H. 1996. Effects of the method of body segment parameter estimation on airborne angular momentum. *Journal of Applied Biomechanics* 12:413-30.

Lamontagne, M., R. Doré, H. Yahia, and J.M. Dorlot. 1985. Tendon and ligament measurement. *Medical Electronics* 6:74-6.

Landjerit, B., Maton, B., and G. Peres. 1988. In vivo muscular force analysis during the isometric flexion on a monkey's elbow. *Journal of Biomechanics* 21:577-84.

Lanshammer, H. 1982a. On practical evaluation of differentiation techniques for human gait analysis. *Journal of Biomechanics* 15:99-105.

Lanshammer, H. 1982b. On precision limits for derivatives calculated from noisy data. *Journal of Biomechanics* 15: 459-70.

Lariviere, C., D. Gagnon, and P. Loisel. 2000. The comparison of trunk muscles EMG activation between subjects with and without chronic low back pain during flexion-extension and lateral bending tasks. *Journal of Electromyography and Kinesiology* 10:79-91.

Laub, A.J., and G.R. Schiflett. 1982. A linear algebra approach to the analysis of rigid body displacement from initial and final position data. *Journal of Applied Mechanics* 49:213-6.

Lee, R.G., P. Ashby, D.G. White, and A.J. Aguayo. 1975. Analysis of motor conduction velocity in the human median nerve by computer simulation of compound muscle action potentials. *Electroencephalography and Clinical Neurophysiology* 39:225-37.

Lehman, G.J., and S.M. McGill. 1999. The importance of normalization in the interpretation of surface electromyography: A proof of principle. *Journal of Manipulative and Physiological Therapeutics* 22:444-6.

Lemaire, E.D., and D.G.E. Robertson. 1989. Power in sprinting. *Track & Field Journal* 35:13-7.

Lemaire, E.D., and D.G.E. Robertson. 1990a. Validation of a computer simulation for planar airborne human motions. *Journal of Human Movement Studies* 18:213-28.

Lemaire, E.D., and D.G.E. Robertson. 1990b. Force-time data acquisition system for sprint starting. *Canadian Journal of Sport Sciences* 15:149-52.

LeVeau, B., and G. Andersson. 1992. Output forms: Data analysis and applications. In *Selected Topics in Surface Electromyography for Use in the Occupational Setting: Expert Perspectives*, ed. G.L. Soderberg. Washington, DC: National Institute for Occupational Safety and Health.

Lexell, J., K. Henriksson-Larsen, B. Winblad, and M. Sjostrom. 1983. Distribution of different fiber types in human skeletal muscles: Effects of aging studied in whole muscle cross sections. *Muscle and Nerve* 6:588-95.

Li, L., and G.E. Caldwell. 1999. Coefficient of cross correlation and the time domain correspondence. *Journal of Electromyography and Kinesiology* 9:385-9.

Lieber, R.L., and J. Friden. 1997. Intraoperative measurement and biomechanical modeling of the flexor carpi ulnaris-to-extensor carpi radialis longus tendon transfer. *Journal of Biomechanical Engineering* 119:386-91.

Liu, M.M., W. Herzog, and H.H. Savelberg. 1999. Dynamic muscle force predictions from EMG: An artificial neural network approach. *Journal of Electromyography and Kinesiology* 9:391-400.

Lutz, G.L., and R.L. Lieber. 1999. Skeletal muscle myosin II structure and function. *Exercise and Sport Sciences Reviews* 27:63-77.

Manal, K., I. McClay, S. Stanhope, J. Richards, and B. Galinat. 2000. Comparison of surface mounted markers and attachment methods in estimating tibial rotations during walking: An in vivo study. *Gait and Posture* 11:38-45.

Mansour, J.M., and M.L. Audu. 1986. The passive elastic moment at the knee and its influence on human gait. *Journal of Biomechanics* 19:369-73.

Marks, R.J., II. 1993. *Advanced Topics in Shannon Sampling and Interpolation Theory*. New York: Springer-Verlag.

Martindale, W.O., and D.G.E. Robertson. 1984. Mechanical energy variations in single sculls and ergometer rowing. *Canadian Journal of Applied Sport Sciences* 9:153-63.

Marzan, T., and H.M. Karara. 1975. A computer program for direct linear transformation of the colinearity condition and some applications of it. In *American Society of Photogrammetry Symposium on Close Range Photogrammetry*, 420-76. Falls Church, VA: American Society for Photogrammetry.

Masuda, T., H. Miyano, and T. Sadoyama. 1985. A surface electrode array for detecting action potential trains of single motor units. *Electroencephalography and Clinical Neurophysiology* 60:435-43.

Masuda, T., H. Miyano, and T. Sadoyama. 1992. The position of innervation zones in the biceps brachii investigated by surface electromyography. *IEEE Transactions on Biomedical Engineering* 32:36-42.

Mathiassen, S.E., J. Winkel, and G.M. Hägg. 1995. Normalization of surface EMG amplitude from the upper trapezius muscle in ergonomic studies: A review. *Journal of Electromyography and Kinesiology* 5:197-226.

Maton, B., and D. Gamet. 1989. The fatigability of two agonistic muscles in human isometric voluntary submaximal contraction: An EMG study. II. Motor unit firing rate and recruitment. *European Journal of Applied Physiology* 58:369-74.

Mayagoitia, R.E., A.V. Nene, and P.H. Veltnik. 2002. Accelerometer and rate gyroscope measurement of kinematics: An inexpensive alternative to optical motion analysis systems. *Journal of Biomechanics* 35:537-42.

McClay, I., and K. Manal. 1997. Coupling parameters in runners with normal and excessive pronation. *Journal of Applied Biomechanics* 13:109-24.

McClay, I., and K. Manal. 1999. Three-dimensional kinetic analysis of running: Significance of secondary planes of motion. *Medicine and Science in Sports and Exercise* 31: 1629-37.

McFaull, S., and M. Lamontagne. 1993. The passive elastic moment about the *in vivo* human knee joint. In *Proceedings of the 14th Annual Conference, International Society of Biomechanics*, ed. S. Bouisset, 848-9. Paris: International Society of Biomechanics.

McFaull, S., and M. Lamontagne. 1998. *In vivo* measurement of the passive viscoelastic properties of the human knee joint. *Human Movement Science* 17:139-65.

McGill, S., and R.W. Norman. 1985. Dynamically and

statically determined low back moments during lifting. *Journal of Biomechanics* 18:877-85.

McMahon, T.A. 1984. Mechanics of locomotion. *International Journal of Robotics Research* 3:4-28.

McMahon, T.A., G. Valiant, and E.C. Frederick. 1987. Groucho running. *Journal of Applied Physiology* 62:2326-37.

Meglan, D., and F. Todd. 1994. Kinetics of human locomotion. In *Human Walking*, eds. J. Rose and J.G. Gamble, 73-99. Baltimore: Williams & Wilkins.

Merletti, R., D. Farina, and A. Granata. 1999. Non-invasive assessment of motor unit properties with linear electrode arrays. *Electroencephalography and Clinical Neurophysiology Supplement* 50:293-300.

Miller, D.I. 1970. A computer simulation of the airborne phase of diving. PhD diss., Pennsylvania State Univiversity.

Miller, D.I. 1973. Computer simulation of springboard diving. In *Medicine and Sport, Volume 8: Biomechanics III*, ed. S. Cerquiglini, A. Venerando, and J.Wartenweiler, 116-9. Basel: Karger.

Miller, D.I., and W.E. Morrison. 1975. Prediction of segmental parameters using the Hanavan human body model. *Medicine and Science in Sports* 7:207-12.

Miller D.I., and R.C. Nelson. 1973. *Biomechanics of Sport*. Philadelphia: Lea & Febiger.

Miller, D.I., and E.J. Sprigings. 2001. Factors influencing the performance of springboard dives of increasing difficulty. *Journal of Applied Biomechanics* 17:217-31.

Miller, N.R., R. Shapiro, and T.M. McLaughlin. 1980. A technique for obtaining spatial kinematic parameters of segments of biomechanical systems from cinematographic data. *Journal of Biomechanics* 13:535-47.

Mills, K.R., and R.T. Edwards. 1984. Muscle fatigue in myophosphorylase deficiency: Power spectral analysis of the electromyogram. *Electroencephalography and Clinical Neurophysiology* 57:330-5.

Milner-Brown, H.S., and R.G. Miller. 1990. Myotonic dystrophy: Quantification of muscle weakness and myotonia and the effect of amitriptyline and exercise. *Archives of Physical Medicine and Rehabilitation* 71:983-7.

Milner-Brown, H.S., R.B. Stein, and R.G. Lee. 1975. Synchronization of human motor units: Possible roles of exercise and supraspinal reflexes. *Electroencephalography and Clinical Neurophysiology* 38:245-54.

Minetti, A.E., and G. Belli. 1994. A model for the estimation of visceral mass displacement in periodic movements. *Journal of Biomechanics* 27:97-101.

Mineva, A., J. Dushanova, and L. Gerilovsky. 1993. Similarity in shape, timing and amplitude of H- and T-reflex potentials concurrently recorded along the broad skin area over soleus muscle. *Electromyography and Clinical Neurophysiology* 33:235-45.

Mirka, G.A. 1991. The quantification of EMG normalization error. *Ergonomics* 34:343-52.

Mitchelson, D.L. 1975. Recording of movement without photograph. *Techniques for the Analysis of Human Movement*. London: Lepus Books.

Mizrahi, J., and Z. Susak. 1982. In-vivo elastic and damping response of the human leg to impact forces. *Journal of Biomechanical Engineering* 104:63-6.

Mochon, S., and T.A. McMahon. 1980. Ballistic walking. *Journal of Biomechanics* 13:49-57.

Monti, R.J., R.R. Roy, J.A. Hodgson, and V.R. Edgerton. 1999. Transmission of forces within mammalian skeletal muscles. *Journal of Biomechanics* 32:371-80.

Morgan, D.L. 1990. Modeling of lengthening muscle: The role of intersarcomere dynamics. In *Multiple Muscle Systems*, ed. J.M. Winters and S.L.-Y. Woo, 46-56. New York: Springer-Verlag.

Morimoto, S. 1986. Effect of length change in muscle fibers on conduction velocity in human motor units. *Japanese Journal of Physiology* 36:773-82.

Moritani, T., and M. Muro. 1987. Motor unit activity and surface electromyogram power spectrum during increasing force of contraction. *European Journal of Applied Physiology* 56:260-5.

Morrenhof, J.W., and H.J. Abbink. 1985. Cross-correlation and cross-talk in surface electromyography. *Electromyography and Clinical Neurophysiology* 25:73-9.

Moss, R.F., P.B. Raven, J.P. Knochel, J.R. Peckham, and J.D. Blachley. 1983. The effect of training on resting muscle membrane potentials. In *Biochemistry of Exercise*, eds. H.G. Knuttgen, J.A. Vogel, and J. Poortmans, 806-11. 3rd ed. Champaign, IL: Human Kinetics.

Mungiole, M., and P.E. Martin. 1990. Estimating segmental inertia properties: Comparison of magnetic resonance imaging with existing methods. *Journal of Biomechanics* 23:1039-46.

Murphy, S.D., and D.G.E. Robertson. 1994. Construction of a high-pass digital filter from a low-pass digital filter. *Journal of Applied Biomechanics* 10: 374-81.

Murtaugh, K., and D.I. Miller. 2001. Initiating rotation in back and reverse armstand somersault tuck dives. *Journal of Applied Biomechanics* 17:312-25.

Nigg, B.M., and W. Herzog. 1994. *Biomechanics of the Musculo-skeletal System*. Toronto: John Wiley & Sons.

Nishimura, S., Y. Tomita, and T. Horiuchi. 1992. Clinical application of an active electrode using an operational amplifier. *IEEE Transactions on Biomedical Engineering* 39:1096-9.

Norman, R., M. Sharratt, J. Pezzack, and E. Noble. 1976. Re-examination of the mechanical efficiency of horizontal treadmill running. In *Biomechanics V-B*, ed. P. Komi, 87-93. International Series on Biomechanics. Baltimore: University Press.

Norman, R.W., G.E. Caldwell, and P.V. Komi. 1985. Differences in body segment energy utilization between world class and recreational cross country skiers. *International Journal of Sports Biomechanics* 1:253-62.

Norman, R.W., and P.V. Komi. 1987. Mechanical energetics of world class cross-country skiing. *International Journal of Sports Biomechanics* 3:353-69.

Oh, S.J. 1993. *Clinical Electromyography: Nerve Conduction Studies*. 2nd ed. Baltimore: Williams & Wilkins.

Ohashi, J. 1995. Difference in changes of surface EMG during low-level static contraction between monopolar and bipolar lead. *Applied Human Science* 14:79-88.

Ohashi, J. 1997. The effects of preceded fatiguing on the relations between monopolar surface electromyogram and fatigue sensation. *Applied Human Science* 16:19-27.

Okada, M. 1987. Effect of muscle length on surface EMG wave forms in isometric contractions. *European Journal of Applied Physiology* 56:482-6.

Onyshko, S., and D.A. Winter. 1980. A mathematical model for the dynamics of human locomotion. *Journal of Biomechanics* 13:361-8.

Padgaonkar, A.J., K.W. Krieger, and A.I. King. 1975. Measurement of angular acceleration of a rigid body using accelerometers. *Transactions of ASME, Journal of Applied Mechanics* 42:552-6.

Pain, M.T.G., and J.H. Challis. 2001. Whole body force distributions in landing from a drop. In *Proceedings of the XVIIIth Congress of the International Society of Biomechanics*, ed. R. Müller, H. Gerber, and A. Stacoff, 202. Zurich: International Society of Biomechanics.

Pan, Z.S., Y. Zhang, and P.A. Parker. 1989. Motor unit power spectrum and firing rate. *Medicine and Biological Engineering and Computing* 27:14-8.

Pandy, M.G., and N. Berme. 1989. Quantitative assessment of gait determinants during single stance via a three-dimensional model. Part 1. Normal gait. *Journal of Biomechanics* 22:717-24.

Pandy, M.G., and F.E. Zajac. 1991. Optimal muscular coordination strategies for jumping. *Journal of Biomechanics* 24:1-10.

Pandy, M.G., F.E. Zajac, E. Sim, and W.S. Levine. 1990. An optimal control model for maximum-height human jumping. *Journal of Biomechanics* 23:1185-98.

Pattichis, C.S., I. Schofield, R. Merletti, P.A. Parker, and L.T. Middleton. 1999. Introduction to this special issue: Intelligent data analysis in electromyography and electroneurography. *Medical Engineering and Physics* 21:379-88.

Pavol, M.J., T.M. Owings, and M.D. Grabiner. 2002. Body segment parameter estimation for the general population of older adults. *Journal of Biomechanics* 35:707-12.

Perreault, E.J., C.J. Heckman, and T.G. Sandercock. 2003. Hill muscle model errors during movement are greatest within the physiologically relevant range of motor unit firing rates. *Journal of Biomechanics* 36:211-8.

Pezzack, J.C. 1976. An approach for the kinetic analysis of human motion. MS thesis, Univ. Waterloo, ON.

Pezzack, J.C., R.W. Norman, and D.A. Winter. 1977. An assessment of derivative determining techniques used for motion analysis. *Journal of Biomechanics* 10:377-82.

Pierrynowski, M.R. 1982. A physiological model for the solution of individual muscle forces during normal human walking. PhD diss., Simon Fraser Univ.

Pierrynowski, M.R., and J.B. Morrison. 1985. Estimating the muscle forces generated in the human lower extremity when walking: A physiological solution. *Mathematical Biosciences* 75:43-68.

Pierrynowski, M.R., R.W. Norman, and D.A. Winter. 1981. Mechanical energy analyses of the human during load carriage on a treadmill. *Ergonomics* 24:1-14.

Pierrynowski, M.R., D.A. Winter, and R.W. Norman. 1980. Transfers of mechanical energy within the total body and mechanical efficiency during treadmill walking. *Ergonomics* 23:147-56.

Plagenhoef, S. 1968. Computer programs for obtaining kinetic data on human movement. *Journal of Biomechanics* 1:221-34.

Plagenhoef, S. 1971. *Patterns of Human Motion: A Cinematographic Analysis*. Englewood Cliffs, NJ: Prentice Hall.

Proakis, J.G., and D.G. Manolakis. 1988. *Introduction to Digital Signal Processing*. New York: Macmillan.

Putnam, C.A. 1991. A segment interaction analysis of proximal-to-distal sequential motion patterns. *Medicine and Science in Sport and Exercise* 23:130-44.

Ramey, M.R. 1973a. A simulation of the running broad jump. In *Mechanics in Sport*, ed. J.L. Bleustein, 101-12. New York: American Society of Mechanical Engineers.

Ramey, M.R. 1973b. Significance of angular momentum in long jumping. *Research Quarterly* 44:488-97.

Ramey, M.R. 1974. The use of angular momentum in the study of long-jump take-offs. In *Biomechanics IV*, ed. R.C. Nelson and C.A. Morehouse, 144-8. Baltimore: University Park Press.

Ramsey, R.W., and S.F. Street. 1940. The isometric length-tension diagram of isolated skeletal muscle fibres of the frog. *Journal of Cellular and Comparative Physiology* 15:11-34.

Rau, G., C. Disselhorst-Klug, and J. Silny. 1997. Noninvasive approach to motor unit characterization: Muscle structure, membrane dynamics and neuronal control. *Journal of Biomechanics* 30:441-6.

Redfern, M.S. 1992. Functional muscle: Effects on electromyographic output. In *Selected Topics in Surface Electromyography for Use in the Occupational Setting: Expert Perspectives*, ed. G.L. Soderberg, 104-20. Washington, DC: National Institute for Occupational Safety and Health.

Reinschmidt, C., A.J. van den Bogert, B.M. Nigg, A. Lundberg, and N. Murphy. 1997. Effect of skin movement on the analysis of skeletal knee joint motion during running. *Journal of Biomechanics* 30:729-32.

Reuleux, F. 1876. *The Kinematics of Machinery: Outlines of a Theory of Machines*. London: Macmillan.

Reynolds, C. 1994. Electromyographic biofeedback evaluation of a computer keyboard operator with cumulative trauma disorder. *Journal of Hand Therapy* 7:25-7.

Risher, D.W., L.M. Schutte, and C.F. Runge. 1997. The use of inverse dynamics solutions in direct dynamic simulations. *Journal of Biomechanical Engineering* 119:417-22.

Robertson, D.G.E., and J.J. Dowling. 2003. Design and responses of Butterworth and critically damped digital filters. *Journal of Electromyography & Kinesiology* 13: 569-73.

Robertson, D.G.E., and D. Fleming. 1987. Kinetics of standing broad and vertical jumping. *Canadian Journal of Sport Science* 12:19-23.

Robertson, D.G.E., and Y.D. Fortin. 1994. Mechanics of rowing. In *Proceedings of the Eighth Conference of the Canadian Society for Biomechanics*, ed. W. Herzog, B. Nigg, and A. van den Bogert, 248-9. Calgary: Canadian Society for Biomechanics.

Robertson, D.G.E., J. Hamill, and D.A. Winter. 1997. Evaluation of cushioning properties of running footwear. In *Proceedings of the XVIth Congress of the International Society of Biomechanics*, ed. M. Miyashita and T. Fukunaga, 263. Tokyo: International Society of Biomechanics.

Robertson, D.G.E., and R.E. Mosher. 1985. Work and power of the leg muscles in soccer kicking. In *Biomechanics IX-B*, ed. D.A. Winter, R.W. Norman, R.P. Wells, K.C. Hayes and A.E. Patla, 533-8. Champaign, IL: Human Kinetics.

Robertson, D.G.E., and V.L. Stewart. 1997. Power production during swim starting. In *Proceedings of the XVIth Congress of the International Society of Biomechanics*, ed. M. Miyashita and T. Fukunaga, 22. Tokyo: International Society of Biomechanics.

Robertson, D.G.E., and D.A. Winter. 1980. Mechanical energy generation, absorption and transfer amongst segments during walking. *Journal of Biomechanics* 13: 845-54.

Rolf, C., P. Westblad, I. Ekenman, A. Lundberg, N. Murphy, M. Lamontagne, and K. Halvorsen. 1997. An experimental *in vivo* method for analysis of local deformation on tibia, with simultaneous measures of ground forces, lower extremity muscle activity and joint motion. *Scandinavian Journal of Medicine and Science in Sports* 7:144-51.

Ronager, J., H. Christensen, and A. Fuglsang-Frederiksen. 1989. Power spectrum analysis of the EMG pattern in normal and diseased muscles. *Journal of the Neurological Sciences* 94:283-94.

Roy, B. 1978. Biomechanical features of different starting positions and skating strides in ice hockey. In *Biomechanics VI-B*, ed. E. Asmussen and K. Jørgensen, 137-41. Baltimore: University Park Press.

Sandberg, A., B. Hansson, and E. Stalberg. 1999. Comparison between concentric needle EMG and macro EMG in patients with a history of polio. *Clinical Neurophysiology* 110:1900-8.

Sanders, D.B., E.V. Stålberg, and S.D. Nandedkar. 1996. Analysis of the electromyographic interference pattern. *Journal of Clinical Neurophysiology* 13:385-400.

Schneider, K., and R.F. Zernicke. 1992. Mass, center of mass and moment of inertia estimates for infant limb segments. *Journal of Biomechanics* 25:145-8.

Seireg, A., and R.J. Arvikar. 1975. The prediction of muscular load sharing and joint forces in the lower extremities during walking. *Journal of Biomechanics* 18:89-102.

Selbie, W.S., and G.E. Caldwell. 1996. A simulation study of vertical jumping from different starting postures. *Journal of Biomechanics* 29:1137-46.

Shapiro, R. 1978. The direct linear transformation method for three-dimensional cinematography. *Research Quarterly* 49:197-205.

Sherif, M.H., R.J. Gregor, and J. Lyman. 1981. Effects of load on myoelectric signals: The ARIMA representation. *IEEE Transactions on Biomedical Engineering* 5:411-6.

Shiavi, R. 1974. A wire multielectrode for intramuscular recording. *Medical and Biological Engineering* 12:721-3.

Shiavi, R., L.Q. Zhang, T. Limbird, and M.A. Edmondstone. 1992. Pattern analysis of electromyographic linear envelopes exhibited by subjects with uninjured and injured knees during free and fast speed walking. *Journal of Orthopedic Research* 10:226-36.

Shorten, M.R., and D.S. Winslow. 1992. Spectral analysis of impact shock during running. *International Journal of Sport Biomechanics* 8:288-304.

Siegel, K.L., T.M. Kepple, and G.E. Caldwell. 1996. Improved agreement of foot segmental power and rate of energy during gait: Inclusion of distal power terms and use of 3D models. *Journal of Biomechanics* 29:823-7.

Siegler, S., H.J. Hillstrom, W. Freedman, and G. Moskowitz. 1985. The effect of myoelectric signal processing on the relationship between muscle force and processed EMG. *Electromyography and Clinical Neurophysiology* 25: 499-512.

Sjøgaard, G., B. Kiens, K. Jorgensen, and B. Saltin. 1986. Intramuscular pressure, EMG and blood flow during low-level prolonged static contraction in man. *Acta Physiologica Scandinavica* 128:475-84.

Smith, G. 1989. Padding point extrapolation techniques for the Butterworth digital filter. *Journal of Biomechanics* 22:967-71.

Smith, R.M. 1996. Distribution of mechanical energy fractions during maximal ergometer rowing. PhD diss., Univ. Woolongong, Australia.

Soest, A.J. van, A.L. Schwab, M.F. Bobbert, and G.J. van Ingen Schenau. 1993. The influence of the biarticularity of the gastrocnemius muscle on vertical-jumping achievement. *Journal of Biomechanics* 26:1-8.

Solomonow, M., R. Baratta, M. Bernardi, B. Zhou, Y. Lu, M. Zhu, and S. Acierno. 1994. Surface and wire EMG crosstalk in neighbouring muscles. *Journal of Electromyography and Kinesiology* 4:131-42.

Solomonow, M., R. Baratta, H. Shoji, and R. D'Ambrosia. 1990. The EMG-force relationships of skeletal muscle;

dependence on contraction rate, and motor units control strategy. *Electromyography and Clinical Neurophysiology* 30:141-52.

Solomonow, M., C. Baten, J. Smit, R. Baratta, H. Hermens, R. D'Ambrosia, and H. Shoji. 1990. Electromyogram power spectra frequencies associated with motor unit recruitment strategies. *Journal of Applied Physiology* 68:1177-85.

Sommer, H.J. 1992. Determination of first and second order instant screw parameters from landmark trajectories. *ASME Journal of Mechanical Design* 114:274-82.

Soutas-Little, R.W., G.C. Beavis, M.C. Verstraete, and T.L. Markus. 1987. Analysis of foot motion during running using a joint co-ordinate system. *Medicine and Science in Sports and Exercise* 19:285-93.

Spoor, C.W., and F.E. Veldpaus. 1980. Rigid body motion calculated from spatial coordinates of markers. *Journal of Biomechanics* 13:391-3.

Stefanyshyn, D.J., and B.M. Nigg. 1998. Contributions of the lower extremity joints to mechanical energy in running vertical and running long jumps. *Journal of Sports Sciences* 16:177-86.

Tang, A., and W.Z. Rymer. 1981. Abnormal force-EMG relations in paretic limbs of hemiparetic human subjects. *Journal of Neurology, Neurosurgery and Psychiatry* 44:690-8.

Thomsen, M., and P.H. Veltink. 1997. Influence of synchronous and sequential stimulation on muscle fatigue. *Medical and Biological Engineering and Computing* 35:186-92.

Thusneyapan, S., and G.I. Zahalak. 1989. A practical electrode-array myoprocessor for surface electromyography. *IEEE Transactions on Biomedical Engineering* 36:295-99.

Trewartha, G., M.R. Yeadon, and J.P. Knight. 2001. Marker-free tracking of aerial movements. In *Proceedings of the XVIIIth Congress of the International Society of Biomechanics*, eds. R. Müller, H. Gerber, and A. Stacoff, 185. Zurich: International Society of Biomechanics.

Tsai, C.-S., and J.M. Mansour. 1986. Swing phase simulation and design of above knee prostheses. *Journal of Biomechanical Engineering* 108:65-72.

Vardaxis, V.G., and T.B. Hoshizaki. 1989. Power patterns of the lower limb during the recovery phase of the sprinting stride of advanced and intermediate sprinters. *International Journal of Sport Biomechanics* 5:332-49.

Vaughan, C.L. 1996. Are joint torques the holy grail of human gait analysis? *Human Movement Science* 15:423-43.

Vaughan, C.L. 1982. Smoothing and differentiating of displacement-time data: An application of splines and digital filtering. *International Journal of Bio-medical Computing* 13:375-86.

Vaughan, C.L., B.L. Davis, and J.C. O'Connor. 1992. *Dynamics of Human Gait*. Champaign, IL: Human Kinetics.

Veeger, H.E., L.S. Meershoek, L.H. van der Woude, and J.M. Langenhoff. 1998. Wrist motion in handrim wheelchair propulsion. *Journal of Rehabilitation Research and Development* 35:305-13.

Veldpaus, F.E., H.J. Woltring, and L.J.M.G. Dortmans. 1988. A least-squares algorithm for the equiform transformation from spatial marker coordinates. *Journal of Biomechanics* 21:45-54.

Vigreux, B., J.C. Cnockaert, and E. Pertuzon. 1979. Factors influencing quantified surface EMGs. *European Journal of Applied Physiology* 41:119-29.

Wakeling, J.M., V. Von Tscharner, and B.M. Nigg. 2001. Muscle activity in the leg is tuned in response to ground reaction forces. *Journal of Applied Physiology* 91:1307-17.

Wallinga-De Jonge, W., F.L. Gielen, P. Wirtz, P. De Jong, and J. Broenink. 1985. The different intracellular action potentials of fast and slow muscle fibres. *Electroencephalography and Clinical Neurophysiology* 60:539-47.

Walton, J.S. 1981. Close-range cine-photogrammetry: A generalized technique for quantifying gross human motion. PhD diss., Pennsylvania State Univ.

Webster, J.G. 1984. Reducing motion artifacts and interference in biopotential recording. *IEEE Transactions on Biomedical Engineering* 31:823-6.

Wells, R.P. 1988. Mechanical energy costs of human movement: An approach to evaluating the transfer possibilities of two-joint muscles. *Journal of Biomechanics* 21:955-64.

White, S.C., and D.A. Winter. 1985. Mechanical power analysis of the lower limb musculature in race walking. *International Journal of Sport Biomechanics* 1:15-24.

Whittlesey, S.N., and J. Hamill. 1996. An alternative model of the lower extremity during locomotion. *Journal of Applied Biomechanics* 12:269-79.

Williams, K.R. 1985. The relationship between mechanical and physiological energy estimates. *Medicine and Science in Sports and Exercise* 17:317-25.

Williams, K.R., and P.R. Cavanagh. 1983. A model for the calculation of mechanical power during distance running. *Journal of Biomechanics* 16:115-28.

Windhorst, U., T.M. Hamm, and D.G. Stuart. 1989. On the function of muscle and reflex partitioning. *Behavioral and Brain Sciences* 12:629-81.

Winter, D.A. 1976. Analysis of instantaneous energy of normal gait. *Journal of Biomechanics* 9:253-7.

Winter, D.A. 1978. Calculation and interpretation of mechanical energy of movement. *Exercise and Sports Science Reviews* 6:183-201.

Winter, D.A. 1979a. *Biomechanics of Human Movement*. Toronto: John Wiley & Sons.

Winter, D.A. 1979b. A new definition of mechanical work done in human movement. *Journal of Applied Physiology* 46:79-83.

Winter, D.A. 1980. Overall principle of lower limb support during stance phase of gait. *Journal of Biomechanics* 13: 923-7.

Winter, D.A. 1983a. Moments of force and mechanical power in jogging. *Journal of Biomechanics* 16:91-7.

Winter, D.A. 1983b. Biomechanical motor patterns in normal walking. *Journal of Motor Behaviour* 15:302-30.

Winter, D.A. 1983c. Energy generation and absorption at the ankle and knee during fast, natural and slow cadences. *Clinical Orthopedics and Related Research* 175: 147-54.

Winter, D.A. 1990. *Biomechanics and Motor Control of Human Movement.* 2nd ed. Toronto: John Wiley & Sons.

Winter, D.A. 1991. *Biomechanics and Motor Control of Human Gait: Normal, Elderly and Pathological.* 2nd ed. Waterloo, ON: Waterloo Biomechanics.

Winter, D.A. 1996. Total body kinetics: Our diagnostic key to human movement. *Proceedings of the International Society of Biomechanics in Sports,* ed. J.M.C.S. Abrantes, 10. Fuchal, Portugal: International Society of Biomechanics in Sports.

Winter, D.A., J.A. Eng, and M.G. Ishac. 1995. A review of kinetic parameters in human walking. In *Gait Analysis: Theory and Application,* ed. R.L. Craik and C.A. Oatis, 252-70. St. Louis, MO: Mosby.

Winter, D.A., A.J. Fuglevand, and S. Archer. 1994. Crosstalk in surface electromyography: Theoretical and practical estimates. *Journal of Electromyography and Kinesiology* 4: 15-26.

Winter, D.A., A.O. Quanbury, and G.D. Reimer. 1976. Analysis of instantaneous energy of normal gait. *Journal of Biomechanics* 9:253-7.

Winter, D.A., and D.G.E. Robertson. 1979. Joint torque and energy patterns in normal gait. *Biological Cybernetics* 29:137-42.

Winter, D.A., and S.E. Sienko. 1988. Biomechanics of below-knee amputee gait. *Journal of Biomechanics* 21:361-7.

Winter, D.A., R.P. Wells, and G.W. Orr. 1981. Errors in the use of isokinetic dynamometers. *European Journal of Applied Physiology* 46:409-21.

Winter, D.A., and S. White. 1983. Moments of force and mechanical power in jogging. *Journal of Biomechanics* 16:91-7.

Winters, J.M., and L. Stark. 1985. Analysis of fundamental movement patterns through the use of in-depth antagonistic muscle models. *IEEE Transactions on Biomedical Engineering* 32:826-39.

Woittiez, R.D., P.A. Huijing, and R.H. Rozendal. 1983. Influence of muscle architecture on the length force diagram of mammalian muscle. *Pfluegers Archiv* 399: 275-9.

Woltring, H.J. 1980. Planar control in multi-camera calibration for 3-D gait studies. *Journal of Biomechanics* 13: 39-48.

Woltring, H.J. 1991. Representation and calculation of 3D joint movement. *Human Movement Science* 10: 603-16.

Woltring, H.J., R. Huiskes, A. De Lange, and F.E. Veldpaus. 1985. Finite centroid and helical axis estimation from noisy landmark measurements in the study of human joint kinematics. *Journal of Biomechanics* 18: 379-89.

Wood, G.A. 1982. Data smoothing and differentiating procedures in biomechanics. *Exercise and Sport Science Reviews* 10:308-62.

Yang, J.F., and D.A. Winter. 1983. Electromyography reliability in maximal and submaximal isometric contractions. *Archives of Physical Medicine and Rehabilitation* 64: 417-20.

Yeadon, M.R. 1990a. The simulation of aerial movement: II. A mathematical inertia model of the human body. *Journal of Biomechanics* 23:67-74.

Yeadon, M.R. 1990b. The simulation of aerial movement: III. The determination of the angular momentum of the human body. *Journal of Biomechanics* 23:75-84.

Yeadon, M.R. 1993. The biomechanics of twisting somersaults. Part I: Rigid body motions. *Journal of Sport Science* 11:187-98.

Yeadon, M.R., and M. Morlock. 1989. The appropriate use of regression equations for the estimation of segmental inertia parameters. *Journal of Biomechanics* 22:683-9.

Yoshihuku, Y., and W. Herzog. 1990. Optimal design parameters of the bicycle-rider system for maximal muscle power output. *Journal of Biomechanics* 23:1069-79.

Yu, B., and J.G. Hay. 1995. Angular momentum and performance in the triple jump: A cross-sectional analysis. *Journal of Applied Biomechanics* 11:81-102.

Zahalak, G.I. 1981. A distribution-moment approximation for kinetic theories of muscular contraction. *Mathematical Biosciences* 55:89-114.

Zajac, F.E. 1989. Muscle and tendon: Properties, models, scaling, and application to biomechanics and motor control. *CRC Critical Reviews in Biomedical Engineering* 17:359-411.

Zajac, F.E., R.R. Neptune, and S.A. Kautz. 2002. Biomechanics and muscle coordination of human walking. Part I: Introduction to concepts, power transfer, dynamics and simulation. *Gait and Posture* 16:215-32.

Zajac, F.E., R.R. Neptune, and S.A. Kautz. 2003. Biomechanics and muscle coordination of human walking. Part II: Lessons from dynamical simulations, clinical implications and concluding remarks. *Gait and Posture* 17:1-17.

Zarrugh, M.Y. 1981. Power requirements and mechanical efficiency of treadmill walking. *Journal of Biomechanics* 14:157-65.

Zatsiorsky, V.M. 2002. *Kinetics of Human Motion.* Champaign, IL: Human Kinetics.

Zatsiorsky, V.M., and V.N. Seluyanov. 1983. The mass and inertia characteristics of the main segments of the human body. *Biomechanics VIII-B,* ed. H. Matsui and K. Kobayashi, 1152-9. Champaign, IL: Human Kinetics.

Zatsiorsky, V.M., and V.N. Seluyanov. 1985. Estimation of the mass and inertia characteristics of the human body by means of the best predictive regression equations. *Biomechanics IX-B,* ed. D.A. Winter et al., 233-9. Champaign, IL: Human Kinetics.

Zipp, P. 1982. Recommendations for the standardization of lead positions in surface electromyography. *European Journal of Applied Physiology* 50:41-54.

INDEX

Note: The italicized *f* and *t* following page numbers refer to figures and tables, respectively.

ABOUT THE AUTHORS

Dr. D. Gordon E. Robertson wrote *Introduction to Biomechanics for Human Motion Analysis* and coauthored *Canadian Foundations of Physical Education, Recreation and Sport Studies.* He has taught undergraduate- and graduate-level biomechanics at the University of British Columbia and currently teaches at the University of Ottawa. He is also Web page editor for the Canadian Society for Biomechanics.

Dr. Graham E. Caldwell (fellow of the Canadian Society for Biomechanics) teaches undergraduate and graduate-level biomechanics at the University of Massachusetts at Amherst and previously held a similar faculty position at the University of Maryland. He is a winner of the Canadian Society for Biomechanics New Investigator Award, and in 1998 he won the Outstanding Teacher Award for the School of Public Health and Health Sciences at the University of Massachusetts at Amherst. Recently, he served as an associate editor for *Medicine and Science in Sports and Exercise.*

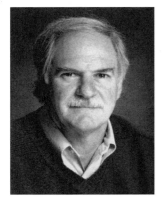

Dr. Joseph Hamill (fellow of the American Alliance for Health, Physical Education, Recreation and Dance; American College of Sports Medicine; and American Academy of Kinesiology and Physical Education) is coauthor of a popular undergraduate textbook, *Biomechanical Basis of Human Movement.* He teaches undergraduate- and graduate-level biomechanics and is director of the exercise science department at the University of Massachusetts at Amherst.

Dr. Gary Kamen (fellow of the American Alliance for Health, Physical Education, Recreation and Dance and American College of Sports Medicine) is author of an undergraduate textbook on kinesiology, *Introduction to Exercise Science.* He is former president of the Research Consortium of AAPHERD and teaches undergraduate and graduate courses in motor behavior and motor control in the exercise science department at the University of Massachusetts at Amherst.

Dr. Saunders (Sandy) N. Whittlesey is a research associate at the University of Massachusetts at Amherst. He has a background in mathematics, engineering, and electronics and works as a technical consultant for FootJoy and Titlest.

*You'll find
other outstanding
biomechanics resources at*

www.HumanKinetics.com

In the U.S. call

1-800-747-4457

Australia.. 08 8277 1555
Canada 1-800-465-7301
Europe...................................... +44 (0) 113 255 5665
New Zealand...................................... 0064 9 448 1207

 HUMAN KINETICS
The Information Leader in Physical Activity
P.O. Box 5076 • Champaign, IL 61825-5076 USA